MW01590999

Composite Materials

Properties, Characterization, and Applications

Edited By

Amit Sachdeva, Pramod Kumar Singh and Hee Woo Rhee

CRC Press
Taylor & Francis Group
Boca Raton London New York

CRC Press is an imprint of the
Taylor & Francis Group, an **informa** business

First edition published 2021
by CRC Press
6000 Broken Sound Parkway NW, Suite 300, Boca Raton, FL 33487-2742

and by CRC Press
2 Park Square, Milton Park, Abingdon, Oxon, OX14 4RN

CRC Press is an imprint of Taylor & Francis Group, LLC

Library of Congress Cataloging-in-Publication Data
Names: Sachdeva, Amit, editor. | Singh, Pramod, 1971- editor. | Rhee, Hee Woo, editor.
Title: Composite materials : properties, characterisation, and applications / [edited] by Amit Sachdeva, Pramod Singh, Hee Woo Rhee.
Other titles: Composite materials (Sachdeva)
Description: First edition. | Boca Raton, FL : CRC Press, 2021. | Includes bibliographical references and index. | Summary: "This book provides an in-depth description of the synthesis, properties, and various characterisation techniques used for the study of composite materials. It also covers applications and simulation tests of these advanced materials. Aimed at industry professionals and researchers, this book offers readers thorough knowledge of the fundamentals as well as advanced level techniques involved in composite material characterization, development, and applications"-- Provided by publisher.
Identifiers: LCCN 2020043362 (print) | LCCN 2020043363 (ebook) | ISBN 9780367490768 (hardback) | ISBN 9781003080633 (ebook)
Subjects: LCSH: Composite materials.
Classification: LCC TA418.9.C6 C554455 2021 (print) | LCC TA418.9.C6 (ebook) | DDC 620.1/18--dc23
LC record available at https://lccn.loc.gov/2020043362
LC ebook record available at https://lccn.loc.gov/2020043363

ISBN: 978-0-367-49076-8 (hbk)
ISBN: 978-1-003-08063-3 (ebk)

Typeset in Times
by SPi Global, India

Contents

Preface

Nanotechnology is defined as technology that deals with particles having dimensions in the range 1–100 nm. Nanoparticles were available in nature before the advent of the human race, in the form of nano-dimensioned particles in the atmosphere, proteins, DNA, RNA, and cells in the human body. All the objects in this size range represent the nano world. In order to realize the nano world, scientists around the globe adopt a "top-down" or a "bottom-up" approach. In the top-down approach, large particles are crushed to nanoscale dimensions. In the bottom-up approach, particles or devices are created by joining single atoms or molecules together via cohesive/adhesive forces.

Major utilizers of such small devices are biotechnologists, physicists, or those working in the electronics field. It started with microelectronics and now we have moved on to nano-electronics. The Nobel Prize in physics was given in 1956 to William Bradford Shockley for the invention of the transistor. This amplification device at that time used a trio whose dimensions can never be reduced, and it consumed a large amount of energy. Today, however, we have developed transistors using nanotechnology that have dimensions of less than 100 nm.

Research at the laboratory scale can go even smaller. The rules obeyed at the atomic scale are quite different from those at the macroscale. For example, when we talk about integrated circuits working at the electronic dimension, quantum tunneling effects dominate on account of the dual wave–particle nature of electrons.

Technological development across the globe started to happen only after the development of various characterization techniques. Initially it was difficult to measure the dimensions of particles in the nanoscale range, but now we have scanning electron microscopy that studies the surface morphology of a sample at nanoscale and transmission electron microscopy to study ultra-thin samples. We can even visualize individual atoms using scanning tunneling microscopy. IBM has written its name using individual atoms of Fe_3O_4 by manipulating individual atoms: they can be moved left and right or up and down.

Now, LEDs are being replaced by quantum dot technology. Quantum dots are particles of nanoscale dimensions (range 4–10 nm) that are made of 10–20 atoms in total. These particles have gained so much popularity in recent times as their loss of energy is almost zero. In the near future, quantum dot technology will rule the television industry based on its picture quality and performance.

Nanoparticles are known for their high surface-area-to-volume ratio or aspect ratio. They can be spread over a large area without any depth, which leads to minimal loss of energy. As per recent studies, nano materials are quantized in particular directions, giving them different shapes like wires, films, dots, or particles. Materials that are not quantized in any direction are free to move along all three dimensions and form nanoparticles. If nanoparticles are quantized from one direction and allowed to move along two directions they form two-dimensional nanoparticles such as thin films. Particles quantized from two directions and allowed to move in only one form one-dimensional nanoparticles, nanowires. Particles that are quantized from all three directions and not allowed to move along any direction form zero-dimensional nanoparticles or quantum dots.

Recently nanotechnology has also made advancements in the field of biotechnology leading to a new merger popularly known as nano biotechnology. It provides advanced and more efficient methods of drug delivery, disease diagnosis, and therapy. Thus, the study of nanotechnology involves a multidisciplinary approach combining various domains.

Editor

Dr. Amit Sachdeva is an Associate Professor at Lovely Professional University, Jalandhar, Punjab, India. He has more than 8 years' teaching experience and his field of specialization is material technology. Dr. Sachdeva has authored around 20 technical research papers in SCI, Scopus, indexed quality journals, and presented at national/international conferences. Dr. Sachdeva is also an editorial member of various journals and is a lifetime member of the International Association of Engineers and Institute For Engineering Research and Publication. Dr. Sachdeva received a Young Scientist Award from the University of Malaya at ICFPAM 2019, Penang Island, Malaysia. He also chaired a session and was selected as a judge for evaluating poster sessions. He has participated in around 15 international conferences and was also part of the organizing committee in 6 international conferences. Dr. Sachdeva also coordinates all the faculty development programs along with an academia–industry interface at Lovely Professional University.

Prof. (Dr.) Pramod Kumar Singh was awarded his Ph.D. from Banaras Hindu University, India, and is currently working as Professor in the Department of Physics at the School of Basic Sciences and Research, Sharda University. As a postdoctoral fellow, Prof. Singh visited South Korea, Norway, and Turkey, spending around 6 years abroad. Prof. Singh has published more than 150 articles in international journals and 5 book chapters. He is currently an editorial board member for five international journals. His areas of research are polymer electrolytes, nanoporous materials for energy devices, dye-sensitized solar cells, and supercapacitors. For more details please visit http://www.materialsresearchlab.net/faculty/prof.-p.k.-singh.html.

Prof. Hee Woo Rhee has a Ph.D. from University of Connecticut, USA, and is currently working as Professor in the Department of Chemical and Biomolecular Engineering, Sogang University, Seoul, South Korea. Prof. Rhee has wide research experience and is additionally appointed as Director of Samsung Display Research Center, South Korea.

Prof. Rhee has published more than 250 international journal articles and 12 book chapters. He is currently editorial board member for many international journals. His areas of research are polymer electrolytes, nanoporous materials for energy devices, dye-sensitized solar cells, and fuel cells.

Contributors

Aseem Acharya
Amity School of Engineering &
  Technology
Department of Mechanical
  Engineering
Amity University
Noida, India

Manoj Singh Adhikari
School of Electronics and Electrical
  Engineering
Lovely Professional University
Phagwara, India

Z. M. A. Ainun
Institute of Tropical Forestry and Forest
  Products (INTROP)
Universiti Putra Malaysia
Selangor, Malaysia

Somjeet Biswas
Light Metals and Alloys Research Lab
Department of Metallurgical and
  Materials Engineering
Indian Institute of Technology
  Kharagpur
West Bengal, India

A. R. Bushroa
AMMP Centre Level 8, Engineering
  Tower
Faculty of Engineering
University of Malaya
Kuala Lumpur, Malaysia

Vijay Chaudhary
Department of Mechanical
  Engineering
Amity School of Engineering &
  Technology
Amity University
Noida, India

Arvind Kumar Chauhan
Department of Physics
Swami Premanand Mahavidyalaya
Hoshiarpur, India

K. L. Chin
Institute of Tropical Forestry and Forest
  Products (INTROP)
Universiti Putra Malaysia
Selangor, Malaysia

Devesh K. Chouhan
Light Metals and Alloys Research Lab
Department of Metallurgical and
  Materials Engineering
Indian Institute of Technology
  Kharagpur
West Bengal, India

Partha Pratim Das
Amity School of Engineering &
  Technology
Department of Mechanical Engineering
Amity University
Noida, India

Sheela Devi
Department of Applied Sciences
Maharaja Surajmal Institute of
  Technology (MSIT)
New Delhi, India

Saurabh Dixit
Department of Biotechnology
Dr. Ambedkar Institute of Technology
  for Handicapped
Kanpur, India

Burhan Gulbahar
Shilpi Jindal
Department of Physics
Chandigarh University
Mohali, India

Zishan H. Khan
Department of Applied Sciences and
    Humanities
Jamia Millia Islamia
New Delhi, India

Deepak Kumar
Electronics and Mechanical Engineering
    School
Affiliated to Gujarat Technological
    University
Vadodara, India

C. H. Lee
Institute of Tropical Forestry and Forest
    Products (INTROP)
Universiti Putra Malaysia
Selangor, Malaysia

S. H. Lee
Institute of Tropical Forestry and Forest
    Products (INTROP)
Universiti Putra Malaysia
Selangor, Malaysia

Gorkem Memisoglu
Kuldeep Mishra
Department of Physics and Materials
    Science
Jaypee University
Anoopshahr, India

Mitali Mishra
Department of Biotechnology
Dr. Ambedkar Institute of Technology
    for Handicapped
Kanpur, India

M. N. F. Norrrahim
Research Centre for Chemical Defence
Universiti Pertahanan Nasional
    Malaysia
Kuala Lumpur, Malaysia

F. N. M. Padzil
Institute of Tropical Forestry and Forest
    Products (INTROP)
Universiti Putra Malaysia
Selangor, Malaysia

A. K. Pandey
Research Centre for Nano-Materials and
    Energy Technology (RCNMET)
School of Science and Technology
Sunway University
Selangor, Malaysia

Ashutosh Kumar Pandey
Department of Biotechnology
Dr. Ambedkar Institute of Technology
    for Handicapped
Kanpur, India

Kritika Pandey
Department of Biotechnology
Dr. Ambedkar Institute of Technology
    for Handicapped
Kanpur, India

Shri Prakash Pandey
Department of Physics
Teerthankar Mahaveer University
Moradabad, India

Rajashri Priyadarshini
Department of Biotechnology
Indian Institute of Technology
Roorkee, India

Nasrudin Abd. Rahim
Higher Institution Centre of Excellence
    (HICoE)
UM Power Energy Dedicated Advanced
    Centre (UMPEDAC), Level 4
Wisma R&D University of Malaya
Kuala Lumpur, Malaysia

Rahul
Department of Applied Sciences and
    Humanities
Jamia Millia Islamia
New Delhi, India
and
Material Research Laboratory
Department of Physics, School of Basic
    Sciences and Research
Sharda University
Greater Noida, India

K. Ramesh
Center for Ionics University of Malaya
Department of Physics
University of Malaya
Kuala Lumpur, Malaysia

S. Ramesh
Center for Ionics University of Malaya
Department of Physics
University of Malaya
Kuala Lumpur, Malaysia

Amit Sachdeva
School of Electronics and Electrical
    Engineering
Lovely Professional University
Phagwara, India

Sobhit Saxena
School of Electronics and Electrical
    Engineering
Lovely Professional University
Phagwara, India

Aparna Seth
Department of Biotechnology
Dr. Ambedkar Institute of Technology
    for Handicapped
Kanpur, India

Aman J. Shukla
Light Metals and Alloys Research Lab
Department of Metallurgical and
    Materials Engineering
Indian Institute of Technology
    Kharagpur
West Bengal, India

P. K. Shukla
Department of Applied Physics
I.T.S Engineering College
Greater Noida, India

Amarjeet Singh
Department of Physics
Himachal Pradesh University
Shimla, India

Kshitij R. B. Singh
Indira Gandhi National Tribal
    University
Department of Biotechnology
Lalpur, India

Ravindra Pratap Singh
Indira Gandhi National Tribal
    University
Department of Biotechnology
Lalpur, India

Pooja Singh
Indira Gandhi National Tribal
    University
Department of Biotechnology
Lalpur, India

Pramod K. Singh
Material Research Laboratory
School of Engineering and Technology
and
Department of Physics
School of Basic Sciences and Research
Sharda University
Greator Noida, India

Sanchita Singh
Department of Biotechnology
Dr. Ambedkar Institute of Technology
    for Handicapped
Kanpur, India

Rakesh K. Sonker
Department of Physics and Astrophysics
University of Delhi
Delhi, India

A. Syafiq
Higher Institution Centre of Excellence
    (HICoE)
UM Power Energy Dedicated Advanced
    Centre (UMPEDAC), Level 4
Wisma R&D University of Malaya
Kuala Lumpur, Malaysia

Suman Lata Tripathi
School of Electronics and Electrical
    Engineering
Lovely Professional University
Phagwara, India

B. Vengadaesvaran
Higher Institution Centre of Excellence
    (HICoE)
UM Power Energy Dedicated
    Advanced Centre (UMPEDAC),
    Level 4
Wisma R&D University of Malaya
Kuala Lumpur, Malaysia

Canan Varlikli
Yogesh Kumar Verma
School of Electronics and Electrical
    Engineering
Lovely Professional University
Phagwara, India

Fatima Zohra
Department of Biotechnology
Dr. Ambedkar Institute of Technology
    for Handicapped
Kanpur, India

1 Introduction to Composite Materials

Nanocomposites and their Potential Applications

Ravindra Pratap Singh and Kshitij R. B. Singh

CONTENTS

1.1 INTRODUCTION

A material made up of more than one component is called a composite. Nanocomposite materials typically comprise a matrix supporting one or more fillers that are particles, sheets, and fibers with dimensions less than 100 nm and with a high surface-to-volume ratio. Nanocomposites are lighter, stiffer, less brittle, more scratch-resistant, more recyclable, more flame retardant, less porous, and better conductors of electricity than normal composite materials. Adding nanofillers such as carbon nanotubes to composite makes the material longer lasting and more resistant to wear and tear and breakage without affecting the surface quality or transparency.

Owing to the high heat resistance and low flammability of some nanocomposites, they are suitable for use as insulators and wire coverings. Nanocomposites containing additives as well as filler components can show substantially improved properties such as decreased permeability to gases, water, and hydrocarbons; improved thermal stability and flame retardancy and reduced smoke emissions; and better chemical resistance, surface appearance, electrical conductivity, and optical properties. Nonporous nanocomposites are used in the packaging of foods and drinks, vacuum packs, and to protect medical instruments: they have high durability, strength, or recyclability (Singh 2019; Caseri 2003).

Nanocomposites have unique properties at the nanoscale level. They are widely utilized in various fields such as in water treatment, supercapacitors, electroconductive scaffolds, anticorrosive/antiballistics, optoelectronic devices, solar cells, hard coatings, biosensors, nanodevices, and green energy generation.

Silver/polyaniline (Ag–Pani) nanocomposites that were prepared via *in situ* reduction of silver in aniline by photolysis at 265 nm not only offer a promising approach for electro-catalytic hydrazine oxidation but could also be utilized in other biosensing applications (Singh et al. The nanocomposites have ecofriendly applications in new technology developments like nanobiosensor-based bioanalysis and biodetection applications (Singh 2016). The exciting and fast-evolving field of multifunctional nanocomposites offers better adsorption capacity, selectivity, and stability than traditionally available materials; they have immense potential for heavy metal detection and removal of pollutants, toxicants, and adulterants from contaminated water. This chapter presents a comprehensive overview of potential applications of nanocomposites in various domains, which will be highly useful to researchers working in this interdisciplinary topic (Singh and Pandey 2011, Singh and Choi 2010).

Nanocomposite materials are used in general automotive and industrial applications. Nanocomposites are revolutionizing the world of materials with high-impact enhanced functionality and broad-spectrum applications. A range of nanostructural nanocomposites, i.e., functional nanomaterials, are used in the construction of biosensors and these nanobiosensors have applications in biological, chemical, and environmental monitoring (Singh et al. 2014). The nanofiller in nanocomposite has enhanced gas-barrier properties and resulted in food packaging applications for processed meats, cheese, confectionery, cereals, fruit juice, dairy products, beer and carbonated drinks bottles, also enhancing the shelf life of foods. Nanoclay polymer composites are using in food packaging materials to check refrigeration and also maintain food freshness due to its excellent gaseous barrier nature. Nanocomposites are high-performance materials that offer unique properties and design applications. Because they are environmentally friendly, nanocomposites offer new technology and business opportunities for aerospace, automotive, electronics, and biotechnology industries.

Nanocomposites are composites with at least one component of nanoscale dimensions. They have been described as the materials of the 21st century and offer new technology and business opportunities for all sectors of industry. Figure 1.1 shows the broad spectrum applications of nanocomposites. Nanocomposite materials may be classified on the basis of their matrix materials: metal matrix nanocomposites ($Fe–Cr/Al_2O_3$, Ni/Al_2O_3, Co/Cr, Fe/MgO, Al/carbon nanotube [CNT], and Mg/CNT), ceramic matrix nanocomposites (Al_2O_3/SiO_2, SiO_2/Ni, Al_2O_3/TiO_2, and Al_2O_3/SiC), and polymer matrix nanocomposites (thermoplastic/thermoset polymer/layered silicates, polyester/TiO_2, polymer/CNT, etc.). Nanocomposite systems have particular sizes at which significant changes in their properties may be expected, such as catalytic property at <5 nm, making hard magnetic materials soft at <20 nm, refractive index changes <50 nm, producing supermagnetism and other electromagnetic phenomenon at <100 nm, strengthening and toughening at <100 nm, and modifying hardness and plasticity at <100 nm (Kamigaito 1991).

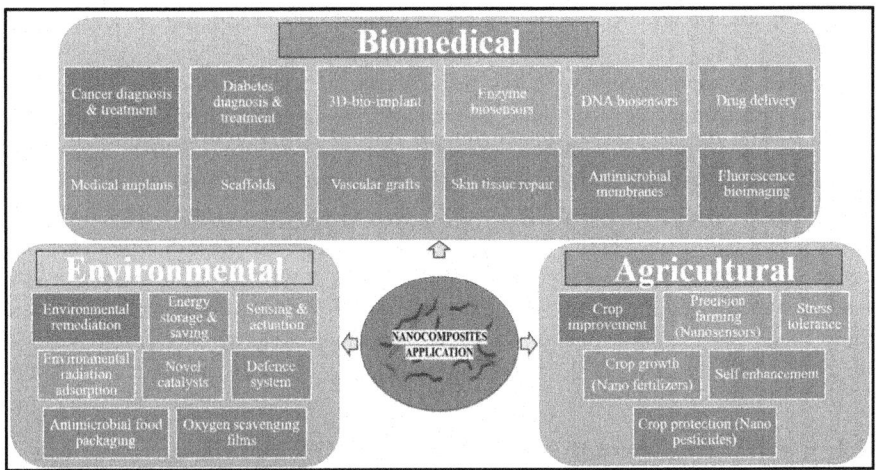

FIGURE 1.1   Potential applications of nanocomposites

Tribological coating of tools for hard and dry cut operations such as drilling, turning, and milling has been made efficient by the development of nanocomposites (Lim et al. 2002, Voevodin and Zabinski 2005). Polymer nanocomposites based on carbon bucky fibers provide lightweight bodies for cars. In aerospace, electronic, and military domains, metal- and ceramic-based nanocomposites are highly utilized in battery cathodes, microelectronics, sensors, catalysts, structural materials, and in electronic, optical, magnetic, and energy conversion devices. Example metal nanocomposites include Fe/MgO for catalysts and magnetic devices, Ni/TiO$_2$ for photoelectrical applications, Al/SiC for aerospace, naval, and automotive structures, Cu/Al$_2$O$_3$ for electronic packaging, and Au/Ag for optical devices and light–energy conversion (Sternitzke, M. 1997, Choa et al. 2003). CNT-based ceramic nanocomposites are utilized in aerospace and sports goods, composite mirrors, and automotive spares, and are also useful for flat-panel displays, gas storage devices, toxic gas sensors, lithium-ion batteries, and conducting paints. They are highly useful in engineering and biomedical applications. However, they are of limited use industrially because of challenges in processing and their cost (Andrews and Weisenberger 2004, Peigney et al. 2000, Alexandre and Dubois 2000).

Polymer-based nanocomposites containing insulating, semiconducting, or metallic nanoparticles are highly used in electronic and food packaging industries; e.g., nylon-6, polypropylene for packaging, and injection-molded articles (nylon-6/surface-modified montmorillonite). They are also useful in tires, fuel systems, seat textiles, mirror housings, door handles, engine covers, timing belt covers, and pollution filters. Polymer/inorganic nanocomposites show good conductivity, permeability, and water management properties for use in water nanofilters (Gangopadhyay and Amitabha. 2000, Giannelis 1996, Fischer 2003 and Pandey et al. 2005).

Ecofriendly nanocomposites for better packaging materials are being developed. For example, the use of nanoclay particles in thermoplastic resins improves barrier properties and package survivability. CNT polymer composites can be used for data storage media, photovoltaic cells and photo diodes, optical limiting devices, drums for printers, etc. (Dresselhaus et al. 2001, Choi and Awaji 2005, and Ajayan et al. 2003). Nanocomposites are speeding healing process for broken bones: e.g., CNT/polymer nanocomposites and these nanocomposites conduct electricity and can be used as a stress sensor on windmill blades to alarm in case of excessive damage and then shut down to save the blade. Epoxy/CNT composite makes a windmill blade that is strong and lightweight. Graphene/epoxy composite is stronger and stiffer than CNT because graphene bonds better to the polymers in the epoxy; this property means this composite is suitable for windmill blades or aircraft components. In the biomedical field, a nanocomposite blend with magnetic and fluorescent nanoparticles makes tumor cells more visible during magnetic resonance imaging (MRI) (Breuer and Sunderraj 2004, Ke and Bai 2005, Presting and Konig 2003, Swearingen et al. 2003).

1.2 APPLICATIONS OF NANOCOMPOSITES IN THE BIOMEDICAL DOMAIN

Yang and Yang (2020) reported nanocomposites based on metal-organic frameworks (MOFs) used in hydrogen/methane storage, catalysis, biological imaging, biosensing, drug delivery, etc. MOF nanocomposites in biomedical fields are used as cargo delivery (drugs, nucleic acids, proteins, and dyes) for cancer therapy, and for bioimaging, biosensing, and biocatalysis; they also show antimicrobial properties. Jones et al. (2020) reported chitin and its derivative chitosan, derived from fungi and crustaceans, respectively, for natural wound treatment. They are hemostatic and antibacterial agents and also support cell proliferation and attachment.

Giliopoulos et al. (2020) reported the use of polymer/MOF nanocomposites in various biomedical applications such as drug delivery and imaging applications. Rahimi et al. (2020) reported carbohydrate polymer-based silver nanocomposites for wound healing, antibacterial, and antifungal effects. Rabiee et al. (2020) reported porphyrin-based nanocomposites: porphyrins are the pigments of life and have a role in photodynamic and sonodynamic therapy and also in magnetic resonance, fluorescence, and photoacoustic imaging, drug delivery, healing and repairing of damaged organs, and cancer theranostics. Maeda (2019) reported thermoresponsive hydrogels showing biocompatibility and degradability. They are composed of linear block copolymers of hydrophilic poly(ethylene glycol) (PEG) and hydrophobic poly(lactic-co-glycolic acid) (PLGA) in biomedical applications such as drug-delivery systems, and in regeneration medicine. Kefeni et al. (2020) reported the biomedical applications and toxicity of spinel ferrite nanoparticles, which are used for the diagnosis and treatment of tumor cells. Wang et al. (2020) reported new possibilities for the construction of glucose-oxidase-based nanocomposites for multimodal synergistic cancer therapy. Glucose oxidase can react with intracellular glucose and oxygen (O_2) to produce hydrogen peroxide and gluconic acid; cancer cells' source of glucose and checks cell proliferation and is recognized as an endogenous oxido-reductase for cancer starvation therapy by increasing hypoxia and acidity. Renu et al. (2020) reported mediated silver/poly-D,L-PLGA nanocomposites extracted from plants for functions such as imaging, biosensors, diagnosis, disease treatment, and antimicrobial-based wound healing. Zhong et al. (2019) reported using MOFs to fabricate polymeric nanocomposites as innovative nanocarriers like dendrimers and mesoporous silica nanoparticles.

Sharma et al. (2019) reported ZnO-based nanocomposites for biomedical applications toward the development of vaccine adjuvants and cancer immunotherapeutics. Yan et al. (2019) reported layered double hydroxide nanostructures for a range of biomedical applications. Fan et al. (2019) reported graphene quantum dot nanocomposites for cancer treatment. Ahmed et al. (2019) reported chitosan nanoparticles as nanostructures of chitosan biopolymer, used in biomedical applications such as tissue engineering, targeted drug delivery, gene delivery, antimicrobial agents against bacteria, fungi, and viral pathogens, in food preservation, as immunomodulatory agents, and in aquaculture and fish farming. Darabdhara et al. (2019) reported the preparation of silver and gold nanoparticles/reduced graphene oxide (rGO) nanocomposites for applications in biomedical (drug delivery and photothermal therapy) and biosensing domains. Qi et al. (2019) reported nanostructured calcium phosphates (NCaPs) as nanocarriers for drug/gene/protein delivery due to their high specific surface area, pH-responsive degradability, high drug/gene/protein loading capacity, and sustained release performance. NCaPs formed from biomolecule/CaP nanocomposites are used in nanomedicine and tissue engineering. Shojaeiarani et al. (2019) reported cellulose nanocrystal hydrogels with high hydrophilic cross-linked polymer networks, used in biomedical, biosensing, and wastewater treatment applications.

Barrios et al. (2019) reported nanomaterial-based aerogels used in lithium-ion batteries, environmental remediation, energy storage, controlled drug delivery, tissue engineering, and biosensing. Castillo and Vallet-Regí (2019) described mesoporous silica nanocomposite used in biomedical and possible biosafety. Simon et al. (2019) reported CNTs as nanocomposite materials for biomedical applications such as diagnostics, tissue engineering, or targeted drug delivery. Graphene is a 2-D nanomaterial with mechanical, optical, electronic, thermal, and electrochemical properties. Graphene oxide and rGO are utilized in the food industry, environmental monitoring, and biomedical fields. Taniselass et al. (2019) reported the detection performances of graphene-based electrochemical biosensors for monitoring biomarkers of non-communicable diseases. Bhat et al. (2019) reported cellulose, a renewable natural fiber, i.e., a ubiquitous and renewable biopolymer resource, as having a broad range of medical applications: tissue engineering, cardiovascular surgery, dental, pharmaceuticals, veterinary, adhesion barriers, and skin therapy. Yadav et al. (2019) reported MoS$_2$-based nanocomposites for applications in sensing, catalysis, therapy, and imaging, drug delivery, gene delivery, phototherapy, combined therapy, bioimaging, theranostics, biosensing, and having potential uses in nanomedicine.

Shin and Choi (2018) reported waterborne polyurethane nanocomposite for biomedical applications. Raza et al. (2018) reported redox-responsive nanocarriers as tumor-targeted drug-delivery systems. Zhang et al. (2019) reported DNA-functionalized nanoparticle technology for several biological and biomedical applications such as cell imaging, cancer therapy, and bioanalytical detection. Vashist et al. (2018) reported CNT-based hybrid hydrogels for biomedical applications for skeletal muscles and cardiac and neural cells, and also diverse applications in regenerative medicines, tissue engineering, drug-delivery devices, implantable devices, biosensing, and biorobotics. Cardoso et al. (2018) reported biomedical applications of magnetic nanoparticles in areas such as hyperthermia, drug release, tissue engineering, theranostics, and lab-on-a-chip. Freag and Elzoghby (2018) reported hybrid protein inorganic nanoparticles for different biomedical applications including bone and cartilage regeneration, imaging tissues, development of antithrombogenic implant biomaterials, and antibacterial wound dressing. Li et al. (2019) reported intelligent polymeric nanogels for the application of stimuli-responsive drug delivery and controlled drug release for activatable theranostics. Yang et al. (2019) reported multifunctional nanocomposite materials or cysteine-rich protein-based biomedical materials such as protein–metal nanohybrids, gold nanoparticle–protein agglomerates, protein-based nanoparticles, and hydrogels for applications in tumor-targeted drug delivery and diagnostics. Ahmad et al. (2017) reported the polyfunctional nature of chitosan and showed it to have antibacterial, mucoadhesive, nontoxic, biodegradable, and biocompatible properties. Chitosan-based self-assembled nanocomposites are used for biomedical applications, specifically in tissue engineering, drug and gene delivery, wound healing, and bioimaging. Wu et al. (2017) reported the use of enzymes as biocatalytic nanocomposites, which act as antimicrobial agents and have been proved to be effective against bacterial pathogens. Gaaz et al. (2017) reported nanotubular clay minerals, composed of aluminosilicate naturally structured in layers known as halloysite nanotubes (HNTs), in biomedical applications. Dykman and Khlebtsov (2016) reported multifunctional gold-based nanocomposites for biomedical applications. Trache et al. (2017) reported cellulose nanocrystals, bio-based nanoscale materials for applications in fields such as biomedical, pharmaceuticals, electronics, barrier films, nanocomposites, membranes, supercapacitors, etc. Dziadek et al. (2017) reported biodegradable polymer composites with ceramic fillers for medical applications; e.g., silica nanocomposites, wollastonite as a composite modifier, and calcium phosphate ceramics, namely hydroxyapatite, tricalcium phosphate, and biphasic calcium phosphate. Motealleh and Kehr (2017) reported nanocomposite hydrogels, organic–inorganic hybrid materials for biomedical applications. Jalili et al. (2016) reported smart thermoresponsive magnetic hydrogels for a range of biomedical applications like therapeutic drug delivery, bioimaging, and regenerative engineering.

Bacterial cellulose (BC) is a biopolymer used in traditional desserts and as a gelling, stabilizing, and thickening agent in the food industry. It is an interesting biocompatible nanomaterial for biomedical applications such as artificial skin, artificial blood vessels and microvessels, and wound dressing. Some examples are BC/collagen, BC/gelatin, BC/fibroin, and BC/chitosan. Ullah et al. (2016) reported BC immobilization of enzymes, bacteria, and fungi, and its use in tissue engineering; heart valve prosthesis; artificial blood vessels, bone, cartilage, cornea, and skin; and dental root treatment.

Deepthi et al. (2016) reported the use of chitin- and chitosan-based nanocomposite scaffolds for bone tissue engineering. Govindhan et al. (2016) reported nanostructured Pt-based materials (electrochemical sensors) for the detection of NO in neuroscience applications. They also utilized nanostructured Pt-based electrode materials such as nanoporous Pt, Pt and PtAu nanoparticles, PtAu nanoparticle/rGO, and PtW nanoparticle/rGO–ionic liquid nanocomposites for the detection of NO in biological and medical applications. Zare and Shabani (2016) reported polymer/

metal nanocomposites for biomedical applications. Gaaz et al. (2015) reported polyvinyl alcohol–HNT nanocomposites for medicinal and biomedical use such as wound dressings, drug delivery, targeted-tissue transportation systems, and soft biomaterial implants. Bolocan et al. (2015) reported three main types of gold dendritic structures (gold–dendrimer nanocomposites, dendrimer-entrapped nanoparticles, and gold monocrystalline dendritic growths) for drug-delivery systems. Gold nanoparticles, due to their optical properties and quantum size effect, are used in sensing, photodynamic therapy, therapeutic agent delivery, and diagnostics.

John et al. (2015) reported polymer block polypeptides and polymer-conjugated hybrid materials for various stimuli-responsive drug and gene delivery applications. Choi et al. (2015) reported nanostructured biomaterial coatings for biomedical and dental clinical applications. Zhou et al. (2014) reported multifunctional magneto-plasmonic ($Au–Fe_xO_y$) nanomaterials for biomedical applications such as biosensors, bioseparation, multimodal imaging, and therapeutics. Vellayappan et al. (2015) reported multifaceted use of nanocomposites across the globe for cardiovascular grafts and stents for the treatment of cardiovascular disease. Tan et al. (2013) reported that polyhedral oligomeric silsesquioxane poly(carbonate-urea) urethane can act as a scaffold for bioartificial organs, nanoparticles for biomedical applications, and a coating for medical devices. Tang et al. (2011) reported metallodendrimers or dendrimer nanocomposites for biomedical applications such as biomimetic catalysts, imaging contrast agents (especially for MRI imaging), biomedical sensors, and therapeutic agents. Oishi and Nagasaki (2010) reported stimuli-responsive PEG-coated (PEGylated) nanogels used for biomedical applications as smart nanomedicines for cancer diagnostics and therapy. Sana et al. (2016) reported the delivery and therapeutic actions of galantamine drug (GAL) against Alzheimer's disease in rat brain by attaching GAL to ceria-containing hydroxyapatite ($GAL/CeO_2/HAp$). Ceria-containing carboxymethyl chitosan-coated hydroxyapatite ($GAL/CeO_2/HAp/CMC$) nanocomposites have also been established.

Cholesterol oxidase/1-fluoro-2-nitro-4-azidobenzene/octadecanethiol/gold (ChOx/FNAB/ODT/Au) nanocomposite film, which has been utilized for estimation of cholesterol in solution using a surface plasmon resonance technique, was developed by Arya et al. (2006). Wu et al. (2016) reported alginate/chitosan nanocomposite particles loaded with S-nitrosoglutathione for oral treatment of cardiovascular diseases. Sabherwal et al. (2016) reported biofunctionalized carbon nanocomposites – mainly bioreceptor-functionalized nonocomposites – that are able to generate detection systems in clinical/environmental monitoring. Han et al. (2016) reported anisotropic yolk/shell or Janus inorganic/polystyrene nanocomposites. These nanocomposites showed stable and strong fluorescence on the introduction of quantum dots for biomedical applications, particularly biodetection. Peng et al. (2016) reported graphene-oxide-based polymer nanocomposites for biomedical applications. Song et al. (2015) reported porous Co nanobead/rGO nanocomposites for glucose sensing, which have become important functional materials in sensors, catalysis, energy conversion, etc. Iordanskii et al. (2016) reported magnetic nanocomposites consisting of polymer matrix and encapsulated functional nanoparticles for controlled drug release. Raj and Prabha (2015) reported cassava starch acetate-polyethylene glycol-gelatin nanocomposites for controlled drug-delivery systems with an anticancer drug.

Wang et al. (2015) reported a luminol electrochemiluminescence sensor strategy based on TiO_2/CNT nanocomposites for the detection of glucose using a glassy carbon electrode. Nguyen et al. (2015) reported that multifunctional Ag/Fe_3O_4–chitosan nanocomposites using chitosan as a stabilizing cross-linking agent showed antibacterial property against *Pseudomonas aeruginosa* and were also used for local hyperthermia treatment of cancers. Mohammed et al. (2016) reported nickel-disulfide-decorated CNT nanocomposites for the detection of the toxic chemical 4-methoxyphenol in environmental and healthcare applications. Filipe et al. (2016) reported Eu^{3+}-doped SiO_2/poly(methyl methacrylate) hybrid nanocomposites used as thermal sensors. Afroze et al. (2016) reported hydroxyapatite-functionalized multiwalled CNT (HA/f-MWCNT) nanocomposites used as a biomaterial for the hyperthermia treatment of bone cancer. Yang et al. (2016) reported a pH-sensitive polymer/lipid nanocomposite for oral colon-targeted drug delivery. Rahn et al. (2016) reported MRI and X-ray imaging of biological tissues with magnetic nanocomposites for cancer treatment. Jafari et al. (2016) reported the anti-inflammatory effect of triamcinolone acetonide-loaded hydroxyapatite nanocomposites in the arthritic rat model and suggested it for treatment of rheumatoid arthritis. Wang et al. (2016) reported superparamagnetic cobalt ferrite/graphene oxide nanocomposites used for MRI and controlled drug delivery for simultaneous cancer diagnosis and chemotherapy.

1.3 APPLICATIONS OF NANOCOMPOSITES IN THE ENVIRONMENTAL DOMAIN

Monitoring of toxicants, contaminants, or pollutants in the air, water, and soil is very important because of the risk they pose to human health and ecosystems. In this context, recent advances in the development and application of biosensor arrays using aptamers for environmental detection are highlighted (Singh et al. 2008).

Progress and development of biosensors will inevitably focus upon the technology of nanocomposites for the real-time detection of pesticides, antibiotics, pathogen, toxins, and biomolecules in food, soil, and water. The current trends and challenges with nanocomposites for various applications focus not only on biosensor development but also on nanobiosensors (Singh 2011); the current trends and challenges with smart nanomaterials pertain to the development of biosensors, nanotechnology, and nanobiotechnology. All these growing areas will have a significant influence on the development of new ultra-biosensing devices (nanobiosensors) to resolve severe pollution problems in the future (Singh et al. 2012).

Gold/3-aminopropyltriethoxysilane/glutaraldehyde/glutathione-s-transferase nanocomposite film was developed (Singh and Choi 2009) for the electrochemical detection of captan in water. Captan is a known harmful chemical and potential carcinogen; its continual use poses an environmental pollution problem.

Biosensors based on conjugated conducting polymers such as polyaniline, polypyrrole, or polythiophene blend with metal or ceramic to form nanocomposites for wide applications, including the detection of toxicants in environmental samples (Singh and Choi 2009). A catalase/polyaniline/indium tin oxide nanocomposite film was developed (Singh and Choi 2009) for the detection of hydrogen peroxide and azide in water or various biological samples. Nano-sized polyaniline possesses a nanofiber morphology and strongly influences the morphology of the composite. As the silver nanoparticles are synthesized in the polyaniline solution, the nanoparticles become embedded into the polyaniline matrix.

Singh and Choi (2010) reported PANi/ClO$_4$ film as a nanocomposite to immobilize DNA as a biosensing platform to detect sanguinarine, an alkaloid present in adulterated edible mustard oil.

Water is the universal solvent for life on earth and an essential resource for living systems, industry, agriculture, and domestic use. Pollutants/toxicants in water can adversely interfere with health, comfort, property, or the environment (Itodo and Itodo 2010, Duncan 2003). Arora et al. (2006) reported DNA/polypyrrole–polyvinyl sulfonate nanocomposite films that have been developed for estimation of o-chlorophenol and 2-aminoanthracine in water and wastewater samples. Singh et al. (2011) reported double-stranded DNA/cytochrome-c/glutathione-self-assembled monolayer/gold film as a nanocomposite for the detection of toxicants like 2-amino anthracine, a poly-aromatic compound that is a pesticide and potent carcinogen; 3-bromobenzanthrone, a dye intermediate that causes hepatic injury; and bisphenol A is a xenoestrogen that is a well-known pollutant in water as well as an endocrine disrupter.

Nanoclay is a well-known natural-fiber nanofiller in nanocomposites. Wypych (2004) reported jute/polyethylene/nanoclay nanocomposites using jute treated with propionic anhydride. Seyedeh and Seyed (2016) reported Ag/Fe$_3$O$_4$ nanocomposites using starch, which showed activity as biocompatible capping agents and reducing agents, superparamagnetic properties, and antibacterial activity. Fatemeh et al. (2016) reported Fe$_2$O$_3$/CuFe$_2$O$_4$/chitosan nanocomposites using onion extract for the detection of the environmental analyte of interest. Magesan and Umapathy (2016) reported TiO$_2$ nanocomposites doped with Al$_2$O$_3$, Bi$_2$O$_3$, CuO, and ZrO$_2$ that showed not only photocatalytic activity to degrade the toxic activity of organic dye pollutants but also antibacterial activity against *Escherichia coli*. Peng et al. (2016) reported zinc ferrite–rGO nanocomposites as an excellent absorbent for removing dye pollutants.

Ibuprofen is a persistent organic pollutant that causes severe adverse effects in humans and wildlife. To remove it from wastewater is a worldwide necessity. Xia and Lo (2016) reported a superparamagnetic Bi$_2$O$_4$/Fe$_3$O$_4$ nanocomposite for the photocatalytic removal of ibuprofen. Heavy metals are one of the most widespread environmental concerns that threaten human health and ecosystems. Giusy et al. (2016) reported polymer-functionalized nanocomposites for the effective removal of heavy metals by an adsorption approach. Rehim et al. (2016) reported a TiO$_2$/sodium alginate nanocomposite that was found to exhibit high photocatalytic activity and antibacterial effect which means it is good agent for waste-water management and can be used in the hygienic and food packaging industries.

Ain et al. (2016) reported rGO/MgFe$_2$O$_4$ nanocomposites and showed photocatalytic activity using methylene blue as a model organic compound in polluted water. Aziz and Gohari (2016) properties. Hong et al. (2016) reported silver nanowire/carbon fiber cloth nanocomposites for electrochemical point-of-use water disinfection. Gui et al. (2015) reported Ag-doped MWCNT titanium dioxide core-shell nanocomposites and showed their application in carbon dioxide photoreduction. Sarah et al.

(2015) reported stable bentonite chitosan nanocomposites used as antibacterial materials for water disinfection.

Aladpoosh and Montazer (2016) reported cellulosic-fabric coated with Ag/ZnO nanocomposites using the ashes of *Seidlitzia rosmarinus* plants, which showed antibacterial activities against *Staphylococcus aureus* and *Escherichia coli*. Bhaumik et al. (2016) reported polypyrrole-wrapped oxidized MWCNT nanocomposites for the effective removal of hexavalent chromium, Cr(VI). Wang et al. (2016) reported montmorillonite nanocomposites as adsorbents for removing Cr(VI) from wastewater. SnO_2 quantum-dot-decorated rGO nanocomposite is used to remove different organic dyes such as methylene blue, methyl orange, and rhodamine B, toxic metal ions such as Co^{2+}, Ni^{2+}, Cu^{2+}, Cd^{2+}, Cr^{3+}, Pb^{2+}, Hg^{2+}, and As^{3+}, and pathogenic bacteria (*Escherichia coli*) from wastewater. However, Dutta et al. (2016) reported quantum dot–rGO nanocomposites that showed higher antibacterial activity against *E. coli* and removal of organic dyes and toxic heavy metals. Fu et al. (2016) reported castor-oil-based waterborne polyurethane/silver–HNT nanocomposites that showed antibacterial activity against *Staphylococcus aureus* and *Escherichia coli*.

Adsorption processes are used for the removal of inorganic and organic micropollutants from aqueous solution for water treatment; e.g., clay–polymer nanocomposite adsorbents treat water by adsorption and flocculation of both inorganic and organic micropollutants from aqueous solutions. Emmanuel and Taubert (2016) reported clay–polymer nanocomposite adsorbents in the removal of micropollutants (inorganic, organic, and biological) from aqueous solutions. Ghosh et al. (2015) reported a manganese-oxide-incorporated ferric oxide nanocomposite as a novel adsorbent for the removal of Cr(VI) from contaminated water. Ghanbari et al. (2015) reported TiO_2 nanoparticle/HNT nanocomposites using nanofillers for forward osmosis applications and also removing the bad smell.

Koli et al. (2016) reported TiO_2–MWCNT nanocomposites as efficient antibacterial agents against a wide range of microorganisms to prevent and control the persistence and spreading of bacterial infections. Xu et al. (2016) reported $Ni/Fe–Fe_3O_4$ nanocomposites for effective remediation of pollution like the herbicide 2,4-dichlorophenol. Pant et al. (2016) reported a multifunctional $Ag–TiO_2/rGO$ nanocomposite that showed superior photocatalytic activities, and antibacterial properties for wastewater treatment. Chen et al. (2016) reported organic–inorganic heterostructured photocatalysts, e.g., a porous graphitic C_3N_4 hybrid with CuS, which could be applied in the field of environmental remediation. Shen et al. (2016) reported $Au–Fe_3O_4$ nanocomposites for the bimodal detection of melamine. Fang et al. (2016) reported CdS/$BiVO_4$ nanocomposites that showed photocatalytic properties in solving environmental pollution issues utilizing solar energy. Deng et al. (2016) reported SnS_2/TiO_2 nanocomposites that were effective photocatalysts in the treatment of Cr(VI) in wastewater.

Song et al. (2015) reported 3-D hierarchical carbon-coated nickel (Ni/C) nanocomposites that showed a high adsorption capacity for heavy metal ions in wastewater treatment. Xu et al. (2016) reported bicomponent $ZnO/ZnCo_2O_4$ nanocomposites used as an anode for lithium-ion batteries. Wei et al. (2013) reported $BiOBr/TiO_2$ nanocomposites that showed photocatalytic performance toward photodegradation of rhodamine B under visible-light irradiation for water purification as well as other environmental applications. Mallakpour and Khadem (2016) reported the combination of CNTs with metal oxides such as aluminum dioxide, titanium dioxide, zinc oxide, and iron oxide, used for applications such as sensors, supercapacitors, and absorbents, and in photocatalytic and photovoltaic functions. Li et al. (2015) reported magnetic CNT adsorbents, CNT/Fe_3O_4 nanocomposites for the removal of bisphenol A in aqueous solution. Rabieh et al. (2016) reported green methods for synthesis of hierarchical ZnO–rGO nanocomposites that showed photocatalyst activity for the photodegradation of azure B dye. Yang et al. (2016) reported a Cu_2O/CNT hierarchical chrysanthemum-like nanocomposite and showed its ability to degrade phenol under visible-light irradiation due to its unique 3-D ordered nanostructure. Kadam et al. (2016) reported ZnO/Ag_2O nanocomposites as photocatalysts for the degradation of methyl orange under UV and visible light. Duangjam et al. (2016) reported $CoFe_2O_4/BiVO_4$ nanocomposites with photocatalytic activities to degrade methylene blue under visible-light irradiation.

Nasrollahzadeh et al. (2016) reported green synthesis of Cu/eggshell, Fe_3O_4/eggshell, and Cu/Fe_3O_4/eggshell nanocomposites via an environmental and economical method using aqueous extract of the leaves of *Orchis mascula* without stabilizer or surfactant. They showed high catalytic activity in the reduction of a variety of dyes in water at room temperature: 4-nitrophenol, methyl orange, Congo red, methylene blue, and rhodamine B. Zango et al. (2016) reported copper (II)–HNT nanocomposites as adsorbents for the removal of 2,4,6-trichlorophenol in water. Czech and Waldemar (2015) reported nanocomposites of MWCNTs and TiO_2/SiO_2 ($MWCNT–TiO_2–SiO_2$) for the removal of bisphenol A and carbamazepine from water solution. Lam et al. (2016) reported graphene/C_3N_4/semiconductor nanocomposites and showed photocatalytic destruction of industrial effluents, i.e., dyestuff effluents, toward environmental conservation. Istratie et al. (2016) reported magnetic iron oxide/carbon nanocomposites as adsorbents for the removal of methyl orange and phenol from aqueous solutions. Ullah et al. (2015) reported AgI/functionalized graphene/TiO_2 nanocomposites and showed photocatalytic activity for hydrogen evolution. Mallakpour et al. (2016) reported poly(vinyl alcohol)/TRIS-functionalized graphene nanosheets (TRIS-GO) films that showed water-absorbent ability.

Chen et al. (2013) reported magnetic hollow Fe_3O_4 nanoparticles that are coated with a polystyrene layer that acts as water repellent and have oil-absorbing surfaces to remove organic contaminants (such as oil spills from water surface) from the environment, and also suggested that they will pave the way for new applications in wastewater treatment. Fonseca et al. (2015) reported poly(lactic acid)/TiO_2 nanocomposites showing biocidal properties against *Escherichia coli*. Zhang et al. (2016) reported HNTs used as ideal templates for immobilizing nanoparticles as heterogeneous catalysts utilized in fuel cells. Jung et al. (2015) reported biochar/Mg–Al assembled nanocomposites and showed them to be excellent adsorbents that can remove phosphate from aqueous solutions. Qutub et al. (2015) reported CdS/ZnS sandwich and core-shell nanocomposites and showed their efficient photocatalytic properties using acid blue dye and p-chlorophenol. Hsieh et al. (2015) reported ZnSe/graphene nanocomposites acting as photocatalysts, tested with methylene blue in the aqueous phase. Allafchian et al. (2015) reported a magnetically responsive three-component nanocomposite $NiFe_2O_4$/poly(acrylonitrile-co-maleic anhydride)/ Ag that showed antibacterial activities and prevents environmental contamination.

Kamal et al. (2016) reported antifouling applications for various nanocomposite membranes in the biomedical industry, food processing, and water treatment. Liao et al. (2020) reported silver nanoparticles of doped TiO_2 for bacterial killing. Further, TiO_2 and Cu-doped TiO_2 nanoparticles in a chitosan or textile matrix formed polymer nanocomposites with antimicrobial properties.

Water and energy are major issues for human beings, with growing pollution problems. Mehta et al. (2019) reported carbon quantum dots for sensor and water treatment applications. Hashemi and Rezania (2019) reported carbon-based adsorbents using CNTs, graphene, fullerenes, activated carbon, carbon nanohorns, carbon nanofibers, and graphitic carbon nitride for the detection of heavy metals in water. Liao et al. (2019) reported silver nanocomposites and bimetallic silver nanocomposites as catalysts for catalytic oxidation, catalytic reduction, photocatalysis, and electrocatalysis. Ali et al. (2019) reported interfacial active nanoparticles and nanocomposites or magnetically responsive nanomaterials and nanocomposites for sustainable wastewater remediation. Fu et al. (2019) reported a metal–semiconductor nanocomposite used as a photocatalyst. Gouzman et al. (2019) reported polyimide-based materials for space applications such as dealing with space environment hazards like solar radiation, energetic particles, vacuum, debris, and space plasma. Durable polyimide-based materials and nanocomposites are developed and tested to resist degradation in space.

Zou et al. (2019) reported carbon-based nanocomposites for the photocatalytic degradation of volatile organic compounds. They are harmful for humans and their removal from ecosystems is mandatory. Lingamdinne et al. (2019) reported nanocomposites based on magnetic graphene oxide for the detection and removal of heavy metals and toxic pollutants, radionuclides, and organic dyes in water bodies. Magnetic graphene-oxide-based nanocomposites are also used as adsorbents for the treatment of water pollutants and organic and agricultural pollutants. Gangu et al. (2019) reported MWCNT-based nanocomposites for the removal and detection of harmful pollutants in water system. They are also used as photocatalysts for the removal of organic pollutants. Chang et al. (2018) reported a quartz crystal microbalance device using affinity ionic liquids for gas sensing, the detection of volatile organic compounds, which is very important in health, safety, military, industry, and the environment.

Most metallic structures degrade due to atmospheric corrosion, which is a major problem for bridges, pipelines, and storage tanks. Valença et al. (2015) reported ZnO nanoparticles/polypyrrole (ZnO–NPs/PPy) hybrid nanocomposites used as additives in an epoxy paint to protect SAE 1020 carbon steel from corrosion due to humidity, pollutants, and temperature. The physicochemical characteristics that may interfere with the corrosive action of the environment are the presence of water, salts, gases, differences in pH, and electrical conductivity. Hammouda et al. (2011) reported the use of zinc as an anticorrosive agent in paints to improve the corrosion resistance of metallic components. Hihara et al. (2013) reported a treatment with chromium and phosphate that can be used to prolong the lifetime of zinc coatings. However, the leaching of such compounds may cause serious ecological damage, and that encourages the development of corrosion inhibitor compounds that could have less impact on the environment. Riccardis and Martina (2014) reported conductive polymers with a good corrosion stability both in contact with solution and in the dry state, to minimize the health risk to humans and damage to the environment. Batool et al. (2012) reported ZnO/PPy hybrid materials that have a variety of applications in optoelectronic devices and good anti-corrosion properties. Furthermore, Lehr and Saidman (2013) reported polypyrrole as an anticorrosive additive for steel coating.

Supercapacitors/ultracapacitors/electrochemical capacitors are energy storage devices. Yang (2011) described electrode materials made up of carbon-based nanomaterials, metal oxides/hydroxides, and conducting polymers; they are the most common electroactive materials for supercapacitors and exhibit very high power output and better cycling ability. Wu et al. (2010) reported that combining nanoscale dissimilar capacitive materials to form nanocomposite electroactive materials is an important approach to control, develop, and optimize the structures and properties of electrode materials for enhancing their performance for supercapacitors. There are still a lot of challenges to be overcome regarding the design and fabrication of nanocomposite electroactive materials for supercapacitor applications. Nanocomposite electroactive materials that have been developed so far have demonstrated huge potential. Lee et al. (2005) reported significant improvement in terms of specific surface area, electrical and ionic conductivities, specific capacitance, cyclic stability, and energy and power density of supercapacitors using nanocomposite electroactive materials.

Ho et al. (2011) reported nanostructures and nanocomposites based on nanoparticles of semiconductor materials that exhibit dielectric nanostructures, using Al_2O_3, SiO_2, and MgO nanoparticles to fabricate optoelectronic and fiber-optic devices. Voevodin and Zabinski (2000) reported that nanostructured and nanocomposite films have nanocrystalline or amorphous structures and have unique physicofunctional properties for the development of hard and superhard films of controlled size (1 to 10 nm) and with high oxidation resistance exceeding 2000°C. The geometry of the grains is responsible not only for the enhanced hardness of the films but also for enhancement of other properties such as magnetic or catalytic behavior.

1.4 APPLICATIONS OF NANOCOMPOSITES IN THE AGRICULTURAL DOMAIN

Pradhan et al. (2015) reported nano SiC chitosan nanocomposites that showed oxygen barrier properties with increased thermal stability, tensile strength, and chemical resistance properties, so may be suitable for food packaging applications. Jamroz et al. (2019) reported biopolymer (e.g., natural chitosan, starch, and synthetic poly(lactic acid)) nanocomposite-containing nanofiller (clay, organic, inorganic, or carbon) nanostructures for use in the food industry and in food packaging systems. Thiruvengadam et al. (2018) reported innovative applications in the food industry, such as food processing technology, packaging materials, and ingredients using encapsulated nanocomposites. Sadeghi et al. (2017) reported nanocomposites' applications in food science and agriculture to improve consumer health and safety, product shelflife and stability, bioavailability, environmental sustainability, efficiency of processing and packaging, and real-time monitoring.

Ghanbarzadeh et al. (2015) reported ecofriendly nanostructured materials utilized in biopolymer-based plastics for food packaging applications. The addition of reinforcing nanofiller forming nanocomposite enhances mechanical and barrier properties. Konwarh et al. (2013) reported diverse applications of electrospun cellulose acetate (CA) nanofibers. CA has been used in applications ranging from high-absorbing diapers to membrane filters. Electrospun CA fibers have potential applications in nutraceutical delivery, bioseparation, biomolecule immobilization, tissue engineering, bioremediation, biosensing, crop protection, anti-counterfeiting, pH-sensitive material, photocatalytic self-cleaning textiles, temperature-adaptable fabric, and antimicrobial mats.

Cellulose is an agricultural product, a biopolymer with the potential to replace petroleum polymers. Mahadeva et al. (2011) reported a cellulose-based nanocomposite material used as a flexible humidity and temperature sensor. Tunc et al. (2011) reported an active antimicrobial packaging material using methyl cellulose as the base material and montmorillonite as reinforcement. Sithique and Alagar (2010) reported bio-based nanocomposites developed from epoxidized soybean oil, diglycidyl ether of bisphenol A, and organically modified montmorillonite. Azeredo et al. (2009) reported the uses of mango-puree-based edible films for nanocomposites and showed potential packaging applications. Tate et al. (2010) reported a soy-based polyurethane that can be used as a matrix for the production of bio-based nanocomposites.

Green nanocomposites are ecofriendly and sustainable in all ways and possess renewability and degradability, reducing and preventing pollution. Natural polymers such as cellulose, chitin, starch, polyhydroxyalkanoates, polylactide, polycaprolactone, collagen, and polypeptides are biodegradable.

Green nanocomposites are used in automotive, construction, packaging, and medical fields. They have unique properties of strength, elastic modulus, dimensional stability, permeability toward gases, water, and hydrocarbons, thermal stability, heat distortion temperature, smoke emissions, chemical resistance, surface appearance, weight, and electrical conductivity at the nanoscale level. They are biodegradable and lead to a reduction of carbon dioxide concentration in the atmosphere (Leja and Lewandowicz 2010). Polymers from natural sources (such as starch, lignin, cellulose acetate, polylactic acid, polyhydroxylalkanoates, polyhydroxylbutyrate, etc.) and some synthetic sources (aliphatic and aromatic polyesters, polyvinyl alcohol, modified polyolefins, etc.) that are degradable are classified as biopolymers (John and Thomas 2008). However, those from synthetic sources are not renewable and therefore do not conform wholly to the concept of renewability and degradability (Siracusa et al. 2008). Thermoplastic starch properties also appear to benefit from silica addition and it has been reported that inclusion of dry powder SiO_2 particles in starch/polyvinyl alcohol films increased tensile strength and improved water barrier properties (Tang et al. 2008).

The improved mechanical properties, transmittance, and water resistance of starch films containing nano-SiO_2 particles have been reported (Xiong et al. 2008). Starch-based resin 11C is a biodegradable and compostable resin based on a blend of thermoplastic starch for the manufacturing of film-type products. It can be used directly in the film blowing process (Billmeyer 1984). The composites were prepared with regular cornstarch plasticized with glycerin and reinforced with hydrated kaolin. A sequence of addition of plasticizers using a solution method, to determine their effect on the mechanical and structural properties, was established. Thermal stability, mechanical properties, and water-absorption studies were conducted to measure the material properties and it was deduced that the sequence of addition of components (starch/plasticizer (glycerol)/clay) had a significant effect on the nature of the composites formed. Starch modified with acetate and biocomposites produced with cellulose fibers have been reported (Guan and Hanna 2006).

Polylactic acid (PLA) is an alternative to petrochemical-based plastics for many applications. Lee et al. (2008) reported the thermal, mechanical, and morphological properties of PLA-based composites. PLA/organically modified clay blends were developed (Ogata et al. 1997). PLA has unique properties such as high mechanical strength, low toxicity, and good barrier properties but is limited in applications due to its low glass transition temperature, weak thermal stability, low ductility and toughness, and low modulus above the glass transition temperature (Harada et al. 2007). Commercially, PLA has been reported as a matrix with cellulose whiskers treated with anionic surfactant as reinforcement (Bondeson and Oksman 2007). Studies on the thermal, mechanical, and morphological properties of PLA-based composites have been reported (Lee et al. 2008).

The blending of two or more polymers to achieve a biodegradable nanocomposite polymer has been reported, e.g., starch/PLA blends, polybutylene succinate/CA blends, starch/modified polyester blends, polycaprolactone/poly(vinyl alcohol) blends, and thermoplastic starch/polyesteramide blends (Averous et al. 2000). Nanocomposites have been used in several applications such as mirror housings on vehicle types, door handles, door panels, trunk liners, instrument panels, parcel shelves, head rests, roofs, upholstery, engine covers, intake manifolds, and timing belt covers. Other applications currently being considered include impellers and blades for vacuum cleaners, power tool housings, mower hoods, and covers for portable electronic equipment such as mobile phones, pagers, etc. Its excellent barrier properties, chemical resistance, and surface appearance make it an excellent material for packaging applications such as beer and carbonated drinks bottles, and paperboard for fruit juice and dairy (Willett and Shogren 2002).

Nanocomposites also have a future in aerospace applications because of their light weight. There is a wide range of possible applications of nanocomposites with just a few problems to be overcome such as poor adhesion of matrix and fiber, difficulty of fiber orientation, the achievement of nanoscale sizes, and the evolution of truly green polymers (Ashori, 2008, Kim et al. 2006, Teixeira et al. 2009).

Honeywell developed commercial clay/nylon-6 nanocomposite products for drink packaging applications. Starch-based resin 11C is a biodegradable and compostable resin based on a blend of thermoplastic starch that is suitable for making compostable bags (shopping bags, green bin liners) and meat liners (overwrap packaging, mulch film, breathable film). The innovation and developments in technologies for green polymer nanocomposites using biodegradable polymer as matrix have shown properties such as renewability and degradability. A series of interesting polymers have been realized through multiple research activities into substances such as thermoplastic starch and its blends, PLA and its modifications, cellulose, gelatin, chitosan, etc. Natural fiber is preferred over synthetic fibers for environmental reasons and has been synthesized from agricultural sources such as kenaf, jute, hemp, flax, banana, bamboo, sisal, coconut coir, etc. There is thus a wide range of possible applications of nanocomposites (Pandey et al. 2007).

1.5   PERSPECTIVES AND CONCLUSIONS

Owing to the outstanding potential applications of nanocomposites, they have generated a great impact on the world economy as well as local business. Polymer/clay nanocomposites have been commercialized in many leading industries for packaging, coating, and automotive use. The high demands of high-performance systems using nanofillers at low costs mean the future is bright for a wide range of applications such as consumer products like electroconductive polymers, nanosmart switches, and sensors for the automotive industry. Further power production and storage devices like hydrogen storage, fuel cells, supercapacitors, and batteries are utilizing metal/ceramic/polymer nanocomposites, improving fire retardancy in interior parts and weatherability in exterior parts. Nanoclays, CNTs for automotive and beverage/food packaging, CNT-based nanocomposites for electrical and electronic industries, and polymer-based nanocomposites for engineered plastics are selling well globally. However, a few outstanding challenges and issues pertaining to nanocomposites mean that future R&D is required for ecofriendly nanocomposites for good health and a cleaner environment.

Nanocomposites are creating macroscopic engineered materials at the nanoscale. Nanocomposites will benefit many domains of our society, in areas such as electronics, chemicals, space, general automobiles, medicine, health care, and the environment. Thus, nanocomposites will improve our quality of life in the near future.

However, social implications of nanocomposites as an emerging technology are a serious cause for concern. Their large surface area, their crystalline structure, and their reactivity may facilitate their easy transport into the environment, allow them to interact with cell constituents, and elicit many harmful effects. Thus, utilization of nanocomposites and their release into the environment is a major health and safety issue. In this context, there is an increasing need for research into the nanotoxicological profiles of nanocomposites before their utilization and applications. The U.S. Environmental Protection Agency is evaluating the potential impacts of such nanomaterials on human health and the environment, not only to minimize their use but also to make a "gold standard" to avoid toxicity to human beings.

Owing to scientific and technological advances in nanocomposites, these are suitable materials to meet new demands. A lot of applications of nanomaterials are already known and new applications are expected to develop because of their unique properties. Thus, nanocomposites provide opportunities and rewards for creating new fields.

ACKNOWLEDGMENT

Authors are thankful to Indira Gandhi National Tribal University, Amarkantak, M.P., India for providing facilities to prepare this chapter.

REFERENCES

Afroze, J.D., M.J. Abden, M.S. Alam, N.M. Bahadur, M.A. Gafur. 2016. Development of functionalized carbon nanotube reinforced hydroxyapatite magnetic nanocomposites. *Mater. Lett.* 169: 24–27.

Ahmad M, Manzoor K, Singh S, Ikram S. 2017. Chitosan centered bionanocomposites for medical specialty and curative applications: A review. *Int. J. Pharm.* 529(1–2):200–217.

Ahmed F, Soliman FM, Adly MA, Soliman HAM, El-Matbouli M, Saleh M. 2019. Recent progress in biomedical applications of chitosan and its nanocomposites in aquaculture: A review. *Res. Vet. Sci.* 126:68–82.

Ain, N., W. Shaheen, B. Bashir, N.M. Abdelsalam, M. F. Warsi, M.A. Khan, and M. Shahid. 2016. Electrical, magnetic and photoelectrochemical activity of rGO/MgFe$_2$O$_4$ nanocomposites under visible light irradiation. *Ceram. Int.* 42(10):12401–12408.

Ajayan, P.M., L. Schadler, and P.V. Braun. 2003. *Nanocomposite science and technology.* Weinheim: Wiley-VCH, Verlag Gmbh & Co. KgaA.

Aladpoosh, R., and M. Montazer. 2016. Nano-photo active cellulosic fabric through in situ phytosynthesis of star-like Ag/ZnO nanocomposites: Investigation and optimization of attributes associated with photocatalytic activity. *Carbohydr. Polym.* 141:116–125.

Alexandre, M., and P. Dubois 2000. Polymer-layered silicate nanocomposites: Preparation, properties and uses of a new class of materials. *Mater. Sci. Eng.* 28:1–63.

Ali N, Zaman H, Bilal M, Shah AA, Nazir MS, Iqbal HMN. 2019. Environmental perspectives of interfacially active and magnetically recoverable composite materials – A review. *Sci. Total Environ.* 670:523–538.

Allafchian, A., H. Bahramian, S.A.H. Jalali, H. Ahmadvand. 2015. Synthesis, characterization and antibacterial effect of new magnetically core-shell nanocomposites. *J. Magn. Magn. Mater.* 394:318–324.

Andrews, R., and M.C. Weisenberger. 2004. Carbon nanotube polymer composites. *Curr. Opin. Solid State Mater. Sci.* 8(1): 31–37.

Arora, K., A. Chaubey, R. Singhal, R.P. Singh, M.K. Pandey, S.B. Samanta, B.D. Malhotra and S. Chand 2006. Application of electrochemically prepared polypyrrole-polyvinylsulphonate films to DNA Biosensor. *Biosens. Bioelectron.* 21(9): 1777–1783.

Arya, S.K., P.R. Solanki, R.P. Singh, M.K. Pandey, M. Datta, and B.D. Malhotra.2006 Application of octadecanethiol self assembled monolayer to cholesterol biosensor based on surface Plasmon Resonance technique. *Talanta* 69(4): 918–926.

Ashori, A. 2008. Wood-plastic composites as promising green-composites for automotive industries! *Bioresour. Biotechnol.* 99: 4661–4667.

Averous, L., N. Fauconnier, and L. Moro 2000. Fringant Blends of thermoplastic starch and polyesteramide: Processing and properties. *J. Appl. Polym. Sci.* 76: 1117–1128.

Azeredo, H.M.C., L.H.C. Mattoso, D. Wood, T.G. Williams, R.J. Avena-Bustillos, and T.H. Mchugh.2009. Nanocomposite edible films from mango puree reinforced with cellulose nanofibers. *J. Food Sci.* 74: 31–35.

Aziz, H.Y., and M.S. Gohari. 2016. Fe$_3$O$_4$/ZnO/Ag$_3$VO$_4$/AgI nanocomposites: Quaternary magnetic photocatalysts with excellent activity in degradation of water pollutants under visible light. *Sep. Purif. Technol.* 166: 63–72.

Barrios E, Fox D, Li Sip YY, Catarata R, Calderon JE, Azim N, Afrin S, Zhang Z, Zhai L.2019.Nanomaterials in advanced, high-performance aerogel composites: A review. *Polymers (Basel).* 11(4). pii: E726.

Batool, A., F. Kanwal, M. Imran, T. Jamil, and S.A. Siddiqi 2012. Synthesis of polypyrrole/zinc oxide composites and study of their structural, thermal and electrical properties. *Synth. Met.* 161(23–24): 2753–2758.

Bhat AH, Khan I, Usmani MA, Umapathi R, Al-Kindy SMZ.2019.Cellulose an ageless renewable green nanomaterial for medical applications: An overview of ionic liquids in extraction, separation and dissolution of cellulose. *Int. J. Biol. Macromol.* 129: 750–777.

Bhaumik, M., S. Agarwal, V.K. Gupta, A. Maity 2016. Enhanced removal of Cr(VI) from aqueous solutions using polypyrrole wrapped oxidized MWCNTs nanocomposites adsorbent. *J. Colloid Interf. Sci.* 470: 257–267.

Billmeyer, F.W.J. 1984. *Textbook of Polymer Science.* 3rd ed. New York: John Wiley & Sons. Inc.

Bolocan A, Mihaiescu DE, Meşterca AR, Spirescu VA, Tote EM, Mogoantă L,Mogoşanu GD, Grumezescu AM. 2015. In vitro and in vivo applications of 3D dendritic gold nanostructures. *Rom. J. Morphol. Embryol.* 56(3): 915–924.

Bondeson, D., and K. Oksman 2007. Dispersion and characteristics of surfactant modified cellulose whiskers nanocomposites. *Compos. Interface* 14, 617–630.

Breuer, O., and Sunderraj U. 2004. Big returns from small fibers: A review of polymer/carbon nanotube composites. *Polym. Compos.* 25(6): 630–645.

Cardoso VF, Francesko A, Ribeiro C, Banobre-Lopez M, MartinsP, Lanceros-Mendez S. 2018. Advances in magnetic nanoparticles for biomedical applications. *Adv. Healthc. Mater.* 7(5): 1700845.

Carvalho, A.J.F., A.A.S. Curvelo, J.A.M.A. Agnelli 2001. First insight on composites of thermoplastic starch and kaolin. *Carbohydr. Polym.* 45: 189–194.

Caseri W. (2003) Nanocomposites. In: Yang P. (ed.), *The Chemistry of Nanostructured Materials.* World Scientifi, Singapore, 359.

Castillo RR, Vallet-Regí M.2019. Functional mesoporous silica nanocomposites: Biomedical applications and biosafety. *Int. J. Mol. Sci.* 20(4): E929.

Chang A, Li HY, Chang IN, Chu YH.2018. Affinity ionic liquids for chemoselective gas sensing. *Molecules* 23(9): E2380.

Chen, M., W. Jiang, F. Wang, P. Shen, P. Ma, J. Gu, J. Mao, and F. Li 2013. Synthesis of highly hydrophobic floating magnetic polymer nanocomposites for the removal of oils from water surface. *Appl. Surf. Sci.* 286: 249–256.

Chen, X., L. Huankun, W. Yuxin, W. Hanshuo, W. Laidi, T. Pengfei, P. Jun, and X. Xiang 2016. Facile fabrication of novel porous graphitic carbon nitride/copper sulfide nanocomposites with enhanced visible light driven photocatalytic performance. *J. Colloid Interface Sci.* 476: 132–143.

Choa, Y.H., J.K. Yang, B.H. Kim, Y.K. Jeong, J.S. Lee, and T. Nakayama 2003. Preparation and characterization of metal: Ceramic nanoporous nanocomposite powders. *J. Magn. Magn. Mater.* 266(1–2): 12–19.

Choi AH, Cazalbou S, Ben-Nissan B. 2015. Nanobiomaterial coatings in dentistry. *Front. Oral Biol.*, 17:49–61.

Choi, S.M., and H. Awaji 2005. Nanocomposites: A new material design concept. *Sci. Technol. Adv. Mat.* 6(1): 2–10.

Czech, B., and B. Waldemar 2015. Photocatalytic treatment of pharmaceutical wastewater using new multiwall-carbon nanotubes/TiO$_2$/SiO$_2$ nanocomposites. *Environ. Res.* 137: 176–184.

Darabdhara G, Das MR, Singh SP, Rengan AK, Szunerits S,Boukherroub R.2019. Ag and Au nanoparticles/reduced graphene oxide composite materials: Synthesis and application in diagnostics and therapeutics. *Adv. Colloid Interface Sci.* 271: 101991.

De Oliveira Barud HG, da Silva RR, da Silva Barud H, Tercjak A, Gutierrez J, Lustri WR, de Oliveira OB Junior, Ribeiro SJL. 2016. A multipurpose natural and renewable polymer in medical applications: Bacterialcellulose. *Carbohydr. Polym.* 153:406–420.

Deepthi S, Venkatesan J, Kim SK, Bumgardner JD, Jayakumar R. 2016. An overview of chitin or chitosan/nano ceramic composite scaffolds for bonetissue engineering. *Int. J. Biol. Macromol.* 93(Pt B): 1338–1353.

Deng, L., L. Hui, G. Xiaoyi, S. Xing, and Z. Zhenfeng 2016. SnS$_2$/TiO$_2$ nanocomposites with enhanced visible light-driven photoreduction of aqueous Cr (VI). *Ceram. Int.* 42: 3808–3815.

Dresselhaus, M.S., G. Dresslhaus, and P. Avouris 2001. *Carbon nanotubes: synthesis, structure, properties and applications*. Berlin: Springer Verlag.

Duangjam, S., K. Wetchakun, S. Phanichphant, and N. Wetchakun 2016. Hydrothermal synthesis of novel $CoFe_2O_4/BiVO_4$ nanocomposites with enhanced visible-light-driven photocatalytic activities. *Mater. Lett.* 181: 86–91.

Duncan M. 2003. *Domestic water treatment in developing countries*. Duncan Mara. Cromwell Press, UK.

Dutta, D., S. Thiyagarajan, D. Bahadur 2016. SnO_2 quantum dots decorated reduced graphene oxide nanocomposites for efficient water remediation. *Chem. Eng. J.* 297: 55–65.

Dykman LA, Khlebtsov NG. 2016. Biomedical applications of multifunctional gold-based nanocomposites. *Biochemistry (Mosc.)* 81(13): 1771–1789.

Dziadek M, Stodolak-Zych E, Cholewa-Kowalska K. 2017. Biodegradable ceramic-polymer composites for biomedical applications: A review. *Korean J. Couns. Psychother.* 71: 1175–1191.

Emmanuel, I. U. and A. Taubert 2016. Clay-polymer nanocomposites (CPNs): Adsorbents of the future for water treatment. *Appl. Clay Sci.* 99: 83–92.

Fan HY, Yu XH, Wang K, Yin YJ, Tang YJ, Tang YL, Liang XH. 2019. Graphene quantum dots (GQDs)-based nanomaterials for improving photodynamic therapy in cancer treatment. *Eur. J. Med. Chem.* 182: 111620.

Fang, S., S. Xue, W. Can, W. Guanqiu, W. Xi, L. Qian, L. Zhongyu, and X. Song 2016. Fabrication and characterization of $CdS/BiVO_4$ nanocomposites with efficient visible light driven photocatalytic activities. *Ceram. Int.* 42: 4421–4428.

Fatemeh, A., S. Azam, and A.S. Masoud 2016. Green synthesis of magnetic chitosan nanocomposites by a new sol-gel auto-combustion method. *J. Magn. Magn. Mater.* 410: 27–33.

Filipe A.d. J., S.T.S. Santos, J.M. A. Caiut, V.H.V. Sarmento 2016. Effects of thermal treatment on the structure and luminescent properties of Eu^{3+} doped SiO^{2}-PMMA hybrid nanocomposites prepared by a sol–gel process. *J.Lumin..* 170: Part 2, 588–593.

Fischer, H. 2003. Polymer nanocomposites: From fundamental research to specific applications. *Mater. Sci. Eng. C* 23:6–8: 763–772.

Freag MS, Elzoghby AO. Protein-inorganic nanohybrids: A potential symbiosis in tissue engineering. *Curr. Drug Targets* 2018;19(16): 1897–1904.

Fu YS, Li J, Li J. 2019. Metal/semiconductor nanocomposites for photocatalysis: Fundamentals, structures, applications and properties. *Nanomaterials (Basel).* 9(3). pii: E359.

Fu, H., W. Yin, L. Xiaoya, C. Weifeng 2016. Synthesis of vegetable oil-based waterborne polyurethane/silver-halloysite antibacterial nanocomposites. *Comp. Sci.Technol.* 126: 86–93.

Gaaz TS, Sulong AB, Akhtar MN, Kadhum AA, Mohamad AB,Al-Amiery AA. 2015. Properties and applications of polyvinyl alcohol, halloysite nanotubes and their nanocomposites. *Molecules* 20(12): 22833–22847.

Gaaz TS, Sulong AB, Kadhum AAH, Al-Amiery AA, Nassir MH, Jaaz AH. 2017. The impact of halloysite on the thermo-mechanical properties of polymer composites. *Molecules.* 22(5). pii: E838.

Gangopadhyay, R., and D. Amitabha 2000. Conducting polymer nanocomposites: A brief overview. *Chem. Mater.* 127: 608–622.

Gangu KK, Maddila S, Jonnalagadda SB.2019. A review on novel composites of MWCNTs mediated semiconducting materials as photocatalysts in water treatment. *Sci. Total Environ.* 646: 1398–1412.

Ghanbari, M., D. Emadzadeh , W.J. Lau , T. Matsuura, M. Davoody, and A.F. Ismail 2015. Super hydrophilic TiO_2/HNT nanocomposites as a new approach for fabrication of high performance thin film nanocomposite membranes for FO application. *Desalin.* 371: 104–114.

Ghanbarzadeh, B., Oleyaei, S.A., Almasi, H. 2015. Nanostructured materials utilized in biopolymer-based plastics for food packaging applications. *Crit. Rev. Food Sci. Nutr.* 55(12):1699–1723.

Ghosh, A., P. Madhubonti, B. Krishna, C.G. Uday, and M. Biswaranjan 2015. Manganese oxide incorporated ferric oxide nanocomposites (MIFN): A novel adsorbent for effective removal of Cr (VI) from contaminated water. *J. Water Process. Eng.* 7: 176–186.

Giannelis, E.P. 1996. Polymer layered silicate nanocomposites. *Adv. Mater.* 8:1, 29–35.

Giliopoulos D, Zamboulis A, Giannakoudakis D, Bikiaris D, Triantafyllidis K. 2020. Polymer/metal organic framework (MOF) nanocomposites for biomedical applications. *Molecules* 25(1): E185.

Giusy, L.M.C., L. Giovanni, F.D. Rute, M. Arjen, D. Luciana, K.G. Ravindra, B. Daniela, R. Marco, K.S. Sanjay, C.C. Mahesh, G. Maurizio, and M. Sureyya 2016. Polymer functionalized nanocomposites for metals removal from water and wastewater: An overview. *Water Res.* 92: 22–37.

Gouzman I, Grossman E, Verker R, Atar N, Bolker A, Eliaz N. 2019. Advances in POLYIMIDE-BASED MATERIALS FOR SPACE APPLICATIONS. *Adv. Mater.*; 31(18): e1807738.

Govindhan M, Liu Z, Chen A. 2016. Design and electrochemical study of platinum-based nanomaterials for sensitive detection of nitric oxide in biomedical applications. *Nanomaterials (Basel).* 6(11). pii: E211.

Guan, J., and M.A. Hanna 2006. Selected morphological and functional properties of extruded acetylated starch-cellulose foams. *Bioresour. Technol.* 97:17161726.

Gui, M.M., W.M.P. Wong, S.P. Chai, and A.R. Mohamed 2015. One-pot synthesis of Ag-MWCNT@TiO$_2$ core–shell nanocomposites for photocatalytic reduction of CO$_2$ with water under visible light irradiation. *Chem. Eng. J.* 278: 272–278.

Hammouda, N., H. Chadli, G. Guillemot and K. Belmokre 2011. The corrosion protection behaviour of zinc rich epoxy paint in 3% NaCl solution. *Adv. Chem. Eng. Sci.* 1:2,51–60.

Han, X., S. Huang, Y. Wang, and D. Shi 2016. Design and development of anisotropic inorganic/polystyrene nanocomposites by surface modification of zinc oxide nanoparticles. *Mater. Sci. Eng. C*, 64: 87–92.

Harada, M., Ohya, T., K. Iida, H. Hayashi, K. Hirano, and H. Fukuda 2007. Increased impact strength of biodegradable poly (lactic acid)/poly (butylenes succinate) blend composites by using isocyanate as a reactive processing agent. *J. Appl. Polym. Sci.* 106: 1813–1820.

Hashemi B, Rezania S. 2019. Carbon-based sorbents and their nanocomposites for the enrichment of heavy metal ions: A review. *Mikrochim. Acta* 186(8):578.

Hihara, L.H., R.P.I. Adler, and R.M. Latanision 2013. editors. *Environmental degradation of advanced and traditional engineering materials*. Boca Raton: CRC Press; 719 p.

Ho, C.H., C.H. Chan, L.C. Tien, Y. S. Huang (2011). Direct optical observation of band-edge excitons, band gap, and fermi level in degenerate semiconducting oxide nanowires In$_2$O$_3$. *J. Phys. Chem. C*, 115, 25088–25096.

Hong, X., J. Wen, X. Xiong, and Y. Hu 2016. Silver nanowire-carbon fiber cloth nanocomposites synthesized by UV curing adhesive for electrochemical point-of-use water disinfection. *Chemosphere* 154: 537–545.

Hsieh S.H., W.J. Chen, T.H. Yeh 2015. Effect of various amounts of graphene oxide on the degradation characteristics of the ZnSe/graphene nanocomposites. *Appl. Surf. Sci.* 358: 63–69.

Iordanskii, A.L., A.V. Bychkova, K.Z. Gumargalieva, and A.A. Berlin 2016. Magnetoanisotropic biodegradable nanocomposites for controlled drug release. *Nanobiomater. Drug Deliv..* 171–196, Chapter 6.

Istratie, R., M. Stoia, C. Păcurariu, C. Locovei 2016. Single and simultaneous adsorption of methyl orange and phenol onto magnetic iron oxide/carbon nanocomposites. *Arab. J. Chem.*, 12,8, 3704–3722.

Itodo A.U., H.U. Itodo 2010. Quantitative specification of potentially toxic metals in expired canned tomatoes found in village markets. *Nat. Sci.* 8,4, 54–59.

Jafari, S., M.D. Nasrin, B. Jaleh, B.J. Mohammad, R. Maryam, A. Khosro 2016. Physicochemical characterization and in vivo evaluation of triamcinolone acetonide-loaded hydroxyapatite nanocomposites for treatment of rheumatoid arthritis. *Colloids Surf. B Biointerfaces* 140: 223–232.

Jalili NA, Muscarello M, Gaharwar AK. 2016. Nanoengineered thermoresponsive magnetic hydrogels for biomedical applications. *Bioeng Transl Med.* 1(3):297–305.

Jamroz E, Kulawik P, Kopel P. 2019. The effect of nanofillers on the functional properties of biopolymer-based films: A review. *Polymers (Basel).* 11(4). pii: E675.

John JV, Johnson RP, Heo MS, Moon BK, Byeon SJ, Kim I. 2015. Polymer-block-polypeptides and polymer-conjugated hybrid materials as stimuli-responsive nanocarriers for biomedical applications. *J. Biomed. Nanotechnol.* 11(1):1–39.

John, M.J., S. Thomas 2008. Biofibres and biocomposites. *Carbohydr. Polym.* 71: 343–364.

Jones M, Kujundzic M, John S, Bismarck A. 2020. Crab vs. Mushroom: A review of crustacean and fungal chitin in wound treatment. *Mar. Drugs.* 18(1). pii: E64.

Jung, K.W., T.U. Jeong, M.J. Hwang, K. Kim, K.H. Ahn 2015. Phosphate adsorption ability of biochar/Mg–Al assembled nanocomposites prepared by aluminum-electrode based electro-assisted modification method with $MgCl_2$ as electrolyte. *Bioresour. Technol.* 198: 603–610.

Kadam, A., R. Dhabbe, A. Gophane, T. Sathe, and K. Garadkar 2016. Template free synthesis of ZnO/Ag_2O nanocomposites as a highly efficient visible active photocatalyst for detoxification of methyl orange. *J. Photochem. Photobiol. B: Biol.*, 154: 24–33.

Kamal T, Ali N, Naseem AA, Khan SB, Asiri AM.2016. Polymer nanocomposite membranes for antifouling nanofiltration. *Recent Pat. Nanotechnol.* 10(3): 189–201.

Kamigaito, O. 1991. What can be improved by nanometer composites? *J. Jap. Soc. Powder Metall.* 38:315–321.

Ke, Z., and Y.P. Bai 2005. Improve the gas barrier property of pet film with montmorillonite by in situ interlayer polymerization. *Mater. Lett.* 59:27, 3348–3351.

Kefeni KK, Msagati TAM, Nkambule TT, Mamba BB(3). 2020. Spinel ferrite nanoparticles and nanocomposites for biomedical applications and their toxicity. *Korean J. Couns. Psychother.* 107:110314.

Kim, J.P., T.H. Yoon, S.P. Mun, J.M. Rhee and J.S. Lee 2006. Wood-polyethylene composites using ethylene-vinyl alcohol copolymer as adhesion promoter. *Bioresour. Biotechnol.* 97: 494–499.

Koli, V.B., G.D. Ananta, V.R. Abhinav, D.T. Nanasaheb, H.P. Shivaji, and D.D. Sagar 2016. Visible light photo-induced antibacterial activity of TiO_2-MWCNTs nanocomposites with varying the contents of MWCNTs. *J. Photochem. Photobiol. A Chem.* 328: 50–58.

Konwarh, R., Karak, N., Misra, M. 2013. Electrospun cellulose acetate nanofibers: The present status and gamut of biotechnological applications. *Biotechnol. Adv.* 31(4):421–437.

Lam, S.M., J.C. Sin, A. R. Mohamed 2016. A review on photocatalytic application of g-C_3N_4/semiconductor (CNS) nanocomposites towards the erasure of dyeing wastewater. *Mater. Sci. Semicond. Process* 47: 62–84.

Lee, J. Y., K. Liang, K.H. An, Y. Lee (2005). Nickel oxide/carbon nanotubes nanocomposite for electrochemical capacitance. *Synth. Met.*, 150–153.

Lee, S., I. Kang, D. Doh, H. Yoon, B. Park, Wu, Q 2008a. Thermal and mechanical properties of wood flour/talc-filled polylactic acid composites: Effect of filler content and coupling treatment. *J. Thermoplast. Compos. Mater.* 21, 209– 223.

Lee, S., I. Kang, G. Doh, H. Yoon, B. Park, and Q. Wu 2008b. Thermal and mechanical properties of wood flour/talc-filled polylactic acid composites: Effect of filler content and coupling treatment. *J. Thermoplast. Compos. Mater.* 21: 209223.

Lehr, I.L., and S.B. Saidman 2013. Anticorrosive properties of polypyrrole films modified with Zinc onto SAE 4140 steel. *Prog. Org. Coat.* 76: 11, 1586–1593.

Leja, K. and G. Lewandowicz 2010. Polymer biodegradation and biodegradable polymers-a review. *Pol J. Environ. Stud.* 19: 255–266.

Li F, Liang Z, Ling D. 2019. Smart organic-inorganic nanogels for activatable theranostics. *Curr. Med. Chem.* 26(8):1366–1376.

Li, S., Y. Gong, Y. Yang, C. He, L. Hu, L. Zhu, L. Sun, and D. Shu 2015. Recyclable CNTs/Fe_3O_4 magnetic nanocomposites as adsorbents to remove bisphenol A from water and their regeneration. *Chem. Eng. J.* 260: 231–239.

Liao C, Li Y, Tjong SC. 2020. Visible-light active titanium dioxide nanomaterials with bactericidal properties. *Nanomaterials (Basel).* 10(1): 124.

Liao G, Fang J , Li Q , Li S , Xu Z , Fang B. 2019. Ag-Based nanocomposites: Synthesis and applications in catalysis. *Nanoscale* 11(15):7062–7096.

Lim, D.S., J.W. An, and H.J. Lee 2002. Effect of carbon nanotube addition on the tribological behavior of carbon/carbon composites. *Wear* 252: 512–517.

Lingamdinne LP, Koduru JR, Karri RR. 2019. A comprehensive review of applications of magnetic graphene oxide based nanocomposites for sustainable water purification. *J. Environ. Manage.* 231:622–634.

Maeda T. 2019. Structures and applications of thermoresponsive hydrogels and nanocomposite-hydrogels based on copolymers with poly (ethylene glycol) and poly (lactide-co-glycolide) blocks. *Bioengineering (Basel).* 6(4). pii: E107.

Magesan, P.G. and M.J. Umapathy 2016. Ultrasonic-assisted synthesis of doped TiO_2 nanocomposites: Characterization and evaluation of photocatalytic and antimicrobial activity. *Opt. Internat. J. Light Elect. Opt.* 127: 5171–5180.

Mahadeva, S.K., S. Yun, and J. Kim 2011. Flexible humidity and temperature sensor based on cellulose-polypyrrole nanocomposite. *Sensor. Actuator. A Phys.* 165: 194199.

Majdzadeh-Ardakani, K., and S. Sadeghi-Ardakani 2010. Experimental investigation of mechanical properties of Starch/natural rubber/clay nanocomposites. *Dig. J. Nanomater. Biostruct.* 5: 307–316.

Mallakpour, S., A. Abdolmaleki, Z. khalesi, and S. Borandeh 2016. Surface functionalization of GO, preparation and characterization of PVA/TRIS-GO nanocomposites. *Polymer* 81: 140–150.

Mallakpour, S., and E. Khadem 2016. Carbon nanotube–metal oxide nanocomposites: Fabrication, properties and applications. *Chem. Eng. J.* 302: 344–367.

Mehta A, Mishra A, Basu S, Shetti NP, Reddy KR, Saleh TA, Aminabhavi TM. 2019. Band gap tuning and surface modification of carbon dots for sustainable environmental remediation and photocatalytic hydrogen production-A review. *J. Environ. Manage.* 250:109486.

Mohammed, M.R., J. Ahmed, A. M. Asiri, I.A. Siddiquey, and M.A. Hasnat 2016. Development of 4-methoxyphenol chemical sensor based on NiS_2-CNT nanocomposites. *J. Taiwan Inst. Chem. Eng.* 64: 157–165.

Motealleh, A, Kehr, NS. 2017. Nanocomposite hydrogels and their applications in tissue engineering. *Adv. Healthc. Mater.* 6(1): 1600938.

Nasrollahzadeh, M., S. M. Sajadi, and H. Arezo 2016. Waste chicken eggshell as a natural valuable resource and environmentally benign support for biosynthesis of catalytically active Cu/eggshell, Fe_3O_4/eggshell and Cu/Fe_3O_4/eggshell nanocomposites. *Appl. Catal. Environ.* 191: 209–227.

Nguyen, N. T., D.L. Tran, D.C. Nguyen, T. L. Nguyen, T.C. Ba, B.H. Nguyen, T.D. Ba, N. H. Pham, D.T. Nguyen, T. H. Tran, G.D. Pham 2015. Facile synthesis of multifunctional Ag/Fe_3O_4-CS nanocomposites for antibacterial and hyperthermic applications. *Curr. Appl. Phys.* 15: 1482–1487.

Ochi, S. 2008. Mechanical properties of Kenaf fibers and Kenaf/PLA composites. *Mech. Mater.* 40: 446–452.

Ogata, N., G. Jimenez, H. Kawai, and T. Ogihara 1997. Structure and thermal/mechanical properties of poly(L-lactide)-clay blend. *J Polym Sci B* 35: 389–396.

Oishi M, Nagasaki Y. 2010. Stimuli-responsive smart nanogels for cancer diagnostics and therapy. *Nanomedicine (Lond.)* 5(3):451–468.

Pandey, J.K., K.R. Reddy, A.P. Kumar, and R.P. Singh 2005. An overview on the degradability of polymer nanocomposites. *Polym. Degrad. Stab.* 88:2, 234–250.

Pandey, J.K., R.P. Singh 2005. Green nanocomposites from renewable resources: Effect of plasticizer on the structure and material properties of clay-filled starch. *Starch/Starke* 57: 8–15.

Pandey, J.K., W.S. Chu, C.S. Lee, S.H. Ahn 2007. Preparation characterization and performance evaluation of nanocomposites from natural fiber reinforced biodegradable polymer matrix for automotive applications. Presented at *the International Symposium on Polymers and the Environment: Emerging Technology and Science, Bioenvironmental Polymer Society (BEPS),* Vancouver, WA, USA.

Pant, B., S.S. Prem, P. Mira, J.P. Soo, and Y.K. Hak. 2016. General one-pot strategy to prepare Ag-TiO$_2$ decorated reduced graphene oxide nanocomposites for chemical and biological disinfectant. *J. Alloys Compd.* 671: 51–59.

Peigney, A., C.H. Laurent, E. Flahaut, and A. Rousset 2000. Carbon nanotubes in novel ceramic matrix nanocomposites. *Ceram. Int.*; 26(6):677–683.

Peng, F., W. Qiang, Z. Ming, and S. Bitao 2016a. Preparation and adsorption properties of enhanced magnetic zinc ferrite-reduced graphene oxide nanocomposites via a facile one-pot solvothermal method. *J. Alloys Compd.* 685: 411–417.

Peng, G., L. Meiying, T. Jianwen, D. Fengjie, W. Ke, X. Dazhuang, L. Liangji, Z. Xiaoyong, and W. Yen 2016b. Improving the drug delivery characteristics of graphene oxide based polymer nanocomposites through the "one-pot" synthetic approach of single-electron-transfer living radical polymerization. *Appl. Surf. Sci.* 378: 22–29.

Pradhan, G.C., S. Dash, and S.K. Swain 2015. Barrier properties of nano silicon carbide designed chitosan nanocomposites. *Carbohydr. Polym.* 134: 60–65.

Presting, H., and U. Konig 2003. Future nanotechnology developments for automotive applications. *Mater. Sci. Eng. C* 23:6–8,737–741.

Qi C, Musetti S , Fu LH , Zhu YJ , Huang L .2019. Biomolecule-assisted green synthesis of nanostructured calcium phosphates and their biomedical applications. *Chem. Soc. Rev.* 48(10):2698–2737.

Qutub, N., B.M. Pirzada, K. Umar, O. Mehraj, M. Muneer, S. Sabir 2015. Synthesis, characterization and visible-light driven photocatalysis by differently structured CdS/ZnS sandwich and core-shell nanocomposites. *Physica E* 74: 74–86.

Rabiee N, Yaraki MT, Garakani SM, Garakani SM, Ahmadi S, Lajevardi A, Bagherzadeh M, Rabiee M, Tayebi L, Tahriri M, Hamblin MR. 2020. Recent advances in porphyrin-based nanocomposites for effective targeted imaging and therapy. *Biomaterials* 232:119707.

Rabieh, S., K. Nassimi, and M. Bagheri 2016. Synthesis of hierarchical ZnO-reduced graphene oxide nanocomposites with enhanced adsorption–photocatalytic performance. *Mater. Lett.* 162: 28–31.

Rahimi M, Noruzi EB, Sheykhsaran E, Ebadi B, Kariminezhad Z, Molaparast M, Mehrabani MG, Mehramouz B, Yousefi M, Ahmadi R, Yousefi B, Ganbarov K, Kamounah FS, Shafiei-Irannejad V, Kafil HS. 2020. Carbohydrate polymer-based silver nanocomposites: Recent progress in the antimicrobial wound dressings. *Carbohydr. Polym.* 231:115696.

Rahn H., W. Robert, H. Michael, E. Diana, F. Kirk, D. Silvio, O. Stefan, S.P. Tim 2016. Calibration standard of body tissue with magnetic nanocomposites for MRI and X-ray imaging. *J. Magnet. Magnetic Mat.* 405: 78–87.

Raj, V., and G. Prabha 2015. Synthesis, characterization and in vitro drug release of cisplatin loaded Cassava starch acetate–PEG/gelatin nanocomposites. *J. Assoc. Arab Univ. Basic Appl. Sci.* 21,1: 10–16.

Raza A, Hayat U, Rasheed T, Bilal M, Iqbal HMN.2018.Redox-responsive nano-carriers as tumor-targeted drug delivery systems. *Eur. J. Med. Chem.* 157:705–715.

Rehim, M.H. A., M.A. El-Samahy, A.A. Badawy, and M.E. Mohram 2016. Photocatalytic activity and antimicrobial properties of paper sheets modified with TiO2/Sodium alginate nanocomposites. *Carbohydr. Polym.* 148: 194–199.

Renu S, Shivashangari KS, Ravikumar V. 2020. Incorporated plant extract fabricated silver/poly-D,l-lactide-co-glycolide nanocomposites for antimicrobial based wound healing. *Spectrochim. Acta A Mol. Biomol. Spectrosc.* 228:117673.

Riccardis, M.F., and V. Martina 2014. Hybrid conducting nanocomposites coatings for corrosion protection. In: Aliofkhazraei M, editor. *Developments in corrosion protection. Rijeka*: InTech; p. 271–317.

Sabherwal, P., R. Mutreja, C.R. Suri 2016. Biofunctionalized carbon nanocomposites: New-generation diagnostic tools. *TrAC Trends Anal. Chem.* 82: 12–21.

Sadeghi R, Rodriguez RJ, Yao Y, Kokini JL. 2017. Advances in nanotechnology as they pertain to food and agriculture: Benefits and risks. *Annu. Rev. Food Sci. Technol.* 8:467–492.

Sana, M.R.W., S.D. Atef, and M.K. Sara 2016. Ceria-containing uncoated and coated hydroxyapatite-based galantamine nanocomposites for formidable treatment of Alzheimer's disease in ovariectomized albino-rat model. *Mater. Sci. Eng. C* 65: 151–163.

Sarah, C.M., S.S. Ray, M.S. Onyango, and M.N.B. Momba 2015. Preparation and antibacterial activity of chitosan-based nanocomposites containing bentonite-supported silver and zinc oxide nanoparticles for water disinfection. *Appl. Clay Sci.* 114: 330–339.

Seyedeh, M. G., and A.S. Seyed 2016. Evaluation of the antibacterial activity of Ag/Fe_3O_4 nanocomposites synthesized using starch. *Carbohydr. Polym.* 144: 454–463.

Sharma P, Jang NY, Lee JW, Park BC, Kim YK, Cho NH. 2019. Application of ZnO-Based Nanocomposites for Vaccines and Cancer Immunotherapy. *Pharmaceutics.* 11(10). pii: E493.

Shin EJ, Choi SM.2018. Advances in waterborne polyurethane-based biomaterials for biomedical applications. *Adv. Exp. Med. Biol.* 1077:251–283.

Shojaeiarani J, Bajwa D, Shirzadifar A.2019. A review on cellulose nanocrystals as promising biocompounds for the synthesis of nanocomposite hydrogels. *Carbohydr. Polym.* 216:247–259.

Shukla, V.K., R.P. Singh, and A. C. Pandey 2010. Black pepper assisted biomimetic synthesis of silver nanoparticles. *J. Alloys Compd.* 507:1 , L13–L16.

Simon J, Flahaut E, Golzio M.2019. Overview of carbon nanotubes for biomedical applications. *Materials (Basel).* 12(4). pii: E624.

Singh, R. P., J.W. Choi, A. Tiwari, and A.C. Pandey 2012a. Biomimetic materials toward application of nanobiodevices, In *Intelligent Nanomaterials: Processes, Properties, and Applications* (eds A. Tiwari, A. K. Mishra, H. Kobayashi and A. P. Turner), John Wiley & Sons, Inc., Hoboken, NJ, USA. chapter 20, pp.741–782, Published Online: 21 FEB 2012.

Singh R.P. 2019. Nanocomposites: Recent trends, developments and Applications. *Advances in nanostructure Nanocomposites.* CRC Press: Taylor and Francis. Editor; Mahmood Aliofkhazraei. (eds Radheshyam Rai), 575 Pages ISBN 9781482236637, Chapter 2.

Singh, R. P., J.W. Choi, A. Tiwari, and A.C. Pandey 2012b. Utility and potential application of nanomaterials in medicine, In *Biomedical Materials and Diagnostic Devices* (eds A. Tiwari, M. Ramalingam, H. Kobayashi and A. P.F. Turner), John Wiley & Sons, Inc., Hoboken, NJ, USA. doi: 10.1002/9781118523025.ch7.

Singh, R. P., J.W. Choi, A. Tiwari, and A.C. Pandey 2014. Functional nanomaterials for multifarious nanomedicine, In *Biosensors Nanotechnology* (eds A. Tiwari and A. P.F. Turner), John Wiley & Sons, Inc., Hoboken, NJ, USA.

Singh, R.P. 2011. Prospects of nanobiomaterials for biosensing. *Internat. J. Electrochem.* Publisher SAGE-Hindawi journal collection. 2011, Vol. 2011, Review article ID 125487, 30 pages, doi:10.4061/2011/125487.

Singh, R.P. 2016. Nanobiosensors: Potentiality towards bioanalysis. *J Bioanal Biomed* 8: e143. doi:10.4172/1948-593X.1000e143.

Singh, R.P., and A.C. Pandey 2011. Silver nano-sieve using 1, 2-benzenedicarboxylic acid as a sensor for detecting hydrogen peroxide. *Anal.l Methods, RSC*. 3: 586–592.

Singh, R.P., and J.W. Choi 2010. Bio-nanomaterials for versatile bio-molecules detection technology. Letter to Editors. *Adv. Mater. Lett*. 1(1): 83–84.

Singh, R.P., A. Tiwari, A.C. Pandey 2011a. Silver/Polyaniline nanocomposite for the electrocatalytic hydrazine oxidation. *J. Inorg. Organomet. Polym. Mat*. 21: 788–792.

Singh, R.P., and J.W. Choi 2009. Biosensors development Based on Potential Target of Conducting Polymers. *Sen.Transd. J*. 104(5): 1–18.

Singh, R.P., B.K. Oh, and J.W. Choi. 2010a. Application of peptide nucleic acid towards development of Nanobiosensor arrays. *Bioelectrochemistry* 79(2): 153–161.

Singh, R.P., B.K. Oh, K.K. Koo, J.Y. Jyoung, S. Jeong, and J.W. Choi 2008. Biosensor arrays for environmental pollutants detection. *Biochem. J*. 2(4): 223–234.

Singh, R.P., D.Y. Kang, and J.W. Choi 2010b. Electrochemical DNA biosensor for the detection of sanguinarine in adulterated mustard oil. *Adv. Mater. Lett*. 1(1): 48–54.

Singh, R.P., D.Y. Kang, and J.W. Choi 2011b. Nanofabrication of bio-self assembled monolayer and its electrochemical property for toxicant detection. *J. Nanosci. Nanotechnol*. 11: 408–412.

Singh, R.P., D.Y. Kang, B.K. Oh, and J.W. Choi 2009a. Polyaniline based catalase biosensor for the detection of Hydrogen peroxide and Azide. *Biotechnol. Bioprocess Eng*. 14: 4, 443–449.

Singh, R.P., J.W. Choi, and A.C. Pandey 2012c. Smart nanomaterials for biosensors, biochips and molecular bioelectronics. Editor(s) S. Li, Y. Ge, H. Li, Bentham Science Publisher (USA), In *Smart Nanomaterials for Sensor Application* Editors: Songjun Li, Jiangsu University, China, Yi Ge, Cranfield University, UK, He Li, University of Jinan, China, Chapter 1, pp. 3–41.

Singh, R.P., K. Kumar, R. Rai, A. Tiwari, J.W. Choi, and A.C. Pandey 2012d. Synthesis, characterization of Metal oxide based nanomaterials and its application in Biosensing, to the upcoming book "*Synthesis, characterization and application of Smart material*". Nova Science Publishers, Inc USA. Chapter 11, pp.225–238.

Singh, R.P., V.K. Shukla, R.S. Yadav, P.K. Sharma, P.K. Singh, and A.C. Pandey 2011c. Biological approach of zinc oxide nanoparticles formation and its characterization. *Adv. Mater. Lett*. 2(4): 313–317.

Singh, R.P., Y.J. Kim, B.K. Oh, and J.W. Choi 2009b. Glutathione-s-transferase based electrochemical biosensor for the detection of captan. *Electrochem. Commun*. 11: 181–185.

Siracusa, V., P. Rocculi, S. Romani, M.D. Rosa 2008. Biodegradable polymers for food packaging: A review. *Trends Food Sci. Technol*. 19: 634–643.

Sithique, M.A., and M. Alagar 2010. Preparation and Properties of Bio-Based Nanocomposites from Epoxidized Soy Bean Oil and Layered Silicate. Malaysian *Polym. J*. 5: 151–161.

Song, Y., C. Wei, J. He, X. Li, X. Lu, L. Wang 2015a. Porous Co nanobeads/rGO nanocomposites derived from rGO/Co–metal organic frameworks for glucose sensing. *Sens. Actuators B* 220: 1056–1063.

Song, Y., T. Qiang, M. Ye, Q. Ma, and Z. Fang 2015b. Metal organic framework derived magnetically separable 3-dimensional hierarchical Ni@C nanocomposites: Synthesis and adsorption properties. *Appl. Surf. Sci*. 359: 834–840.

Sternitzke, M. 1997. Review: Structural ceramic nanocomposites. *J. Eur. Ceram. Soc*. 17(9): 1061–1082.

Swearingen, C., S. Macha, and A. Fitch 2003. Leashed ferrocenes at clay surfaces: Potential applications for environmental catalysis. *J. Mol. Catal. A - Chem*. 199(1–2): 149–160.

Tan A, Farhatnia Y, Seifalian AM. 2013. Polyhedral oligomeric silsesquioxane poly(carbonate-urea) urethane (POSS-PCU):Applications in nanotechnology and regenerative medicine. *Crit. Rev. Biomed. Eng*. 41(6): 495–513.

Tang S.Z. P., H. Xiong, H. Tang 2008. Effect of nano-SiO2 on the performance of starch/polyvinyl alcohol blend films. *Carbohydr. Polym.* 72: 521–526.

Tang YH, Huang AY, Chen PY, Chen HT, Kao CL. 2011. Metallodendrimers and dendrimer nanocomposites. *Curr. Pharm. Des.*17(22): 2308–2330.

Taniselass S, Arshad MKM, Gopinath SCB. 2019. Graphene-based electrochemical biosensors for monitoring non-communicable disease biomarkers. *Biosens. Bioelectron.* 130: 276–292.

Tate, J.S., A.T. Akinola, and D. Kabakov 2010. Bio-based nanocomposites: An alternative to traditional composites. *J. Technol. Stud.* 1: 25–32.

Teixeira, E., D. Pasquini, A.S. Antonio, C.E. Corradini, M.N. Belgacem, and A. Dufresne 2009. Cassava baggasse cellulose nanofibrils reinforced thermoplastic cassava starch. *Carbohydr. Polym.* 78: 422–431.

Thiruvengadam M, Rajakumar G, Chung IM. 2018. Nanotechnology: Current uses and future applications in the food industry. *Australas. Biotechnol.* 8(1): 74.

Tiwari, A., A. Tiwari, and R.P. Singh 2012a. Bionanocomposite Matrices in Electrochemical Biosensors, In *Biomedical Materials and Diagnostic Devices* (eds A. Tiwari, M. Ramalingam, H. Kobayashi and A. P.F. Turner), John Wiley & Sons, Inc., Hoboken, NJ. doi: 10.1002/9781118523025.ch10.

Tiwari, A., D. Terada, H. Kobayashi, R.P. Singh, and R. Rai 2012b. Bionanomaterials for emerging biosensors technology, In *Synthesis, characterization and application of Smart materials*, Ed. Radheshyam Rai, Nova Publishers, Hauppauge, New York, Chapter 7, pp.137–154.

Tiwari, A., R. P. Singh, and R. Rai 2012c. Vinyls modified guar gum biodegradable plastics, In *Synthesis, characterization and application of Smart materials*, Ed. Radheshyam Rai, Nova Publishers, Hauppauge, New York. Chapter 6, pp.125–136.

Trache D, Hussin MH, Haafiz MK, Thakur VK. 2017. Recent progress in cellulose nanocrystals: Sources and production. *Nanoscale* 9(5): 1763–1786.

Tunç, S., and O. Duman 2011. Preparation of active antimicrobial methyl cellulose/carvacrol/montmorillonite nanocomposite films and investigation of carvacrol release. *Food Sci. Technol.* 44: 465–472.

Ullah H, Wahid F, Santos HA, Khan T. 2016. Advances in biomedical and pharmaceutical applications of functional bacterial cellulose-based nanocomposites. *Carbohydr. Polym.* 150: 330–352.

Ullah, K., A. Ullah, A. Aldalbahi, J. Chung, W.C. Oh 2015. Enhanced visible light photocatalytic activity and hydrogen evolution through novel heterostructure AgI–FG–TiO$_2$ nanocomposites. *J. Mol. Catal. A Chem.* 410: 242–252.

Valença, D.P., K.G.B. Alves, C.P. Melo, and N. Bouchonneau 2015. Study of the efficiency of Polypyrrole/ZnO nanocomposites as additives in anticorrosion coatings. *Mater. Res.* 18: 2–10.

Vashist A, Kaushik A, Vashist A, Sagar V, Ghosal A, Gupta YK, Ahmad S, Nair M. 2018. Advances in carbon nanotubes-hydrogel hybrids in nanomedicine for therapeutics. *Adv. Healthc. Mater.* 7(9): e1701213.

Vellayappan MV, Balaji A, Subramanian AP, John AA, Jaganathan SK, Murugesan S, Supriyanto E, Yusof M. 2015. Multifaceted prospects of nanocomposites for cardiovascular grafts and stents. *Int. J. Nanomedicine* 10: 2785–2803.

Voevodin, A.A., and J.S. Zabinski 2005. Nanocomposite and nanostructured tribological materials for space applications. *Compos. Sci. Technol.* 65: 741–748.

Voevodin, A.A., J.S. Zabinski 2000. *Thin Solid Films* 37: 223.

Wang G., M. Yingying, W. Zhiyong, Q. Min 2016a. Development of multifunctional cobalt ferrite/graphene oxide nanocomposites for magnetic resonance imaging and controlled drug delivery. *Chem. Eng. J.* 289: 150–160.

Wang M, Wang D, Chen Q, Li C, Li Z, Lin J. 2019. Recent advances in glucose-oxidase-based nanocomposites for tumor therapy. *Small* 15(51): e1903895.

Wang, G., Y. Ma, L. Zhang, J. Mu, Z. Zhang, X. Zhang, H. Che, Y. Bai, and J. Hou 2016b. Facile synthesis of manganese ferrite/graphene oxide nanocomposites for controlled targeted drug delivery. *J. Magn. Magn. Mater.* 401: 647–650.

Wang, G., Y. Hua, X. Su, S. Komarneni, S. Ma, Y. Wang 2016c. Cr(VI) adsorption by montmorillonite nanocomposites. *Appl.Clay Sc.Appl. Clay Sci.* 124–125: 111–118.

Wang, Y.H., F.L. Li, Y.Q. Wang, S. Wu, X.X. He, K.M. Wang 2015. A TiO₂/CNTs nanocomposites enhanced luminol electrochemiluminescence assay for glucose detection. *Chin. J. Anal. Chem.* 43: 1682–1687.

Wei, X.X., H. Cui, S. Guo, L. Zhao, and W. Li 2013. Hybrid BiOBr-TiO₂ nanocomposites with high visible light photocatalytic activity for water treatment. *J. Hazard. Mater.* 263: 650–658.

Willett, J.L., and R.L. Shogren 2002. Processing and properties of extruded starch/polymer foams. *Polymer* 43: 5935–5947.

Wu X, Kwon SJ, Kim J, Kane RS, Dordick JS. 2017. Biocatalytic nanocomposites for combating bacterial pathogens. *Annu Rev Chem Biomol Eng.* 8: 87–113.

Wu, Q., Y. Xu, Z. Yao, A. Liu (2010). Supercapacitors based on flexible graphene/polyaniline nanofiber composite films. *ACS Nano* 4(4): 1963–1970.

Wu, W.C.P.S., H. Ming, I. Lartaud, P. Maincent, X.M. Hu, A.S. Minet, and C. Gaucher 2016. Polymer nanocomposites enhance S-nitrosoglutathione intestinal absorption and promote the formation of releasable nitric oxide stores in rat aorta. *Nanomed. Nanotechnol. Biol. Med.* 12(7): 1795–1803.

Wypych F. Chemical modification of clay surfaces. In: Wypych F, Satyanarayana KG, 2004. *Clay surfaces: fundamentals, applications.* Amsterdam: Academic Press.

Xie P., S. Xue, J. Wei, J. Han, W. Zhou, and R. Zou 2016. Morphology-controlled synthesis of grass-like GO-CdSe nanocomposites with excellent optical properties and field emission properties. *J. Solid State Chem.* 234: 63–71.

Xiong H. T.S., H. Tang, P. Zou, 2008. The structure and properties of a starch-based biodegradable film. *Carbohydr. Polym.* 2008: 263–268.

Xu, C., L. Rui, C. Lvjun, and T. Jialu 2016a. Enhanced dechlorination of 2,4-dichlorophenol by recoverable Ni/Fe-Fe₃O₄ nanocomposites. *J. Environ. Sci.* 48: 92–101.

Xu, J. L., H.Y. Wang, C. Zhang, and Y. Zhang 2016b. Preparation of bi-component ZnO/ZnCo₂O₄ nanocomposites with improved electrochemical performance as anode materials for lithium-ion batteries. *Electrochim. Acta* 191: 417–425.

Yadav V, Roy S, Singh P, Khan Z, Jaiswal A.2019. 2D MoS₂ - based nanomaterials for therapeutic, bioimaging, and biosensing applications. *Small.* 15(1): e1803706.

Yadav, R. S., R.P. Singh, P. Verma, A. Tiwari, and A.C. Pandey 2012. Smart Nanomaterials for Space and Energy Applications, In *Intelligent Nanomaterials: Processes, Properties, and Applications* (eds A. Tiwari, A. K. Mishra, H. Kobayashi and A. P. Turner), John Wiley & Sons, Inc., Hoboken, NJ. Chapter 6, pp. 213–250.

Yan L, Gonca S , Zhu G , Zhang W , Chen X .2019.Layered double hydroxide nanostructures and nanocomposites for biomedical applications. *J. Mater. Chem. B* 7(37): 5583–5601.

Yang G, Lu Y, Bomba HN, Gu Z. 2019. Cysteine-rich proteins for drug delivery and diagnosis. *Curr. Med. Chem.* 26(8): 1377–1388.

Yang J, Yang YW. 2020. Metal-organic frameworks for biomedical applications. *Small.* 16(10): e1906846.

Yang, C., G.Y. Deng, P. Deng, K.L. Xin, W. Xia, A.S.W. Bligh, and R.W. Gareth 2016a. Electrospun pH-sensitive core-shell polymer nanocomposites fabricated using a tri-axial process. *Acta Biomater.* 35: 77–86.

Yang, D. (2011). Pulsed laser deposition of manganese oxide thin films for supercapacitor Applications. *J. Power Sources* 196: 8843.

Yang, L., D. Chu, L. Wang, X. Wu, and J. Luo 2016b. Synthesis and photocatalytic activity of chrysanthemum-like Cu₂O/Carbon Nanotubes nanocomposites. *Ceram. Int.* 42(2 Part A): 2502–2509.

Zabihzadeh, S.M. 2010. Water uptake and flexural properties of natural Filler/HDPE composites. *BioResource* 5: 316–323.

Zango, Z.U., N. G. Zaharaddeen, N.H.H.A. Bakar, W.L. Tan, M. A. Bakar 2016. Adsorption studies of Cu^{2+}–Hal nanocomposites for the removal of 2,4,6-trichlorophenol. *Appl. Clay Sci.* 2016: 8–17.

Zare Y, Shabani I. 2016. Polymer/metal nanocomposites for biomedical applications. *Korean J. Couns. Psychother.* 60: 195–203.

Zhang X, Wang F, Sheng JL, Sun MX. 2019. Advances and application of DNA-functionalized Nanoparticles. *Curr. Med. Chem.* 26(40): 7147–7165.

Zhang, Y., A. Tang, H. Yang, and J. Ouyang 2016. Applications and interfaces of halloysite nanocomposites. *Appl. Clay Sci.* 119: 8–17.

Zhong J, Kankala RK, Wang SB, Chen AZ. 2019. Recent advances in polymeric nanocomposites of Metal-Organic Frameworks (MOFs). *Polymers (Basel).* 11(10): E1627.

Zhou H, Zou F, Koh K, Lee J. 2014. Multifunctional magnetoplasmonic nanomaterials and their biomedical applications. *J. Biomed. Nanotechnol.* 10(10): 2921–2949.

Zou W, Gao B, Ok YS, Dong L. 2019. Integrated adsorption and photocatalytic degradation of volatile organic compounds (VOCs) using carbon-based nanocomposites: A critical review. *Chemosphere* 218: 845–859.

2 Biocomposites and Nanocomposites

C. H. Lee, S. H. Lee, F. N. M. Padzil,
Z. M. A. Ainun, M. N. F. Norrrahim and K. L. Chin

CONTENTS

2.1   INTRODUCTION

The continuing record-breaking global temperatures (19 of the 20 warmest years on record occurred after 2001; a 0.98°C increment was observed in 2019) (Shaftel, 2020) have raised the awareness of the effects of using petroleum-based plastic products, including synthetic polymer composites. According to a report from Lisa and Steven (2019), a total of 850 million metric tons of greenhouse gases were released into the atmosphere by plastic production and incineration processes in 2019, equivalent to 189 times the emission rate of a 500 MW coal plant (Lisa and Steven 2019). This scenario is projected to increase to 2,800 million metric tons of greenhouse gases produced in 2050. Light, cheap, but high-strength products (plastics and synthetic polymer composites) have caused an increase in various applications. Ineffective plastic waste management means that only 9% and 12% of the 8.3 billion metric tons of plastic accumulated since 1950 have been recycled and incinerated, respectively (Laura 2018). The remaining plastic waste has gathered in landfills or accumulated in the natural environment.

Long-lasting petroleum-based products create multiple issues to the environment and living organisms. Numerous marine wildlife fatalities due to swallowed plastic have been reported. Microplastics have impacted every ecosystem in the world, from the deepest seabed to the most remote wilderness. A human consumes an estimated 74,000 to 121,000 pieces of microplastic annually, which may possibly cause severe illness (Carly 2019).

To reduce the global burden from these products, natural fibers have been introduced. These are readily available, low-cost materials with properties comparable to those of synthetic fibers: high strength-to-weight ratio, relative low consumption of manufacturing energy, longer tool life, CO_2 neutrality. Most importantly, introducing natural fibers promotes some biodegradability properties. One study has found that a biocomposite specimen's weight (contributed by biodegradable natural fibers) reduced after 50 days buried in the soil, because of holes forming on the surface (Vu et al., 2018). Natural fiber can also be inserted into biopolymer to produce a fully biodegradable polymer composite.

Natural fibers can be categorized into three main groups: plant, animal, and mineral fibers, as shown in Figure 2.1. In this chapter, only natural plant fibers are discussed. Animal and mineral natural fibers have been reviewed in detail in previous studies (Bhattacharyya et al., 2015).

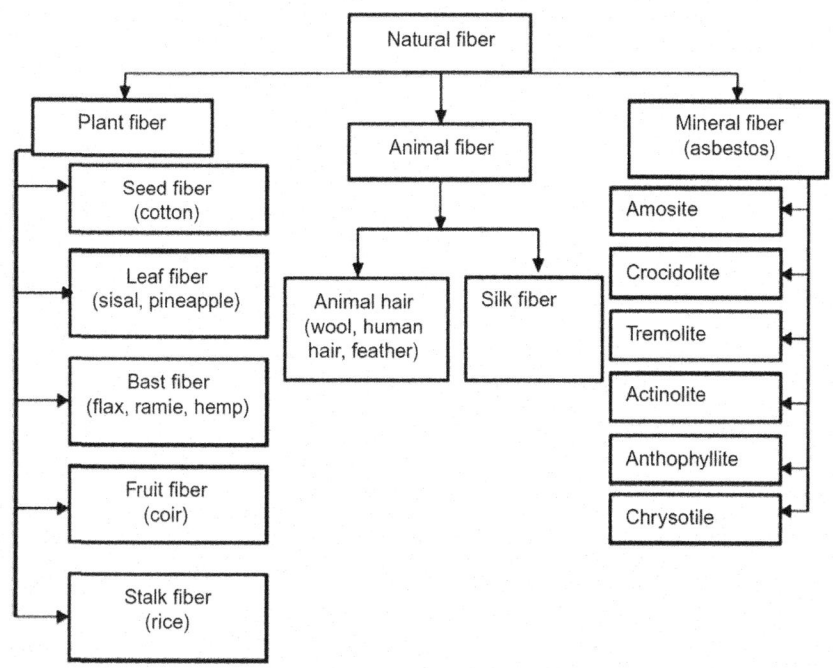

FIGURE 2.1 Classification of natural fibers (Bhattacharyya et al., 2015) (permission granted)

Natural plant fibers are non-woody and include some of the oldest materials found in human history. Ropes for hunting, nets for fishing, textiles for clothing, and roofing materials for shelter are some examples of the use of natural fibers before synthetic fibers were introduced to the world (Kılınç et al., 2017). Natural fiber reinforcements in a matrix composite enhanced tensile, compressive, and impact properties (Hristozov, Wroblewski, and Sadeghian 2016, Rajesh et al. 2019). Unfortunately, the hydrophilic nature of natural fibers is not always compatible with hydrophobic polymers, which results in poor interfacial bonding and the formation of voids, thereby reducing the material's strength (Ayu et al. 2018, Ayu et al. 2020, Dashtizadeh et al. 2019). Also, the natural constituents of fibers (cellulose, hemicellulose, and lignin components) can make the composite low in thermal and flame resistance (Lee, Salit, and Hassan 2014, Lee, Sapuan, and Hassan 2017, Lee, Sapuan, and Hassan 2018).

Surface treatments can improve the fiber surface for better binding. Furthermore, additive fillers like compatibilizers (Baley et al. 2018) and flame-retardant fillers can be incorporated into the biocomposite system to create active sites for greater interfacial bonding and to improve thermal properties. The hybrid fiber reinforcement concept preserves the superior properties of both fibers, making the composite stronger, cheaper, and/or lighter with the same fiber–matrix ratio (Song et al. 2018, Aisyah et al. 2018).

To achieve complete biodegradability, biopolymers have been synthesized. Like natural fibers, biopolymers can be differentiated by the methods used to synthesize them, shown in Figure 2.2. The most widespread biopolymers used in the packaging

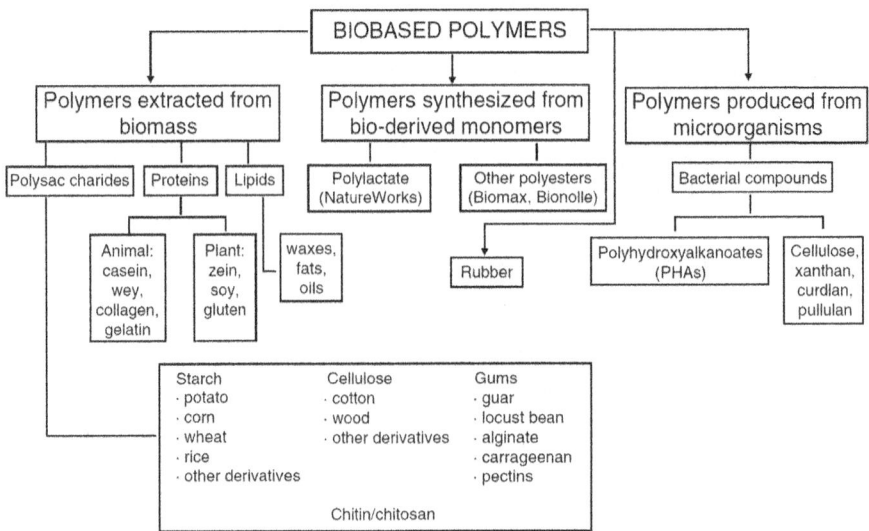

FIGURE 2.2   Biopolymer synthesis methods (Chiellini et al. 2008) (permission granted)

industry are polysaccharides comprising cellulose, starch, and chitin. The packaging industry alone has used 161 million tons of plastics and is the world's biggest consumer of plastic. Polysaccharides consist of monosaccharides (sugars) linked together by O-glycosidic linkages. Differences in monosaccharide composition, linkage types and patterns, chain shapes, and molecular weight dictate physical properties such as solubility, viscosity, gelling potential, and surface and interfacial properties. However, plasticizer fillers are always added to increase the biopolymer's flexibility, workability, or distensibility by reducing its glass transition temperature (Vieira et al. 2011).

In the framework of environmentally friendly processes and products, polylactide (PLA) represents the best polymeric substitute for various petropolymers because of its renewability, biodegradability, biocompatibility, good thermomechanical properties, and relatively low cost. PLA is an eco-friendly product that is non-toxic; hence, the medical industry uses PLA biopolymer for its products (da Silva et al. 2018). However, polyhydroxyalkanoate (PHA), a biopolymer produced from renewable carbon resources by many microorganisms, was found to be highly suitable for fabrication of resorbable medical devices such as tissue engineering scaffolds and bioimplant products (Wang and Chen 2019, Rodríguez-Contreras et al. 2019, Zhang et al. 2018). Over 300 species of Gram-positive and Gram-negative bacteria can be used to produce these polymers under nutrient-depletion conditions such as limited oxygen, phosphorous, or nitrogen with an excess carbon source. The carbon sources normally used are by-product biomass with almost zero commercial value (Kumar 2019, Amaro et al. 2019).

The bacteria convert carbon into PHA granules (Laycock et al. 2013). The PHA polymer is extracted and purified until the shape of the final product is reached, dry and solid (Santos et al. 2017). Figure 2.3 illustrates the industrial production of PHA biopolymer.

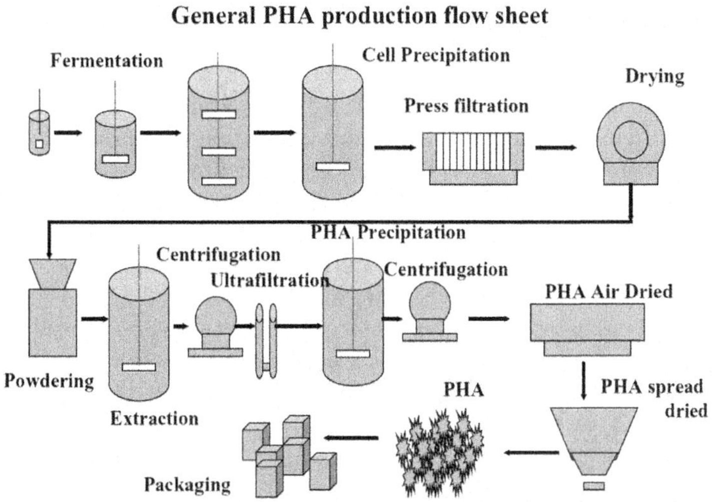

General PHA production flow sheet

FIGURE 2.3 Industrial production flow of PHA biopolymer (Chen 2009) (permission granted)

Nanotechnology is the engineering of materials at the nanoscale, which includes current research and advanced trends. The insertion of even a very low concentration of nanofibers into polymer composite has provided distinct strength improvements (Hosseini 2017, Saba et al., 2014).

Mathematic modeling has enhanced properties further by exploring nanoscale fillers (Kundalwal 2018). Nanofibers have higher surface-area-to-weight ratio, which improves fiber wetting by polymer. However, aggregation phenomena remain a major subject in composites containing spherical nanoparticles. Nanofibers, nanocellulose, and nanocrystal whiskers can be produced from natural fibers by different processing methods and morphologies. Most interestingly, nanocomposites can be synthesized and fabricated in ways similar to conventional polymer composites. In this chapter, natural plant fibers will be discussed in detail. After that, biopolymers extracted from biomass, synthesized from bio-derived monomers, and produced from microorganisms are described as well as nanofillers produced from natural fibers. Last but not least, natural fiber products – nanocellulose and cellulose nanofiber – are discussed. This chapter aims to give readers a general idea about biocomposites, nanocomposites, and the sources of biocomposite materials.

2.2 CATEGORIES OF NATURAL FIBER REINFORCEMENTS

Natural fibers have long gained attention as reinforcing agents for polymer composites owing to their sustainability, readily availability, and satisfactory mechanical strength. According to Ramamoorthy et al. (2015), plant fibers can be generally classified into six types: bast fibers, seed fibers, leaf fibers, straw fibers, bass fibers, and wood fibers (Ramamoorthy et al., 2015). The following section discusses some of these fibers and their use as reinforcing agents for polymer composites.

2.2.1 BAST FIBER

2.2.1.1 Flax

Bast fiber is a fibrous material from a plant, in particular the inner bark of a tree. Examples are flax, hemp, jute, and kenaf. Bast fibers are relatively easy to extract, making them the most widely used non-wood lignocellulosic fibers (Peças et al. 2018). Flax (*Linum usitatissimum* L.) is normally grown in moderate climates such as those found in France, China, and Belarus (Ramesh 2019). According to the Food and Agriculture Organization of the United Nations (FAO), in 2018, a total area of 240,293 hectares (0.24 billion m²) of flax was harvested worldwide, generating 868,374 ton of flax (FAO 2019). Flax fibers have reinforced various polymer composites but the most suitable polymer for flax fibers is polypropylene, owing to its low density, low thermal expansion, superior water resistance, and ability to be recycled (Van de Velde and Kiekens 2001).

2.2.1.2 Hemp

Hemp (*Cannabis sativa*) is normally grown in Asia and Europe. Owing to high yield and versatility, the fibers extracted from this plant are used in the production of ropes, clothing, and paper (Carus et al. 2013, Ranalli and Venturi 2004). In 2018, the hemp tow waste, or the beaten stalks of the hemp plant, was reported as 60,657 ton worldwide (FAO 2019). Hemp is undoubtedly a very useful fiber. However, due to the presence of noncellulosic components such as hemicelluloses, lignin, pectin, fat, and waxes, it needs to be treated or processed for further use (Pejić et al. 2020).

Hemp fibers have long been identified as a reinforcing agent for polymer composites. Neves et al. (2019) reinforced epoxy and polyester with 10, 20, and 30 vol% of continuous and aligned hemp fibers and the mechanical properties of the resultant polymer composites were assessed (Neves et al. 2019). The results revealed that epoxy composites performed better than polyester composites in flexural and tensile strength when the content of hemp fibers was higher, i.e., 30 vol%.

2.2.1.3 Jute

Jute has reportedly been used for textile fabrication in the Indus valley civilization since the third millennium BC. Jute fiber, also known as the golden fiber of Bangladesh, is the most important fiber in Bangladesh and Eastern India (Gupta et al., 2015). India, Bangladesh, and China are the main countries producing jute. In 2018, 1,546,953 ha (1.55 billion m²) of jute were harvested worldwide and the production was recorded as 3,633,550 ton. Jute fiber has a high aspect ratio, superior strength-to-weight ratio, and good insulation properties (Rohit and Dixit 2016). Gupta et al. (2015) reported in a review that jute fiber has reinforced a variety of composites such as epoxy, polyester, vinyl ester, low-density polyethylene, high-density polyethylene, and polypropylene and the resultant composites showed good mechanical properties (Gupta et al., 2015). Different chemical treatment might be required to enhance the mechanical properties of the jute fiber.

2.2.1.4 Kenaf

Kenaf (*Hibiscus cannabinus* L.) is an herbaceous plant native to southern Asia. It is also called Deccan hemp and Java jute. The fibers from kenaf can be extracted from two parts of the stalks: the bark (bast) and the core. Bast fibers make up to 40% of the stalk's dry weight while core fibers make up to 60% of the stalk's dry weight (Thyavihalli Girijappa et al. 2019). The interesting features of kenaf fiber are its low cost, light weight, renewability, biodegradability, and high specific mechanical properties (Rohit and Dixit 2016). Kenaf was introduced to Malaysia in 2010 as a replacement for tobacco. In only a few years, kenaf has become Malaysia's third largest industrial crop after palm oil and rubber. The National Kenaf and Tobacco Board of Malaysia has allocated 2,000 ha (2 million m^2) of land to smallholders for establishment of kenaf plantations in order to cater to the expanding global market (Thomas 2019). Kenaf has been used in the fabrication of various fiber-reinforced polymer composites.

2.2.2 LEAF FIBER

2.2.2.1 Pineapple

Pineapple (*Ananas comosus*) is planted abundantly in tropical regions and the pineapple leaves are a source of pineapple leaf fiber (PALF). Leao et al. (2010) stated that around 15 ton of PALF could be produced from a hectare of pineapple plantation (Leao et al. 2010). In 2018, a total of 27,924,287 ton of pineapples were produced from an area of 1,111,372 ha (1.11 billion m^2) (FAO 2019). In Malaysia, 10,556 ha (10.556 million m^2) of pineapple were harvested in 2019 and 329,365 ton of pineapple produced. PALF is superior to other types of fiber. For instance, PALF has specific modulus and specific strength properties that are very close to those of glass fiber and its aspect ratio is four times that of jute (Todkar and Patil 2019). Odusote and Oyewo (2016) compared pineapple leaf polyester and epoxy composite with glass fiber polyester composite (Odusote 2016), revealing that pineapple leaf epoxy composites with 40% fiber loading have superior mechanical properties compared with the glass fiber polyester composite. This indicates that pineapple leaf fiber is promising as a substitute glass fiber for the manufacturing of polymer composite.

2.2.2.2 Abaca

Abaca (*Musa textilis*), also known as Manila hemp, is an economically important inedible banana species native to the Philippines, which is cultivated commercially in the Philippines, Ecuador, and Costa Rica. The fibers of abaca are extracted from the leaf stems. In 2018, 108,131 ton of abaca were produced worldwide (FAO 2019). The banana can be harvested three times a year and therefore contributes significantly to the continuous supply of abaca fibers. The advantages of abaca fiber include its long fiber length; it is strong, flexible, and durable (ABC 2020). Abaca has been used in the fabrication of fiber-reinforced polymer composite. Sinha et al. (2018) evaluated the flexural properties of single- and double-layer abaca-fiber-reinforced epoxy composites. The results showed that double-layer abaca-fiber-reinforced epoxy composites possessed higher flexural properties and could potentially be used for structural applications.

2.2.2.3 Sisal

Sisal (*Agave sisalana*) is a plant native to southern Mexico. Sisal is easily cultivated with short renewal times and has been planted in several countries outside Mexico.

Sisal production worldwide in 2018 is recorded as 198,309 ton (FAO 2019). Sisal has been traditionally used for agricultural twine due to its superior strength. According to Rohit and Dixit (2016), sisal fiber has several advantages such as high tenacity and tensile intensity, abrasion resistance, salt water resistance, and acid and alkali resistance (Rohit and Dixit 2016). Sisal-fiber-reinforced epoxy composites have been fabricated by Gupta and Srivastava (2016) using fiber weight fractions of 15, 20, 25, and 30% (Gupta and Srivastava 2016). Better thermal and mechanical properties were observed with higher water absorption, when fiber weight fraction increased.

2.2.3   FRUIT FIBER

2.2.3.1   Coir

Coconut fiber (*Cocos nucifera*), also called coir, is a natural fiber extracted from the internal shell and the outer coat of a coconut. In 2018, the production of coir worldwide reached 1,238,725 ton. Malaysia alone produced 21,521 ton of coir in 2018 (FAO 2019). Coconut fiber is regarded as the thickest among all types of fiber (Thyavihalli Girijappa et al. 2019). In addition, coconut coir fibers have superior mechanical strength and weather resistance in comparison to the other types of fiber, owing to their higher lignin and microfibrillar angle but lower cellulose and hemicellulose content (Thyavihalli Girijappa et al. 2019). de Olveira (2018) fabricated polymer composite using epoxy resin reinforced with coir fiber and found that composites reinforced with 35 vol% had the highest flexural modulus and impact resistance (de Olveira et al. 2018). On the other hand, composites reinforced with 30 vol% coir fibers exhibited the highest flexural strength.

2.2.4   STRAW FIBER

Straw fibers such as corn and wheat have been used as reinforcing fillers in several polymers. Biological degradation of the natural fibers is a drawback for composites' applications. Mold growth depends on the water content of natural fibers, with a high moisture content in natural fiber promoting the formation of mold (Bouasker et al. 2014). Straw fiber has the advantage of a low water content. It is assumed to be relatively stable and typically exhibits little evidence of microbial respiration. Thus, straw fibers are strong candidates as bio- and nanocomposites.

2.2.4.1   Corn

Corn (*Zea mays*) or maize is widely cultivated in many countries. Corn is abundant and classified as waste in many countries such as Brazil, China, and the United States, which are the top maize producers in the world (Ranum et al., 2014). Incorporation of corn in polymer could have economic advantages and low environmental impact (Husseinsyah, Mostapha, and Selvi 2014). Corn has also been converted into nanocellulose (Andrade et al. 2019). Corn stover, such as husks, stalks, and leaves, was found to be a good source of nanocellulose because it has a high cellulose content (Costa et al. 2015). The unique characteristics of corn husk fiber, such as flexible, low density, moderate strength, good elongation, and durability, will positively contribute to properties of corn fiber products including composites (Ibrahim et al. 2019).

2.2.4.2 Wheat

Wheat is one of most common staple cereals in the world. The most common species is bread wheat, *Triticum aestivum* (USDA 2020). The annual worldwide production of this fiber has been estimated to be around 529 million tons (Kapoor, Panwar, and Kaira 2016). After harvest, most of the wheat straw is left on the ground to decompose (Zou et al., 2010). In some countries, wheat straw is burned in open fields, which can cause air pollution. Even though a small portion of wheat straw is used as animal feedstock and bedding, industrial applications of wheat straw are still under investigation. As agricultural waste, wheat straw is very cheap and easily obtained (Yu et al. 2016). To the best of best our knowledge, few studies have investigated the use of wheat straw fibers for composites. However, the potential of wheat straw has been discovered for numerous applications such as composites, anion exchangers, and panelboards (Panthapulakkal et al., 2006). Moreover, wheat straw fiber has also been used for nanocellulose production. It was found that the nanocellulose produced from wheat straw by a chemi-mechanical method had diameters in the range of 10–80 nm (Alemdar and Sain 2008).

2.2.5 SEED FIBER

Seed *fibers* are collected from the *seeds* of various plants. Commonly available seed fibers used in bio- and nanocomposites are cotton, kapok, and coconut husk. Cotton is the most common seed fibers used for textiles all over the world. The development of a blend of seed fibers and its application in thermoset composite manufacture has introduced a new class of cellulose-based reinforcement material.

2.2.5.1 Cotton

Cotton is an important agricultural crop that belongs to the genus *Gossypium*, subtribe *Hibisceae,* family *Malvaceae* (Chand and Fahim 2008). It is mainly used for the production of clothing and many other daily products for a large proportion of the world's population. Cotton grows around the seeds of the plant. As it has a high cellulose content, cotton fiber is a common source of cellulose nanostructures (Morais et al. 2013). Several researchers have discovered the synthesis of nanocellulose from cotton. According to Morais et al. (2013), cotton cellulose nanocrystal (CNC) produced from their study had a diameter of 177 nm and a width of 12 nm, while the aspect ratio was 19 (Morais et al. 2013). Composites based on plant fibers, usually cotton fibers recycled from textiles, have been used extensively for thermo-acoustic insulation. The material mainly comprises cotton fibers, up to 80% by weight. Because cotton is usually used in fabrics, several efforts have attempted to incorporate antimicrobial agents, such as argentum/argentum bromide–titanium oxide and copper oxide with cotton, to improve the technical value of fabrics (Perelshtein et al. 2009, Rana et al. 2016). The results obtained showed that the integration of these compounds onto cotton fabrics exhibited excellent antibacterial properties and improved mechanical properties, selective oil adsorption behavior from oil–water mixtures, and ultraviolet-blocking activities. Therefore, the value of cotton fabrics can be significantly enhanced for various applications.

In addition, polyester has been extensively being used for cotton-fiber-reinforced polymer composites. Polyester is inexpensive and easily available as a liquid, has good processing properties, and has good mechanical properties when reinforced with fibers. Polyesters are suitable for various application and also suitable for the fabrication of complex structures and intricate shapes.

2.2.5.2   Kapok

Kapok (*Ceiba pentandra*) is the lightest natural fiber in the world (Mani, Rayappan, and Bisoyi 2012). Kapok has another name, silk cotton, because of its natural luster. This fiber consists of a very thin wall and a huge hollow region full of air. Of natural fibers, kapok circular cross-section cellulose has the largest degree of hollowness, around 80 to 90% (Macedo et al. 2019). This makes this fiber very suitable to produce low-density polymer composites. Kapok has a high content of cellulose fiber. However, compared with cotton fiber, it has the lower content of cellulose and higher content of lignin.

Nanocellulose has also been successfully extracted from kapok (Jayaweera et al. 2017). Scanning electron microscopy has revealed that the diameter thickness of kapok nanocellulose is in between 100 nm and 1000 nm. It has also been shown that a small amount of nanocellulose greatly enhanced the mechanical performance of composites (Jayaweera et al. 2017).

2.2.6   Cane, Grass, and Reed Fiber

Cane, grass and reed are classified into a similar class of fiber. These fibers are more obvious analogs to the synthetic fibers currently in use in composite materials and other applications. Grass fibers are elongated sclerenchyma cells, which can be found in different parts of plants. They can be found in the ground and vascular tissues, which usually use them for mechanical support, but sometimes they occur in dermal tissues as well. Numerous studies have utilized these fibers as reinforcement in a polymer matrix for making partially biodegradable green composites.

2.2.6.1   Bamboo

Bamboo has been reported as good potential renewable woody biomass resource of the future. It has good strength and a short growth cycle (Han et al. 2018). Around 1000 bamboo species are known, all around the world. Interestingly, new species are still discovered almost every day (Md Shah et al. 2016). Bamboo fiber has superior mechanical properties, making it a promising alternative for conventional fibers, such as glass and carbon, in composite materials. According to Han et al. (2018), bamboo consists of approximately 47% cellulose, 26% lignin, and other substrates (Han et al. 2018). Bamboo fiber has a considerably high percentage of lignin compared with other natural fibers, resulting in its high strength (Md Shah et al. 2016). Han et al. (2018) successfully produced nanocellulose from bamboo, with a fiber diameter size of approximately 15 nm (Han et al. 2018). Meanwhile, Chang et al. (2012) also managed to produce nanocellulose from bamboo with a diameter of a few tenths of a nanometer (Chang et al. 2012).

2.2.6.2   Sugarcane Bagasse

Sugarcane bagasse usually does not meet the criteria to be classified as long fibers (Verma 2012). Because of this, the fiber is usually abundantly left over, so their use as a reinforcement of composite is strongly considered as a future possibility. Brazil is the top producer of sugarcane, with approximately 743.0 million metric ton per year (Teboho et al. 2018). There is about 640–660 million ton of sugarcane that could be converted into 28,500 million liters of alcohol. This could lead to the production of 160 million metric ton of sugarcane bagasse (Teboho et al. 2018). Among the ligno-cellulosic biomass available, sugarcane bagasse is classified as the most abundantly available fiber waste worldwide (Panicker et al., 2017). Generally, it consists of 41.0–55.0 wt% of cellulose, 20.0–27.5 wt% of hemicellulose, 18.0–26.3 wt% of lignin, and ~7.0 wt% of other compounds such as inorganic materials (Teboho et al. 2018).

Sugarcane bagasse has also been used for the production of nanocellulose. According to Ghaderi et al. (2014), nanocellulose was successfully synthesized from sugarcane bagasse, with a diameter of 39 nm (Ghaderi et al. 2014). Sugarcane-bagasse-based composites have numerous applications in automotive and railway coaches/buses. This fiber bagasse has been extensively investigated as a reinforcement agent for polymers (Teboho et al. 2018). There is an excellent opportunity in producing bagasse-based composites for a variety of applications in building and construction, such as board panels and blocks as reconstituted wood, flooring tiles, and more.

2.3 CATEGORIES OF BIOPOLYMERS

Biopolymers are formed via linkages of molecules of biological origin. There are several categories of biopolymers, such as polysaccharides, proteins, and lipids. For instance, biopolymers can be a heterodimeric protein with a heavy chain (395 kDa) and a light chain (25 kDa) linked by a single disulfide bond (Elzoghby et al., 2015). The terminology varies: for example, in food applications, biopolymers are called starch. Owing to biological properties such as biodegradability and biocompostability, these biopolymers are always selected as one of the main ingredients in composites. Such production allows the composites to be described as biodegradable or renewable, depending on the percentages incorporated.

2.3.1 BIOPOLYMERS EXTRACTED FROM BIOMASS

Biopolymers can be extracted from agricultural plant or animal products or can be synthetic.

2.2.3.1 Polysaccharides

Polysaccharides are biological polymers composed of monosaccharide units with glycosidic linkages, known as long chains of carbohydrate molecules. The chain can be linear or branched, which may influence its reaction to water. The functions of polysaccharides in living organisms are structure-related like cellulose and chitin or storage-related like starch and glycogen. A polysaccharide that contains all the same type of monosaccharide repeating units is called homopolysaccharide or homogly-can but if more than one type of monosaccharide is present then it is named hetero-polysaccharide or heteroglycan. The chemical formulae of monosaccharide and polysaccharide are $(CH_2O)_n$ and $C_x(H_2O)_y$, respectively. Glucose, fructose, and glyc-eraldehyde are examples of monosaccharides.

Cellulose exists in plant or organism cell walls, which are known be the most common organic molecules on earth. It has applications in pulp, paper, and textiles industries. Chitin has a similar structure to cellulose but contains nitrogen side branches that enhance its strength. Unlike cellulose, chitin is found in arthropod exoskeletons and in the cell walls of some fungi.

A green solvent called deep eutectic solvent is used as an alternative ionic liquid for polysaccharide treatment. It is expected to be a good solvent to solubilize polysaccharides and is beneficial to delignify biomass, extract cellulose, and fibrillate or crystallize nanocellulose (Zdanowicz et al., 2018). This green solvent is expected to help the initiatives to develop industrial scale production.

Hericium erinaceus (Bull.) Pers. is a kind of polysaccharide, edible and medicinal higher fungi which is also known as yamabushitake, houtou, and lion's mane. The presence of this polysaccharide can nourish the stomach, tranquilizing the mind, fighting cancer, and fortifying the spleen (He et al. 2017). Most *H. erinaceus* polysaccharides are heteropolysaccharides that have good potential in industry. Mushrooms are also polysaccharide known for their medicinal and nutritional value (Wang et al., 2017a). In addition, marine life such as algae (macro and microalgae) has unique chemical components in basic and complex components. Polysaccharides present as hemicelluloses, starch, or pectin. Such polysaccharides can be extracted via conventional and innovative techniques. Factors like low yield, inefficient time, and extract quality influence the isolation and purification methods. One potential method is enzyme-aided extraction, which applies enzymes in the recovery of polyphenols, oils, polysaccharides, and more. In order to enhance the recovery, conventional methods also are introduced such as high pressure, supercritical carbon dioxide, microwave, and ultrasound. Magnetic nanoparticles are considered as potential carriers for the enzymes (Nadar et al., 2018).

Polysaccharides have wide anti-tumor, anti-virus, and anti-inflammatory applications due to their low toxicity and therapeutic effects. Various methods are applied in order to isolate and purify polysaccharides related to their massive structural heterogeneity (Shi 2016). Qiu et al. (2017) focused on sorghum bran, bagasse, and biomass to isolate and characterize three carbohydrate-rich fractions known as hemicellulose A, hemicellulose B, and cellulose-rich residue. Monosaccharide composition, insoluble dietary fiber, soluble dietary fiber, total dietary fiber, water-holding capacity, and emulsion stability are tests that need to be considered in any manufacturing, especially for food products (Qiu et al., 2017). The bran of sorghum showed a highly branched structure while sorghum bagasse has the highest insoluble dietary fiber. All types of sorghum exhibited a high water-holding capacity and had good emulsion stability.

2.2.3.2 Proteins

Proteins, also called polypeptides, are formed of one or more long, linear chains of amino-acid residues connected by peptide bonds. Different proteins have different sequences of amino acids. A short polypeptide may contain 20 to 30 residues; these are commonly known as peptides or oligopeptides. Proteins also contain non-peptide groups that are called prosthetic groups or cofactors. The main function of proteins is to be molecular transporters. Proteins are essential parts of organisms and participate in all processes within cells. There are several purification methods for proteins: ultra-filtration (Jang, Liceaga, and Yoon 2016), ultracentrifugation, precipitation, gel filtration chromatography (Agrawal et al., 2016), electrophoresis, and chromatography (Jin et al. 2016). One study carried out by Wang et al. (2017a) gradually isolated microalgal components in order to obtain lipid-rich, protein-rich, and carbohydrate-rich components using the pyrolysis mechanism of the microalgae (Wang et al. 2017b). The three components of lipid, protein, and carbohydrate have specific pyrolysis pathways as shown in Figure 2.4.

Bioactive peptides are reported to positively influence body health; they are famously utilized in the manufacturing of functional foods and dietary interventions. One major challenge in their manufacture comes from biomolecular purification related to scale-up. Agyei et al. (2016) suggested "-omics" techniques and chromatographic purification using monolithic adsorbents to overcome these challenges (Agyei et al. 2016).

2.2.3.3 Lipids

A lipid is defined as a hydrophobic or amphiphilic small molecule. It is soluble in nonpolar solvents like hydrocarbons and has functions in energy storage, signals, and structural components. Nanotechnology and the cosmetic and food industries always involve lipids in their manufacturing processes.

Microbial lipid production is an environmentally sustainable method. Such microbial lipid production involves screening of yeast strains in order to seek efficient oleaginous yeasts that are affordable and accumulate large amounts of lipid when cultivated on lignocellulosic sugars (Poontawee et al., 2017). *Cryptococcus*, *Lipomyces*, *Rhodotorula*, *Rhodosporidium*, and *Trichosporon* are well-known oleaginous yeasts that can generate lipids exceeding 60 to 70% of biomass (Ratledge and Wynn 2002). A well-studied oleaginous yeast species is *Yarrowia lipolytica*, which can build up lipids up to 40% of cell mass (Beopoulos et al. 2009). The yeast is often chosen to be a host organism in overproducing single-cell oil in metabolic

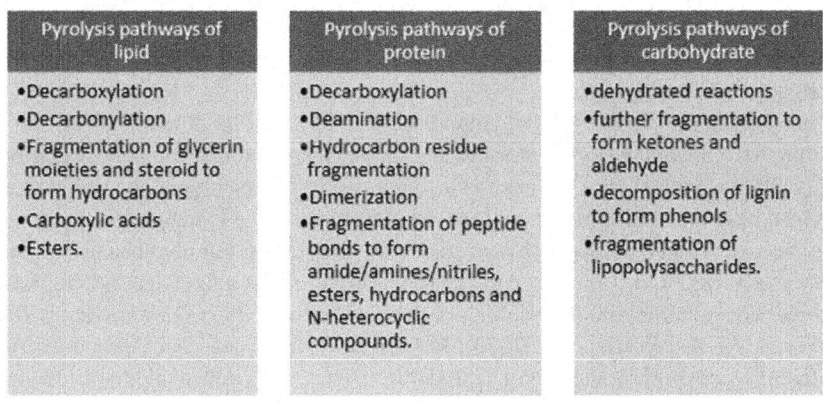

FIGURE 2.4 Pyrolysis pathways of lipids, proteins, and carbohydrates (adapted from Wang et al. 2017b)

engineering (Blazeck et al. 2014, Qiao et al. 2015, Tai and Stephanopoulos 2013). Cell growth (growth at abundance and low-cost biomass feedstock), lipid accumulation, expansion of genetically tractable diversity, inhibitor tolerance, lipid production kinetics, and fatty acid distribution have all been studied (Josh et al., 2017).

2.3.2 BIOPOLYMERS SYNTHESIZED FROM BIO-DERIVED MONOMERS

Bio-derived monomers are manufactured from naturally occurring molecules. They can also be synthesized using chemical and biochemical processes to disintegrate naturally occurring molecules. Owing to the wide variability during the disintegration processes of monomers, the resultant biopolymers also show exceptionally versatile chemical structures. Bio-derived monomers are realistically plausible to be introduced into the current manufacturing system of petroleum-derived polymers and subsequently replace them.

2.3.2.1 Polylactide

Polylactic acid (PLA), a recognized biodegradable polymer, has been widely investigated for regenerative medicine and drug delivery systems. The earliest research into low-molecular-weight PLA was produced by Carothers in 1932. Following research in 1954, DuPont patented a higher-molecular-weight PLA.

Two approaches have been applied to synthesize lactic acid (LA), the main backbone of PLA: (1) chemical processes from petrochemical feedstock and (2) fermentation of carbohydrates such as biomass. Although the first method is more common, the natural decomposition of organic materials containing carbohydrates has been the leading method applied to synthesize LA in recent years due to concerns about biodegradability and carbon emissions. More than 90% of LA production comes from the fermentation of carbohydrate.

Datta and Henry studied the chemical synthesis and purification of lactic acid containing both enantiomers of LA, L- and D-lactic acid (Datta and Henry 2006). A combination of both in equal measure is named DL-lactic acid or racemic lactic acid. Fermentation-derived lactic acid presents almost solely as L-lactic acid. The possibility of obtaining pure L-lactic acid significantly affects the chemical properties and process-structure-property relationship of polymers produced from LA (Hamad et al. 2015).

Biodegradable polymers have gained substantial attention in medical utilizations as they have advantages over non-biodegradable polymers that include biocompatibility and the exclusion of the necessity to remove implants. Lately, the utilization of PLA in these medical applications was not merely based on biodegradability properties, because it was produced from renewable feedstocks, but also because it functioned very well and contributed exceptional properties at a low cost in comparison to other conventional biodegradable polymers used in medicine (Ivanov et al. 2019, Ma et al. 2018, Pandele et al. 2020). Numerous devices have been produced from different types of PLA, including degradable sutures, drug-releasing micro/nanoparticles, and porous scaffolds for cellular applications (Lasprilla et al. 2012). Furthermore, the exceptional properties of PLA, including physico-mechanical features and the possibility to produce it using conventional manufacturing methods such as extrusion, injection molding, compression molding, and blow molding, expanded its applications.

2.3.2.2 Succinic Polymers

Biopolymers based on succinic acid, a bio-based chemical star, are becoming more commercially available (Thakker et al. 2012, Zeikus et al., 1999). Poly(butylene succinate) (PBS), a popular succinate polymer, is formed via immediate polycondensation of succinic acid and butanediol (Xu and Guo 2010). Until recently, commercially available PBS was only manufactured using succinic acid and butanediol from non-renewable fuel sources. With the increased interest in biopolymers from renewable resources, numerous studies have led to the finding that succinic acid and butanediol can be derived from processed biomass. In Japan, a commercial biodegradable PBS polymer was launched in 1993 by Showa Denko under the trade name Bionolle, with a production capacity of 3000–10,000 tons per year. In 2013, Succinity GmbH, a joint venture between Corbion and BASF with a plant located in Spain, began processing bio-based succinic acid for the global market in commercial quantities with an annual capacity of 10,000 metric tons.

2.3.2.3 Bio-polyethylene

The sharp rise in crude oil prices during the 1970s increased the fuel industry's interest in producing bio-ethanol from sugar-containing feedstocks, for use as a transportation fuel. Bio-ethanol is also chemically used to produce bio-ethylene, which is the main component of bio-based polyethylene (bio-PE) (Morschbacker 2009). The stunning fall in oil prices weakened the bio-PE market, but the biopolymer remains in development by key industry players such as Brazil's Braskem because of fluctuations in fuel prices and increasing environmental consciousness (Matsumoto et al. 2009). In fact, the main benefit of bio-PE is that it is chemically similar to conventional PE so they carry the same properties and can be applied directly in commercial fossil-based polymer processing and recycling infrastructure. However, bio-PE production faces stiff competition from the fossil-based polymer, with the heavy fluctuation of fossil price (Hess and Johnson, 2014). Bio-based PE is fully recyclable; however, it is not biodegradable.

2.3.2.4 Bio-based Poly(Ethylene Terephthalate) and Poly(Trimethylene Terephthalate)

Poly(ethylene terephthalate) (PET) is the most common thermoplastic polymer resin of the polyester family. It is a high-performance engineered plastic extensively applied in products such as bottles, fibers, moldings, and sheets due to its high impact and tensile stress, transparency, workability, chemical resistance, and thermal stability (Avila-Orta et al. 2003, Kong and Hay 2003). However, the increasing consumption of PET creates a serious environmental pollution issue owing to its non-biodegradable nature (Edge et al. 1991).

PET waste management with upcycling has become an important social issue. Nevertheless, the properties of the final recycled products are always dominated by deterioration due to thermo-oxidative, chemical, and thermal reactions applied during the recycling process. Chemical recycling of PET has been more successful, further optimizing the PET recycling process including sorting, granulation, washing, and extrusion. The industry is searching for breakthrough technology on PET recycling due to PET's relatively high chemical and thermal stability even under typical hydrolysis, alcoholysis, or breakdown procedures (Nakajima et al., 2017). PET is produced commercially from ethylene glycol (EG) with terephthalic acid (TPA), mostly from petroleum-derived sources. To create sustainability in PET manufacturing, it is necessary to develop bio-based PET (bio-PET) synthesized from monomers derived from viable resources such as biomass. The Coca-Cola Company accelerated the development of bio-PET and launched the "PlantBottle" in 2009, which was made from 100% bio-based EG (bio-EG) and petroleum-derived terephthalic acid (TPA) (Anonymous 2019). To further enhance the viability of PET by producing bio-PET from 100% natural biomass feedstock, bio-based terephthalic acid (bio-TPA) from naturally derived sustainable biomass feedstock is being developed. GEVO, one of the world's largest renewable chemical producers, is currently the leading global player in the development of bio-TPA (Nakajima et al., 2017).

The unique chain conformation of PTT creates a remarkably elastic recovery property that makes it suitable for applications in textile and carpet productions. Bio-based 1,3-propanediol (bio-PDO) is applied in the production of bio-based poly(trimethylene terephthalate) (PTT). The joint venture DuPont Tate & Lyle BioProducts has successfully developed bio-PDO with the brand name "Susterra" by fermenting sugars from starches (Nakajima et al., 2017). Susterra is applied as backbone in the manufacturing of PTT (brand name Sorona®), a polymer that consists of 37 wt% of renewable material. Poly(butylene terephthalate) (PBT) is a high-volume-usage commercial engineering plastic with chemical functional groups similar to PET and PTT. PBT will be one of the potential bio-based polyesters available in the market soon, as bio-based manufacture of the monomer component butanediol is under development (Nakajima et al., 2017).

2.3.2.5   Bio-based Polyamides

According to Kyulavska, Toncheva-Moncheva, and Rydz (2017), polyamides (PAs) possess several promising characteristics such as high resistance to wear and abrasion as well as chemical properties. In addition, PAs have superior mechanical properties, dimensional stability, and low permeability to gases. Because of the increase in environmental awareness, researchers are actively trying to synthesize green polyamides based on biorenewable resources by replacing conventional manufacturing of petrochemical-based chemicals with bio-based chemicals. It is important to continuously put research efforts into developing better bio-based polyamides because petroleum resources will eventually be depleted. There are many resources for synthesizing PAs, for example from crude oil (Jiang and Loos 2016). PAs based on castor oil have the longest history in man-made polymers from renewable raw materials (Salimon et al., 2012, Montero de Espinosa and Meier 2011). However, the most common resource in producing PAs is biomass (Ogunniyi 2006). Kyulavska, Toncheva-Moncheva, and Rydz (2017) have demonstrated in their works several processes for synthesizing PAs, worth exploring for further information.

2.3.3   Biopolymers Produced from Microorganisms

2.3.3.1   Polyhydroxyalkanoates

Polyhydroxyalkanoates (PHAs) are members of a family of polyesters that consists of hydroxyalkanoate monomers. PHAs are naturally or synthetically collected from microorganisms under specific conditions as water-insoluble granules of pure polymer. PHAs are commercially generated by feeding the bacteria with fatty acids, which are converted from carbohydrate-rich resources. In the industrial manufacture of PHAs, cells are isolated and lysed after few cycles of periodic feast–famine conditions. The polymer is then isolated from the bacterial cells, purified, and processed into powder or pellets (Madison and Huisman 1999). A dynamic development is foreseen for the PHAs, which is a big family of different polymers possessing characteristics ranging from thermoplastic to elastomers (Koller 2018). PHAs are considered a favorable replacement to petrochemical polymers. The most commonly commercialized PHAs are polyhydroxybutyrate (PHB) and KANEKA PHBHTM. The short-chain polymers (PHB) are more crystalline; hard but brittle. The medium-chain polymers (KANEKA PHBHTM) are tougher and more resilient. The distinctive properties that make PHAs safer to the environment compared with petrochemical polymers lie in their environmental biodegradability and biocompatibility, which are strong assets (Muhammadi et al. 2015).

As reported by Gerhart et al. (2002), the PHA chemical structure is mainly regulated by the microbial strains, carbon source, and also by the extension of precursors. The focus of recent research has been on constrained bioproduction of PHAs utilizing various substrates and using different microorganisms, with the intention of producing economically feasible concentrations of PHAs in fermentors (Koller et al. 2017). Nevertheless, the manufacturing costs of conventional non-biodegradable plastics are still much lower compared with PHAs, meaning that PHAs fail to compete in many major polymer industrial sectors. Processing PHAs is also challenging due to limited thermomechanical stability, environmental instability, inadequate global manufacture facilities, and stringent processing conditions for sterile fermentation conditions, which hinders larger industrial applications (Kovalcik et al. 2017). Thus, several researchers have conducted intensive research into producing PHA under non-sterile conditions and proved it can be done with extreme care (Anjum et al. 2016, Kourmentza et al. 2017).

2.3.3.2   Poly-glutamic Acid

Poly-γ-glutamic acid (γPGA) is a naturally existing biopolymer. It is produced mainly by Gram-positive bacteria. γPGA, a poly-amino acid, is an unusual anionic homopolyamide, a water-soluble, biodegradable, non-hazardous, non-immunogenic, edible biopolymer made from repeating units of L-glutamic acid, D-glutamic acid, or both (Bajaj and Singhal 2011, Ogunleye et al. 2015). γPGA has already been developed on the industrial scale via microbial fermentation, mainly from the genus *Bacillus* (Bajaj and Singhal 2011, Chettri et al., 2016, Ogunleye et al. 2015). γPGA was initially discovered in 1937 by Bruckner and co-workers when a capsule of *Bacillus anthracis* was discharged into the medium during autoclaving (Shih and Van 2001). Japan's traditional fermented soybean food, natto, is another source of γPGA. The natto mucilage contains a mixture of γPGA and fructan formed by *Bacillus natto*

Sawamura (Candela et al. 2014, Shih and Van 2001). Even though γPGA microbial production is well established, the manufacturing cost remains high. The latest research into γPGA is therefore concentrated on optimization of culture conditions to produce high yields and alter the enantiomeric composition and molecular weight. Most marketable and viable γPGA is synthesized via microbial fermentation from biomass.

2.4 TYPES OF NANO FILLER REINFORCEMENTS FROM NATURAL FIBER

In the past few years, bio-based materials have garnered tremendous interest in the research world due to their vast potential in producing various high-end products with fewer environmental issues. Annual global production of lignocellulosic is estimated to be approximately 1.3×10^{10} metric tons, which can be considered as one of the most abundant biopolymers available for the reinforcement of composites (Saratale and Oh, 2012). Cellulose is known for its abundant availability, biodegradability, low cost, high strength-to-weight ratio, renewability, and low density for natural reinforcing materials (Mat Zubir et al. 2016). Cellulose fibers are able to enhance mechanical as well as physical properties when used to reinforce polymer matrix composites. Nevertheless, high moisture absorption and a lack of compatibility with the hydrophobic polymer matrix make cellulose less suitable as a reinforcing agent in polymer matrices, especially for microscale materials (Gabr et al. 2013). Thus, nanocellulose is newly under research in order to find alternatives to solve the limitation and incompatibility issues of cellulose.

Nanocellulose is produced from natural fiber sources (Gan et al. 2020). It has a large surface area, which leads to great interfacial fiber/fiber and fiber/matrix adhesion that eventually reduces the swelling and water uptake of composite (Abdul Khalil et al., 2012). Only a small amount of well-dispersed nanofiller is necessary to achieve mechanical improvement. In other words, nanocellulose can be considered an excellent material for the development of high-performance nanocomposites. CNC and cellulose nanofiber (CNF) are both nanocellulose and commonly extracted from plant cell walls. The abundance of biomass sources and great mechanical properties make nanocellulose a promising reinforcement material for nanocomposites. Nanocellulose has plenty of hydroxyl groups, which assist in surface modifications and allow hydrogen bonding (Xu et al. 2013). CNF is composed of both crystalline and amorphous areas where it is a micrometer long with a diameter of less than 100 nm; it is also known for its rice-like structure, usually produced via mechanical processes. CNC is known as cellulose nanowhiskers (CNW), with a diameter that ranges from 5 to 40 nm. It is a highly crystalline nanocellulose commonly produced through acid hydrolysis (Abdul Khalil et al. 2014).

2.4.1 Cellulose Nanocrystal

It is commonly known that CNC is produced by acid hydrolysis with sulfuric acid or acidic deep eutectic solvent, hydrochloric acid hydrolysis, ionic liquid, or ultrasonication. CNC production involves the removal of amorphous regions; what remains is the CNC. CNC isolation from natural sources is highly interesting because of its superior mechanical properties in terms of specific strength and modulus, and its high aspect ratio including specific surface area (Gan et al. 2020). Favier and his co-workers (1995) showed an exceptional reinforcing effect within nanocomposite materials when tunicin whiskers are well dispersed in a styrene butyl acrylate copolymer matrix (Favier et al., 1995). With as little as 1.5 wt% of nanosized whisker, a percolation threshold could be achieved by the hydrogen-bonded system that contributes to a significant improvement in mechanical properties of the nanocomposite obtained. Therefore, CNW has been integrated in many biopolymer matrices, such as PLA, through solution casting or extrusion compounding (Bondeson and Oksman 2007, Petersson et al., 2007). Despite its excellent mechanical properties, its aggregation and poor dispersion were apparent in most of the subsequent composite structures, leading to minimal enhancement in mechanical strength. In many cases, the common problem found in nanocomposites when using CNW is their low thermal stability following acid treatment (Petersson and Oksman 2006).

Kamal and Khoshkava reported a spray freeze-drying method that can produce porous CNW agglomerate structures, which improves CNW dispersibility in melts of PLA and polypropylene (Kamal and Khoshkava 2015). Besides that, electrospinning PLA fibers can successfully improve adhesion between CNW and PLA; the CNW film is subsequently hot-pressed in order to melt and homogenize the PLA phase (Martínez-Sanz et al., 2013). A significant decrease in oxygen and water permeability, including a great increase in the mechanical properties of these PLA/CNW films, was observed. It is comparable to or better than conventional packaging plastics like PET (Oksman et al. 2006). In more hydrophilic biopolymers such as polyvinyl alcohol (PVA) (Fortunati et al. 2013, Pereira et al. 2014), ethylene vinyl alcohol copolymer (Martínez-Sanz et al. 2011), agar (Atef et al., 2015), regenerated cellulose (Ma et al. 2011), gelatin (Mondragon et al. 2015), hydroxypropyl methyl cellulose (HPMC) (George et al. 2014), thermoplastic starch (TPS) (Gonzalez et al. 2015, Slavutsky and Bertuzzi 2014), and natural rubber (Visakh et al. 2012), the nanocomposites with CNW were mostly prepared by a film-casting technique from aqueous suspension. A high reinforcing effect and excellent interaction between CNW with these polymer matrices are the results from less than 5 wt% addition of well-dispersed CNW.

Santos and co-workers have found that a sonication method could enhance CNW dispersibility in the gelatin matrix with higher CNW loading, such as 10 wt%. Thus, thermal, gas barrier, and mechanical properties could be enhanced and good transparency retained in these film materials. However, some CNW-reinforced films as well as ductile polymer and HPMC/CNW films have a higher degree of water vapor permeability, which leads them to be more brittle (Soykeabkaew et al. 2017).

2.4.2 CELLULOSE NANOFIBER

The first CNF produced from wood used a very energy-intensive process: high-pressure homogenization without any pretreatment. Introduction of chemical treatments such as 2,2,6,6-tetramethylpiperidine-1-oxyl radical (TEMPO) oxidation, combined with mechanical treatments, has reduced the energy intensity of production. As well as TEMPO oxidation, enzymatic hydrolysis treatment shows high effectiveness and is environmentally friendly in producing CNF when combined with high-pressure homogenization and mechanical shearing (Gan et al. 2020). Previously, CNF has been incorporated into hydrophilic biopolymers such as pectin and carboxymethyl cellulose, PVA, TPS, and wood fibers. The superior mechanical strength of these nanocomposites, due to homogeneous dispersion among the filler and the matrix, was reported. However, less-hydrophilic biopolymers like PLA, chitosan, and cellulose acetate butyrate exhibited difficulty in dispersion. To solve this dispersion issue, a more complicated and well-designed preparation process is required.

Previously, Nakagaito et al. prepared PLA/CNF films using a papermaking-like process including compression molding (Nakagaito et al. 2009). With 10 wt% and more CNF content, nanocomposites have displayed a significant enhancement in their thermal and mechanical properties. On the other hand, PLA-based nanocomposite incorporated with CNF has improved its toughness and reduced oxygen transmission. CNF is believed to increase the crystallization rate by acting as a nucleating agent. However, some nanocomposites produced through compounding and melt mixing agglomerate due to poor dispersion. Therefore, a few steps were added in the processing; for example, master batch preparation of PLA/CNF, extrusion compounding the master batch and PLA, and injection molding of the nanocomposites when some material properties have been improved. Herrera et al. previously reported on the use of liquid feeding of CNF including a plasticizer, which successfully prepared well-dispersed CNF in PLA-based films by using a co-rotating twin-screw extruder and then compression molding (Herrera et al., 2015). Interestingly, a positive effect on this film's toughness was also found in the presence of CNF together with the plasticizer. It was suggested that the combination of slippage of the nanofiber–matrix interface and a massive crazing effect led to the PLA toughening (Soykeabkaew et al. 2017).

2.5 CONCLUSION

As the record-breaking global temperatures continue, awareness of the use of petroleum-based plastic products has been raised. Little plastic waste is recycled or incinerated, with the majority going to landfill or into the sea. Thus, natural-fiber-reinforced biocomposites have been introduced.

Natural fibers are readily available, low-cost materials with properties comparable with synthetic fibers such as high strength-to-weight ratios, relative low consumption of manufacturing energy, longer tool life, and CO_2 neutrality. Most importantly, introduction of natural fibers promotes some biodegradability properties. Natural plant fibers are considered as non-woody fibers and have been used for many years.

Natural fiber reinforcements enhance the strength of biocomposites. However, poor interfacial bonding caused by incompatibility of hydrophilic natural fiber and hydrophobic matrix can result in undesirable performances. Bast, leaves, fruit, straw, seeds, and grass fibers are natural fibers commonly used in biocomposite reinforcements.

To achieve complete biodegradability, biopolymers have been synthesized. They can be classified according to synthetic methods: extracted from biomass, synthesized from bio-derived monomers, and produced by microorganisms. Each type has distinct advantages and disadvantages. Last but not least, nanosize cellulose has a higher surface area and leads to better interfacial adhesion between fiber/fiber and fiber/matrix, which reduces the water uptake and the swelling of composite. Both CNC and CNF can be isolated from plant cell walls, using different processing methods. No doubt nanocomposites will be the future trend in advanced composite materials section.

REFERENCES

ABC. *Abaca Natural Fiber.* 2020 [cited 16th Feb 2020. Available from https://www.abc-oriental-rug.com/abaca-natural-fiber.html

Abdul Khalil, H. P., Y. Davoudpour, M. N. Islam, A. Mustapha, K. Sudesh, R. Dungani, and M. Jawaid. 2014. "Production and modification of nanofibrillated cellulose using various mechanical processes: a review." *Carbohydrate Polymers* no. 99:649–665. doi:10.1016/j.carbpol.2013.08.069.

Abdul Khalil, H. P. S., A. H. Bhat, and A. F. Ireana Yusra. 2012. "Green composites from sustainable cellulose nanofibrils: A review." *Carbohydrate Polymers* no. 87 (2):963–979. doi: https://doi.org/10.1016/j.carbpol.2011.08.078.

Agrawal, H., R. Joshi, and M. Gupta. 2016. "Isolation, purification and characterization of antioxidative peptide of pearl millet (Pennisetum glaucum) protein hydrolysate." *Food Chem* no. 204:365–372. doi:10.1016/j.foodchem.2016.02.127.

Agyei, Dominic, Clarence M. Ongkudon, Chan Yi Wei, Alan S. Chan, and Michael K. Danquah. 2016. "Bioprocess challenges to the isolation and purification of bioactive peptides." *Food and Bioproducts Processing* no. 98:244–256. doi: https://doi.org/10.1016/j.fbp.2016.02.003.

Aisyah, H.A., M.T. Paridah, A. Khalina, S.M. Sapuan, M.S. Wahab, O.B. Berkalp, C.H. Lee, and S.H. Lee. 2018. "Effects of fabric counts and weave designs on the properties of laminated woven kenaf/carbon fibre reinforced epoxy hybrid composites." *Polymers* no. 10 (12):1320.

Alemdar, Ayse, and Mohini Sain. 2008. "Biocomposites from wheat straw nanofibers: Morphology, thermal and mechanical properties." *Composites Science and Technology* no. 68 (2):557–565. doi: https://doi.org/10.1016/j.compscitech.2007.05.044.

Amaro, Tiago M. M. M., Davide Rosa, Giuseppe Comi, and Lucilla Iacumin. 2019. "Prospects for the use of whey for polyhydroxyalkanoate (PHA) production." *Frontiers in Microbiology* no. 10 (992). doi: 10.3389/fmicb.2019.00992.

Andrade, de Nery, de Santana, Leal Rodrigues De Reis, Druzian, and Machado. 2019. "Effect of cellulose nanocrystals from different lignocellulosic residues to chitosan/glycerol films." *Polymers* no. 11:658. doi: 10.3390/polym11040658.

Anjum, A., M. Zuber, K. M. Zia, A. Noreen, M. N. Anjum, and S. Tabasum. 2016. "Microbial production of polyhydroxyalkanoates (PHAs) and its copolymers: A review of recent advancements." *Int J Biol Macromol* no. 89:161–174. doi: 10.1016/j.ijbiomac.2016.04.069.

Anonymous. 2019. *Coca-Cola Expands Access to PlantBottle IP* [cited 13th FEB 2020. Available from https://www.coca-colacompany.com/news/coca-cola-expands-access-to-plantbottle-ip.

Atef, Maryam, Masoud Rezaei, and Rabi Behrooz. 2015. "Characterization of physical, mechanical, and antibacterial properties of agar-cellulose bionanocomposite films incorporated with savory essential oil." *Food Hydrocolloids* no. 45:150–157. doi: https://doi.org/10.1016/j.foodhyd.2014.09.037.

Avila-Orta, Carlos A., Medellı, X., Rodrı N. X., Francisco J. Guez, Zhi-Gang Wang, Rodrı Navarro, X., Dámaso Guez, Benjamin S. Hsiao, and Fengji Yeh. 2003. "On the nature of multiple melting in poly(ethylene terephthalate) (PET) and its copolymers with cyclohexylene dimethylene terephthalate (PET/CT)." *Polymer* no. 44 (5):1527–1535. doi: https://doi.org/10.1016/S0032-3861(02)00832-7.

Ayu, Rafiqah S., Abdan Khalina, Ahmad Saffian Harmaen, Khairul Zaman, Mohammad Jawaid, and Ching Hao Lee. 2018. "Effect of modified tapioca starch on mechanical, thermal, and morphological properties of pbs blends for food packaging." *Polymers* no. 10 (11):1187.

Ayu, Rafiqah S., Abdan Khalina, Ahmad Saffian Harmaen, Khairul Zaman, N. Mohd Nurrazi, Tawakkal Isma, and Ching Hao Lee. 2020. "Effect of empty fruit brunch reinforcement in polybutylene-succinate/modified tapioca starch blend for agricultural mulch films." *Scientific Reports* no. 10 (1):1166. doi: 10.1038/s41598-020-58278-y.

Bajaj, I., and R. Singhal. 2011. "Poly (glutamic acid) – an emerging biopolymer of commercial interest." *Bioresour Technol* no. 102 (10):5551–5561. doi: 10.1016/j.biortech.2011.02.047.

Baley, Christophe, Marine Lan, Alain Bourmaud, and Antoine Le Duigou. 2018. "Compressive and tensile behaviour of unidirectional composites reinforced by natural fibres: Influence of fibres (flax and jute), matrix and fibre volume fraction." *Materials Today Communications* no. 16:300–306. doi: https://doi.org/10.1016/j.mtcomm.2018.07.003.

Beopoulos, Athanasios, Julien Cescut, Ramdane Haddouche, Jean-Louis Uribelarrea, Carole Molina-Jouve, and Jean-Marc Nicaud. 2009. "Yarrowia lipolytica as a model for bio-oil production." *Progress in Lipid Research* no. 48 (6):375–387. doi: https://doi.org/10.1016/j.plipres.2009.08.005.

Bhattacharyya, Debes, Aruna Subasinghe, and Nam Kyeun Kim. 2015. "Chapter 4 – Natura fibers: Their composites and flammability characterizations." In *Multifunctionality of Polymer Composites*, edited by Klaus Friedrich and Ulf Breuer, 102–143. Oxford: William Andrew Publishing.

Blazeck, John, Andrew Hill, Leqian Liu, Rebecca Knight, Jarrett Miller, Anny Pan, Peter Otoupal, and Hal S. Alper. 2014. "Harnessing Yarrowia lipolytica lipogenesis to create a platform for lipid and biofuel production." *Nature Communications* no. 5 (1):3131. doi: 10.1038/ncomms4131.

Bondeson, Daniel, and Kristiina Oksman. 2007. "Polylactic acid/cellulose whisker nanocomposites modified by polyvinyl alcohol." *Composites Part A: Applied Science and Manufacturing* no. 38 (12):2486–2492. doi: https://doi.org/10.1016/j.compositesa.2007.08.001.

Bouasker, Marwen, Naima Belayachi, Dashnor Hoxha, and Muzahim Al-Mukhtar. 2014. "Physical characterization of natural straw fibers as aggregates for construction materials applications." *Materials* no. 7. doi: 10.3390/ma7043034.

Braunegg Gerhart, Rodolfo Bona, Florian Schellauf, and Elisabeth Wallner. 2002. "Polyhydroxyalkanoates (PHAs): Sustainable biopolyester production." *Polimery/Polymers* no. 47:479–484. doi: 10.14314/polimery.2002.479.

Candela, T., S. Balomenou, V. Aucher, V. Bouriotis, J. P. Simore, A. Fouet, and I. G. Boneca. 2014. "N-acetylglucosamine deacetylases modulate the anchoring of the gamma-glutamyl capsule to the cell wall of Bacillus anthracis." *Microb Drug Resist* no. 20 (3): 222–230. doi: 10.1089/mdr.2014.0063.

Carly, Cassella. 2019. *Here's How at Least 74,000 Microplastic Particles End Up in Your Diet in a Single Year.* Scinecealert [cited 13th FEB 2020. https://www.sciencealert.com/humans-consume-over-74-000-microplastic-particles-each-year-without-knowing-it

Carus, Michael, Stefan Karst, Alexandre Kauffmann, John Hobsonand, and Sylvestre Bertucelli. 2013. *The European Hemp Industry: Cultivation, Processing and Applications for Fibres, Shivs and Seeds.* March 1–9, Brussels, Belgium, European Industrial Hemp Association.

Chand, Navin, and Mohammed Fahim. 2008. "*Cotton Reinforced Polymer Composites.*" In, 129-161.

Chang, Fuxiang, Seung-Hwan Lee, Keisuke Toba, Asahiro Nagatani, and Takashi Endo. 2012. "Bamboo nanofiber preparation by HCW and grinding treatment and its application for nanocomposite." *Wood Science and Technology* no. 46 (1):393–403. doi: 10.1007/s00226-011-0416-0.

Chen, Guo-Qiang. 2009. Industrial Production of PHA. In: Chen GQ (eds) *Plastics from Bacteria, Microbiology Monographs*, vol 14. Springer, Berlin, Heidelberg. https://doi.org/10.1007/978-3-642-03287-5_6.

Chettri, R., M. O. Bhutia, and J. P. Tamang. 2016. "Poly-gamma-Glutamic Acid (PGA)-producing bacillus species isolated from kinema, indian fermented soybean food." *Front Microbiol* no. 7:971. doi: 10.3389/fmicb.2016.00971.

Chiellini, E., A. Barghini, P. Cinelli, and V. I. Ilieva. 2008. "15 – Overview of environmentally compatible polymeric materials for food packaging." In *Environmentally Compatible Food Packaging*, edited by Emo Chiellini, 371–395. Cambridge: Woodhead Publishing.

Costa, Larissa A. S., Denilson de J. Assis, Gleice V. P. Gomes, Jania B. A. da Silva, Ananda F. Fonsêca, and Janice I. Druzian. 2015. "Extraction and characterization of nanocellulose from corn stover." *Materials Today: Proceedings* no. 2 (1):287–294. doi: https://doi.org/10.1016/j.matpr.2015.04.045.

Da Silva, Dana, Maya Kaduri, Maria Poley, Omer Adir, Nitzan Krinsky, Janna Shainsky-Roitman, and Avi Schroeder. 2018. "Biocompatibility, biodegradation and excretion of polylactic acid (PLA) in medical implants and theranostic systems." *Chemical Engineering Journal* no. 340:9–14. doi: https://doi.org/10.1016/j.cej.2018.01.010.

Dashtizadeh, Zahra, Abdan Khalina, Francisco Cardona, and Ching Hao Lee. 2019. "Mechanical characteristics of green composites of short kenaf bast fiber reinforced in cardanol." *Advances in Materials Science and Engineering* no. 2019:6. doi: 10.1155/2019/8645429.

Datta, Rathin, and Michael Henry. 2006. "Lactic acid: recent advances in products, processes and technologies — a review." *Journal of Chemical Technology & Biotechnology* no. 81 (7):1119–1129. doi: 10.1002/jctb.1486.

De Olveira, Lívia Ávila, Júlio César dos Santos, Túlio Hallak Panzera, Rodrigo Teixeira Santos Freire, Luciano Machado Gomes Vieira, and Juan Carlos Campos Rubio. 2018. "Investigations on short coir fibre–reinforced composites via full factorial design." *Polymers and Polymer Composites* no. 26(7):391–399. doi: 10.1177/0967391118806144.

Dien, Bruce S., J. Y. Zhu, Patricia J. Slininger, Cletus P. Kurtzman, Bryan R. Moser, Patricia J. O'Bryan, Roland Gleisner, and Michael A. Cotta. 2016. "Conversion of SPORL pre-treated Douglas fir forest residues into microbial lipids with oleaginous yeasts." *RSC Advances* no. 6 (25):20695–20705. doi: 10.1039/C5RA24430G.

Edge, M., M. Hayes, M. Mohammadian, N. S. Allen, T. S. Jewitt, K. Brems, and K. Jones. 1991. "Aspects of poly(ethylene terephthalate) degradation for archival life and environmental degradation." *Polymer Degradation and Stability* no. 32(2):131–153. doi: https://doi.org/10.1016/0141-3910(91)90047-U.

Elzoghby, A. O., M. M. Elgohary, and N. M. Kamel. 2015. "Implications of protein- and peptide-based nanoparticles as potential vehicles for anticancer drugs." *Advances in Protein Chemistry and Structural Biology* no. 98:169–221. doi: 10.1016/bs.apcsb.2014.12.002.

FAO. 2019. *Food and Agriculture Organization of the United Nations* [cited 16th February 2020]. Available from http://www.fao.org/faostat/en/#data/QC

Favier, V., H. Chanzy, and J. Y. Cavaille. 1995. "Polymer nanocomposites reinforced by cellulose whiskers." *Macromolecules* no. 28 (18):6365–6367. doi: 10.1021/ma00122a053.

Fortunati, E., D. Puglia, F. Luzi, C. Santulli, J. M. Kenny, and L. Torre. 2013. "Binary PVA bio-nanocomposites containing cellulose nanocrystals extracted from different natural sources: Part I." *Carbohydrate Polymers* no. 97 (2):825–836. doi: https://doi.org/10.1016/j.carbpol.2013.03.075.

Gabr, Mohamed H., Nguyen T. Phong, Mohammad Ali Abdelkareem, Kazuya Okubo, Kiyoshi Uzawa, Isao Kimpara, and Toru Fujii. 2013. "Mechanical, thermal, and moisture absorption properties of nano-clay reinforced nano-cellulose biocomposites." *Cellulose* no. 20 (2):819–826. doi: 10.1007/s10570-013-9876-8.

Gan, P. G., S. T. Sam, Muhammad Faiq bin Abdullah, and Mohd Firdaus Omar. 2020. "Thermal properties of nanocellulose-reinforced composites: A review." *Journal of Applied Polymer Science* no. 137 (11):48544. doi: 10.1002/app.48544.

George, J., R. Kumar, V. A. Sajeevkumar, K. V. Ramana, R. Rajamanickam, V. Abhishek, S. Nadanasabapathy, and Siddaramaiah. 2014. "Hybrid HPMC nanocomposites containing bacterial cellulose nanocrystals and silver nanoparticles." *Carbohydrate Polymers* no. 105:285–292. doi:10.1016/j.carbpol.2014.01.057.

Ghaderi, Moein, Mohammad Mousavi, Hossein Yousefi, and Mohsen Labbafi. 2014. "All-cellulose nanocomposite film made from bagasse cellulose nanofibers for food packaging application." *Carbohydrate Polymers* no. 104:59–65. doi: https://doi.org/10.1016/j.carbpol.2014.01.013.

Gonzalez, K., A. Retegi, A. Gonzalez, A. Eceiza, and N. Gabilondo. 2015. "Starch and cellulose nanocrystals together into thermoplastic starch bionanocomposites." *Carbohydrate Polymers* no. 117:83–90. doi: 10.1016/j.carbpol.2014.09.055.

Gupta, Manoj, and Ravindra Srivastava. 2016. "Properties of sisal fibre reinforced epoxy composite." *Indian Journal of Fibre and Textile Research* no. 41:235–241.

Gupta, Manoj, Ravindra Srivastava, and Himanshu Bisaria. 2015. "Potential of jute fibre reinforced polymer composites: A review." *International Journal of Fiber and Textile Research* no. 5:30–38.

Hamad, K., M. Kaseem, H. W. Yang, F. Deri, and Y. G. Ko. 2015. "Properties and medical applications of polylactic acid: A Review." *Express Polymer Letters* no. 9 (5):435–455.

Han, Shenjie, Qiufang Yao, Chunde Jin, Bitao Fan, Huanhuan Zheng, and Qingfeng Sun. 2018. "Cellulose nanofibers from bamboo and their nanocomposites with polyvinyl alcohol: Preparation and characterization." *Polymer Composites* no. 39 (8):2611–2619. doi: 10.1002/pc.24249.

He, X., X. Wang, J. Fang, Y. Chang, N. Ning, H. Guo, L. Huang, X. Huang, and Z. Zhao. 2017. "Structures, biological activities, and industrial applications of the polysaccharides from Hericium erinaceus (Lion's Mane) mushroom: A review." *International Journal of Biological Macromolecules* no. 97:228–237. doi: 10.1016/j.ijbiomac.2017.01.040.

Herrera, Natalia, Aji P. Mathew, and Kristiina Oksman. 2015. "Plasticized polylactic acid/cellulose nanocomposites prepared using melt-extrusion and liquid feeding: Mechanical, thermal and optical properties." *Composites Science and Technology* no. 106:149–155. doi: https://doi.org/10.1016/j.compscitech.2014.11.012.

Hess, Glenn, and Jeff Johnson. 2014. *Deconstructing Inherently Safer Technology*. Chemical and Engineering News [cited 13th FEB 2020. Available from https://cen.acs.org/articles/92/i10/Deconstructing-Inherently-Safer-Technology.html.

Hosseini, S. Behnam. 2017. "A Review: nanomaterials as a filler in natural fiber reinforced composites." *Journal of Natural Fibers* no.14(3):311–325.doi:10.1080/15440478.2016.1212765.

Hristozov, Dimo, Laura Wroblewski, and Pedram Sadeghian. 2016. "Long-term tensile properties of natural fibre-reinforced polymer composites: Comparison of flax and glass fibres." *Composites Part B: Engineering* no. 95:82–95. doi: https://doi.org/10.1016/j.compositesb.2016.03.079.

Husseinsyah, Salmah, Marliza Mostapha, and E. Selvi. 2014. "Bio composites from polypropylene and corn cob: effect maleic anhydride polypropylene." *Advanced Materials Research* no. 3:129–137. doi: 10.12989/amr.2014.3.3.129.

Ibrahim, Mohamed, S. Sapuan, E. S. Zainudin, and Mohd Zuhri Mohamed Yusoff. 2019. "Potential of using multiscale corn husk fiber as reinforcing filler in cornstarch-based biocomposites." *International Journal of Biological Macromolecules* no. 139. doi: 10.1016/j.ijbiomac.2019.08.015.

Ivanov, Evgeni, Rumiana Kotsilkova, Hesheng Xia, Yinghong Chen, Ricardo K. Donato, Katarzyna Donato, Anna Paula Godoy, Rosa Di Maio, Clara Silvestre, Sossio Cimmino, and Verislav Angelov. 2019. "PLA/Graphene/MWCNT Composites with Improved Electrical and Thermal Properties Suitable for FDM 3D Printing Applications." *Applied Sciences* no. 9 (6):1209.

Jang, Hye Lim, Andrea M. Liceaga, and Kyung Young Yoon. 2016. "Purification, characterisation and stability of an antioxidant peptide derived from sandfish (Arctoscopus japonicus) protein hydrolysates." *Journal of Functional Foods* no. 20:433–442. doi: https://doi.org/10.1016/j.jff.2015.11.020.

Jayaweera, Denisha, D. W. T. S. Karunaratne, S. T. S. Bandara, and Shantha Walpalage. 2017. *Investigation of the Effectiveness of Nanocellulose Extracted from Sri Lankan Kapok, as a FILLER IN POLYPROPYLENE POLymer matrix*. Moratuwa, Sri Lanka: Institute of electronic and Electrical Engineers.

Jiang, Yi, and Katja Loos. 2016. "Enzymatic synthesis of biobased polyesters and polyamides." *Polymers* no. 8 (7):243.

Jin, D. X., X. L. Liu, X. Q. Zheng, X. J. Wang, and J. F. He. 2016. "Preparation of antioxidative corn protein hydrolysates, purification and evaluation of three novel corn antioxidant peptides." *Food Chemistry* no. 204:427–436. doi: 10.1016/j.foodchem.2016.02.119.

Josh, Quarterman, Cletus Kurtzman, Stephanie Thompson, and Bruce Dien. 2017. "A survey of yeast from the Yarrowia clade for lipid production in dilute acid pretreated lignocellulosic biomass hydrolysate." *Applied Microbiology and Biotechnology* no. 101. doi: 10.1007/s00253-016-8062-y.

Kamal, Musa R., and Vahid Khoshkava. 2015. "Effect of cellulose nanocrystals (CNC) on rheological and mechanical properties and crystallization behavior of PLA/CNC nanocomposites." *Carbohydrate Polymers* no. 123:105–114. doi: https://doi.org/10.1016/j.carbpol.2015.01.012.

Kapoor, M., D. Panwar, and G. S. Kaira. 2016. "Chapter 3 - Bioprocesses for Enzyme Production Using Agro-Industrial Wastes: Technical Challenges and Commercialization Potential." In *Agro-Industrial Wastes as Feedstock for Enzyme Production*, edited by Gurpreet Singh Dhillon and Surinder Kaur, 61–93. San Diego: Academic Press.

Kılınç, Ahmet Çağrı, Cenk Durmuşkahya, and M. Özgür Seydibeyoğlu. 2017. "10 - Natural fibers." In *Fiber Technology for Fiber-Reinforced Composites*, edited by M. Özgür Seydibeyoğlu, Amar K. Mohanty and Manjusri Misra, 209–235. Duxford: Woodhead Publishing.

Koller, M. 2018. "Biodegradable and biocompatible polyhydroxy-alkanoates (PHA): Auspicious microbial macromolecules for pharmaceutical and therapeutic applications." *Molecules* no. 23 (2). doi: 10.3390/molecules23020362.

Koller, M., L. Marsalek, M. M. de Sousa Dias, and G. Braunegg. 2017. "Producing microbial polyhydroxyalkanoate (PHA) biopolyesters in a sustainable manner." *New Biotechnology* no. 37 (Pt A):24–38. doi: 10.1016/j.nbt.2016.05.001.

Kong, Y., and J. N. Hay. 2003. "Multiple melting behaviour of poly(ethylene terephthalate)." *Polymer* no. 44(3):623–633. doi: https://doi.org/10.1016/S0032-3861(02)00814-5.

Kourmentza, C., J. Placido, Venetsaneas Nikolaos, Burniol-figols Anna, Varrone Cristiano, N Gavala Hariklia, and A M Reis. Maria. 2017. "Recent Advances and Challenges towards Sustainable Polyhydroxyalkanoate (PHA) Production." no. 4(2). doi: 10.3390/bioengineering4020055.

Kovalcik, A., K. Meixner, M. Mihalic, W. Zeilinger, I. Fritz, W. Fuchs, P. Kucharczyk, F. Stelzer, and B. Drosg. 2017. "Characterization of polyhydroxyalkanoates produced by Synechocystis salina from digestate supernatant." *International Journal of Biological Macromolecules* no. 102:497–504. doi: 10.1016/j.ijbiomac.2017.04.054.

Kumar, Muthu. 2019. "Wood Waste -Carbon Source for Polyhydroxyalkanoates (PHAs) production." *International Journal of Forestry and Wood Science*.

Kundalwal, Shailesh I. 2018. "Review on micromechanics of nano- and micro-fiber reinforced composites." *Polymer Composites* no. 39 (12):4243–4274. doi: 10.1002/pc.24569.

Kyulavska, Mariya, Natalia Toncheva-Moncheva, and Joanna Rydz. 2017. "Biobased polyamide ecomaterials and their susceptibility to biodegradation." In *Handbook of Ecomaterial*, 1–34. Cham, Switzerland: Springer.

Lasprilla, Astrid J. R., Guillermo A. R. Martinez, Betânia H. Lunelli, André L. Jardini, and Rubens Maciel Filho. 2012. "Poly-lactic acid synthesis for application in biomedical devices — A review." *Biotechnology Advances* no. 30 (1):321–328. doi: https://doi.org/10.1016/j.biotechadv.2011.06.019.

Laura, Parker. 2018. *Here's How Much Plastic Trash Is Littering The Earth*. National Geographic [cited 13th FEB 2020. https://www.nationalgeographic.com/news/2017/07/plastic-produced-recycling-waste-ocean-trash-debris-environment/

Laycock, Bronwyn, Peter Halley, Steven Pratt, Alan Werker, and Paul Lant. 2013. "The chemomechanical properties of microbial polyhydroxyalkanoates." *Progress in Polymer Science* no. 38 (3):536–583. doi: https://doi.org/10.1016/j.progpolymsci.2012.06.003.

Leao, Alcides, Sivoney Souza, Bibin Cherian, Elisabete Frollini, Seemon Thomas, Laly Pothan, and Kottai Samy. 2010. "Agro-based biocomposites for industrial applications." *Molecular Crystals and Liquid Crystals* no. 522:18/[318]–27/[327]. doi:10.1080/15421401003719852.

Lee, C. H., Mohd Sapuan Salit, and M. R. Hassan. 2014. "A review of the flammability factors of kenaf and allied fibre reinforced polymer composites." *Advances in Materials Science and Engineering* no. 2014:8. doi: 10.1155/2014/514036.

Lee, C. H., S. M. Sapuan, and M. R. Hassan. 2017. "Mechanical and thermal properties of kenaf fiber reinforced polypropylene/magnesium hydroxide composites." *Journal of Engineered Fibers and Fabrics* no. 12 (2). doi: 10.1177/155892501701200206.

Lee, C.H., S.M. Sapuan, and M.R. Hassan. 2018. "Thermal analysis of kenaf fiber reinforced floreon biocomposites with magnesium hydroxide flame retardant filler." *Polymer Composites* no. 39 (3):869–875. doi: 10.1002/pc.24010.

Lindner, Paul. 1922. "Das Problem der biologischen Fettbildung und Fettgewinnung." *Angewandte Chemie* no. 35 (19):110–114. doi: 10.1002/ange.19220351903.

Lisa, Anne Hamilton, and Feit Steven. 2019. *Plastic & Climate- The hidden costs of a plastic planet*. Center for International Environmental Law (CIEL), Washington, DC.

Ma, B., J. Han, S. Zhang, F. Liu, S. Wang, J. Duan, Y. Sang, H. Jiang, D. Li, S. Ge, J. Yu, and H. Liu. 2018. "Hydroxyapatite nanobelt/polylactic acid Janus membrane with osteoinduction/barrier dual functions for precise bone defect repair." *Acta Biomater* no. 71:108–117. doi: 10.1016/j.actbio.2018.02.033.

Ma, Hao, Bo Zhou, Hong-Sheng Li, Yi-Qun Li, and Shi-Yi Ou. 2011. "Green composite films composed of nanocrystalline cellulose and a cellulose matrix regenerated from functionalized ionic liquid solution." *Carbohydrate Polymers* no. 84 (1):383–389. doi: https://doi.org/10.1016/j.carbpol.2010.11.050.

Macedo, Murilo J. P., Giovanna S. Silva, Michele C. Feitor, Thércio H. C. Costa, Edson N. Ito, and José D. D. Melo. 2019. "Composites from recycled polyethylene and plasma treated kapok fibers." *Cellulose*. doi: 10.1007/s10570-019-02946-4.

Madison, L. L., and G. W. Huisman. 1999. "Metabolic engineering of poly(3-hydroxyalkanoates): from DNA to plastic." *Microbiology and Molecular Biology Reviews* no. 63 (1):21–53.

Mani, Dr Ganesh Kumar, John Bosco Balaguru Rayappan, and Dillip Bisoyi. 2012. "Synthesis and characterization of kapok fibers and its composites." *Journal of Applied Sciences* no. 12:1661–1665. doi: 10.3923/jas.2012.1661.1665.

Martínez-Sanz, Marta, Amparo Lopez-Rubio, and Jose M. Lagaron. 2013. "High-barrier coated bacterial cellulose nanowhiskers films with reduced moisture sensitivity." *Carbohydrate Polymers* no. 98 (1):1072–1082. doi: https://doi.org/10.1016/j.carbpol.2013.07.020.

Martínez-Sanz, Marta, Richard T. Olsson, Amparo Lopez-Rubio, and Jose M. Lagaron. 2011. "Development of electrospun EVOH fibres reinforced with bacterial cellulose nanowhiskers. Part I: Characterization and method optimization." *Cellulose* no. 18 (2):335–347. doi: 10.1007/s10570-010-9471-1.

Mat Zubir, Nurul Hani, Sung Ting Sam, Ragunathan Santiagoo, N. Z. Noimam, and Jing Wang. 2016. "Tensile properties of rice straw fiber reinforced poly(lactic acid) biocomposites." *Advanced Materials Research* no. 1133:598–602. doi: 10.4028/www.scientific.net/AMR.1133.598.

Matsumoto, K., T. Murata, R. Nagao, C. T. Nomura, S. Arai, Y. Arai, K. Takase, H. Nakashita, S. Taguchi, and H. Shimada. 2009. "Production of short-chain-length/medium-chain-length polyhydroxyalkanoate (PHA) copolymer in the plastid of Arabidopsis thaliana using an engineered 3-ketoacyl-acyl carrier protein synthase III." *Biomacromolecules* no. 10 (4):686–690. doi: 10.1021/bm8013878.

Md Shah, Ain, Mohamed Sultan, Mohammad Jawaid, Francisco Cardona, and Abd Rahim Abu Talib. 2016. "A review on the tensile properties of bamboo fiber reinforced polymer composites." *Bioresources* no. 11. doi: 10.15376/biores.11.4.Shah.

Mondragon, G., C. Peña-Rodriguez, A. González, A. Eceiza, and A. Arbelaiz. 2015. "Bionanocomposites based on gelatin matrix and nanocellulose." *European Polymer Journal* no. 62:1–9. doi: https://doi.org/10.1016/j.eurpolymj.2014.11.003.

Montero de Espinosa, Lucas, and Michael A. R. Meier. 2011. "Plant oils: The perfect renewable resource for polymer science?!" *European Polymer Journal* no. 47 (5):837–852. doi: https://doi.org/10.1016/j.eurpolymj.2010.11.020.

Morais, João Paulo Saraiva, Morsyleide de Freitas Rosa, Men de sá Moreira de Souza Filho, Lidyane Dias Nascimento, Diego Magalhães do Nascimento, and Ana Ribeiro Cassales. 2013. "Extraction and characterization of nanocellulose structures from raw cotton linter." *Carbohydrate Polymers* no. 91 (1):229–235. doi: https://doi.org/10.1016/j.carbpol.2012.08.010.

Morschbacker, Antonio. 2009. "Bio-ethanol based ethylene." *Polymer Reviews* no. 49 (2):79–84. doi: 10.1080/15583720902834791.

Muhammadi, Shabina, Muhammad Afzal, and Shafqat Hameed. 2015. "Bacterial polyhydroxyalkanoates-eco-friendly next generation plastic: Production, biocompatibility, biodegradation, physical properties and applications." *Green Chemistry Letters and Reviews* no. 8 (3–4):56–77. doi: 10.1080/17518253.2015.1109715.

Nadar, Shamraja S., Priyanka Rao, and Virendra K. Rathod. 2018. "Enzyme assisted extraction of biomolecules as an approach to novel extraction technology: A review." *Food Research International* no. 108:309–330. doi: https://doi.org/10.1016/j.foodres.2018.03.006.

Nakagaito, Antonio, Akihiro Fujimura, Toshiaki Sakai, Yoshiaki Hama, and Hiroyuki Yano. 2009. "Production of microfibrillated cellulose (MFC)-reinforced polylactic acid (PLA) nanocomposites from sheets obtained by a papermaking-like process." *Composites Science and Technology - COMPOSITES SCI TECHNOL* no. 69:1293–1297. doi: 10.1016/j.compscitech.2009.03.004.

Nakajima, Hajime, Peter Dijkstra, and Katja Loos. 2017. "The recent developments in bio-based polymers toward general and engineering applications: Polymers that are upgraded from biodegradable polymers, analogous to petroleum-derived polymers, and newly developed." *Polymers* no. 9 (10):523.

Neves, Anna, Lázaro Rohen, Dhyemila Mantovani, Juliana Carvalho, Carlos Vieira, F. P. D. Lopes, Noan Simonassi, Fernanda Luz, and Sergio Monteiro. 2019. "Comparative mechanical properties between biocomposites of Epoxy and polyester matrices reinforced by hemp fiber." *Journal of Materials Research and Technology.* doi: 10.1016/j.jmrt.2019.11.056.

Odusote, Jamiu. 2016. "Mechanical properties of pineapple leaf fibre reinforced polymer composites for application as prosthetic socket." *Journal of Engineering Technology* no. 7:125–139. doi: 10.21859/jet-06011.

Ogunleye, A., A. Bhat, V. U. Irorere, D. Hill, C. Williams, and I. Radecka. 2015. "Poly-gamma-glutamic acid: production, properties and applications." *Microbiology* no. 161 (Pt 1):1–17. doi: 10.1099/mic.0.081448-0.

Ogunniyi, D. S. 2006. "Castor oil: a vital industrial raw material." *Bioresour Technol* no. 97 (9):1086–1091. doi: 10.1016/j.biortech.2005.03.028.

Oksman, K., A. P. Mathew, D. Bondeson, and I. Kvien. 2006. "Manufacturing process of cellulose whiskers/polylactic acid nanocomposites." *Composites Science and Technology* no. 66 (15):2776–2784. doi: https://doi.org/10.1016/j.compscitech.2006.03.002.

Pandele, Andreea Madalina, Andreea Constantinescu, Ionut Cristian Radu, Florin Miculescu, Stefan Ioan Voicu, and Lucian Toma Ciocan. 2020. "Synthesis and characterization of PLA-micro-structured hydroxyapatite composite films." *Materials* no. 13 (2):274.

Panicker, Arun. M., K. A. Rajesh, and T. O. Varghese. 2017. "Mixed morphology nanocrystalline cellulose from sugarcane bagasse fibers/poly(lactic acid) nanocomposite films: synthesis, fabrication and characterization." *Iranian Polymer Journal* no. 26 (2):125–136. doi: 10.1007/s13726-017-0504-6.

Panthapulakkal, S., A. Zereshkian, and M. Sain. 2006. "Preparation and characterization of wheat straw fibers for reinforcing application in injection molded thermoplastic composites." *Bioresource Technology* no. 97 (2):265–272. doi: https://doi.org/10.1016/j.biortech.2005.02.043.

Peças, Paulo, Hugo Carvalho, Hafiz Salman, and Marco Leite. 2018. "Natural Fibre Composites and Their Applications: A Review." *Journal of Composites Science* no. 2 (4):66.

Pejić, Biljana M., Ana D. Kramar, Bratislav M. Obradović, Milorad M. Kuraica, Andrijana A. Zekić, and Mirjana M. Kostić. 2020. "Effect of plasma treatment on chemical composition, structure and sorption properties of lignocellulosic hemp fibers (*Cannabis sativa* L.)." *Carbohydrate Polymers*:116000. doi: https://doi.org/10.1016/j.carbpol.2020.116000.

Pereira, A. L., D. M. do Nascimento, S. Souza Filho Mde, J. P. Morais, N. F. Vasconcelos, J. P. Feitosa, A. I. Brigida, and F. Rosa Mde. 2014. "Improvement of polyvinyl alcohol properties by adding nanocrystalline cellulose isolated from banana pseudostems." *Carbohydrate Polymers* no. 112:165–172. doi: 10.1016/j.carbpol.2014.05.090.

Perelshtein, Ilana, G. Applerot, N. Perkas, Eva Sigl, Andrea Hasmann Heinzle, Guebitz Gm, and A. Gedanken. 2009. "CuO-cotton nanocomposite: Formation, morphology, and antibacterial activity." *Surface and Coatings Technology* no. 204:54–57. doi: 10.1016/j.surfcoat.2009.06.028.

Petersson, L., I. Kvien, and K. Oksman. 2007. "Structure and thermal properties of poly(lactic acid)/cellulose whiskers nanocomposite materials." *Composites Science and Technology* no. 67 (11):2535–2544. doi: https://doi.org/10.1016/j.compscitech.2006.12.012.

Petersson, L., and K. Oksman. 2006. "Biopolymer based nanocomposites: Comparing layered silicates and microcrystalline cellulose as nanoreinforcement." *Composites Science and Technology* no. 66 (13):2187–2196. doi: https://doi.org/10.1016/j.compscitech.2005.12.010.

Poontawee, Rujiralai, Wichien Yongmanitchai, and Savitree Limtong. 2017. "Efficient oleaginous yeasts for lipid production from lignocellulosic sugars and effects of lignocellulose degradation compounds on growth and lipid production." *Process Biochemistry* no. 53:44–60. doi: https://doi.org/10.1016/j.procbio.2016.11.013.

Qiao, K., S. H. Imam Abidi, H. Liu, H. Zhang, S. Chakraborty, N. Watson, P. Kumaran Ajikumar, and G. Stephanopoulos. 2015. "Engineering lipid overproduction in the oleaginous yeast Yarrowia lipolytica." *Metabolic Engineering* no. 29:56–65. doi: 10.1016/j.ymben.2015.02.005.

Qiu, S., M. P. Yadav, and L. Yin. 2017. "Characterization and functionalities study of hemicellulose and cellulose components isolated from sorghum bran, bagasse and biomass." *Food Chemistry* no. 230:225–233. doi: 10.1016/j.foodchem.2017.03.028.

Rajesh, S., B. Vijayaramnath, C. Elanchezhian, S. Vivek, M. Hari Prasadh, and M. Kesavan. 2019. "Experimental investigation of tensile and impact behavior of aramid-natural fiber composite." *Materials Today: Proceedings* no. 16:699–705. doi: https://doi.org/10.1016/j.matpr.2019.05.148.

Ramamoorthy, Sunil Kumar, Mikael Skrifvars, and Anders Persson. 2015. "A review of natural fibers used in biocomposites: Plant, animal and regenerated cellulose fibers." *Polymer Reviews* no. 55 (1):107–162. doi: 10.1080/15583724.2014.971124.

Ramesh, M. 2019. "Flax (*Linum usitatissimum* L.) fibre reinforced polymer composite materials: A review on preparation, properties and prospects." *Progress in Materials Science* no. 102:109–166. doi: https://doi.org/10.1016/j.pmatsci.2018.12.004.

Rana, Masud, Bin Hao, Lei Mu, Lin Chen, and Peng-Cheng Ma. 2016. "Development of multifunctional cotton fabrics with Ag/AgBr-TiO$_2$ nanocomposite coating." *Composites Science and Technology* no. 122:104–112. doi: https://doi.org/10.1016/j.compscitech.2015.11.016.

Ranalli, Paolo, and Gianpietro Venturi. 2004. "Hemp as a raw material for industrial applications." *Euphytica* no. 140 (1):1–6. doi: 10.1007/s10681-004-4749-8.

Ranum, P., J. P. Pena-Rosas, and M. N. Garcia-Casal. 2014. "Global maize production, utilization, and consumption." *Annals of the New York Academy of Sciences* no. 1312: 105–112. doi: 10.1111/nyas.12396.

Ratledge, C., and J. P. Wynn. 2002. "The biochemistry and molecular biology of lipid accumulation in oleaginous microorganisms." *Advances in Applied Microbiology* no. 51:1–51. doi: 10.1016/s0065-2164(02)51000-5.

Rodríguez-Contreras, Alejandra, Jordi Guillem-Marti, Oscar Lopez, José María Manero, and Elisa Ruperez. 2019. "Antimicrobial PHAs coatings for solid and porous tantalum implants." *Colloids and Surfaces B: Biointerfaces* no. 182:110317. doi: https://doi.org/10.1016/j.colsurfb.2019.06.047.

Rohit, Kiran, and Savita Dixit. 2016. "A Review - Future aspect of natural fiber reinforced composite." *Polymers from Renewable Resources* no. 7 (2):43–59. doi: 10.1177/204124791600700202.

Saba, Naheed, Paridah M. Tahir, and Mohammad Jawaid. 2014. "A Review on potentiality of nano filler/natural fiber filled polymer hybrid composites." *Polymers* no. 6:2247–2273. doi: 10.3390/polym6082247.

Salimon, Jumat, Nadia Salih, and Emad Yousif. 2012. "Industrial development and applications of plant oils and their biobased oleochemicals." *Arabian Journal of Chemistry* no. 5 (2):135–145. doi: https://doi.org/10.1016/j.arabjc.2010.08.007.

Santos, Antonio, Luiz Dalla Valentina, Andrey Schulz, and Marcia Duarte. 2017. "From obtaining to degradation of phb:material properties. Part I." *Ingeniería y Ciencia* no. 13:269–298. doi: 10.17230/ingciencia.13.26.10.

Saratale, G. D., and Oh, S. E. 2012. "Lignocellulosics to ethanol: The future of the chemical and energy industry." *African Journal of Biotechnology* no. 11 (5): 1002–1013.

Shaftel, Holly. 2020. *Global Temperature*. NASA's Jet Propulsion Laboratory [cited 13th FEB 2020. California Institute of Technology, Pasadena, United States.

Shi, L. 2016. "Bioactivities, isolation and purification methods of polysaccharides from natural products: A review." *Int J Biol Macromol* no. 92:37–48. doi: 10.1016/j.ijbiomac.2016.06.100.

Shih, I. L., and Y. T. Van. 2001. "The production of poly-(gamma-glutamic acid) from microorganisms and its various applications." *Bioresour Technol* no. 79 (3):207–225. doi: 10.1016/s0960-8524(01)00074-8.

Sinha, Kumar, Agnivesh, H. K. Narang, and S. Bhattacharya. 2018. "Evaluation of Bending Strength of Abaca Reinforced Polymer Composites." *Materials Today: Proceedings* no. 5 (2, Part 2):7284–7288. doi: https://doi.org/10.1016/j.matpr.2017.11.396.

Slavutsky, Aníbal M., and María A. Bertuzzi. 2014. "Water barrier properties of starch films reinforced with cellulose nanocrystals obtained from sugarcane bagasse." *Carbohydrate Polymers* no. 110:53–61. doi: https://doi.org/10.1016/j.carbpol.2014.03.049.

Slininger, P. J., B. S. Dien, C. P. Kurtzman, B. R. Moser, E. L. Bakota, S. R. Thompson, P. J. O'Bryan, M. A. Cotta, V. Balan, M. Jin, C. Sousa Lda, and B. E. Dale. 2016. "Comparative lipid production by oleaginous yeasts in hydrolyzates of lignocellulosic biomass and process strategy for high titers." *Biotechnol Bioeng* no. 113 (8):1676–1690. doi: 10.1002/bit.25928.

Song, Zhenhua, Shiqiang Li, Jianyin Lei, Robert L. Noble, and Zhihua Wang. 2018. "Dynamic tensile properties of ROP/OCC natural hybrid fibers reinforced composites." *Composite Structures* no. 185:600–606. doi: https://doi.org/10.1016/j.compstruct.2017.11.077.

Soykeabkaew, Nattakan, Nattaya Tawichai, Chuleeporn Thanomsilp, and Orawan Suwantong. 2017. "Nanocellulose-reinforced "green" composite materials." *Walailak Journal of Science and Technology* no. 14:353–368.

Tai, M., and G. Stephanopoulos. 2013. "Engineering the push and pull of lipid biosynthesis in oleaginous yeast Yarrowia lipolytica for biofuel production." *Metab Eng* no. 15:1–9. doi: 10.1016/j.ymben.2012.08.007.

Teboho, Mokhena, Mokgaotsa Mochane, Motaung Tshwafo, Linda Linganiso, Oriel Thekisoe, and Sandile Songca. 2018. *Sugarcane Bagasse and Cellulose Polymer Composites.* London, UK: Sugarcane - Technology and Research.

Thakker, C., I. Martinez, K. Y. San, and G. N. Bennett. 2012. "Succinate production in Escherichia coli." *Biotechnology Journal* no. 7 (2):213–224. doi: 10.1002/biot.201100061.

Thomas, J. 2019. *Kenaf: Nature's Little-Known Wonder.* [cited 16th Feb 2020. Available from https://theaseanpost.com/article/kenaf-natures-little-known-wonder

Thyavihalli Girijappa, Yashas Gowda, Sanjay Mavinkere Rangappa, Jyotishkumar Parameswaranpillai, and Suchart Siengchin. 2019. "Natural Fibers as Sustainable and Renewable Resource for Development of Eco-Friendly Composites: A Comprehensive Review." *Frontiers in Materials* no. 6 (226). doi: 10.3389/fmats.2019.00226.

Todkar, Santosh Sadashiv, and Suresh Abasaheb Patil. 2019. "Review on mechanical properties evaluation of pineapple leaf fibre (PALF) reinforced polymer composites." *Composites Part B: Engineering* no. 174:106927. doi: https://doi.org/10.1016/j.compositesb.2019.106927.

Tsigie, Y. A., C. Y. Wang, C. T. Truong, and Y. H. Ju. 2011. "Lipid production from Yarrowia lipolytica Po1g grown in sugarcane bagasse hydrolysate." *Bioresource Technology* no. 102 (19):9216–9222. doi: 10.1016/j.biortech.2011.06.047.

USDA. 2020. *Cereal, Grasses, and Grains.* U.S. FOREST SERVICE [cited 16th FEB 2020. Available from https://www.fs.fed.us/wildflowers/ethnobotany/food/grains.shtml.

Van de Velde, K., and P. Kiekens. 2001. "Thermoplastic polymers: overview of several properties and their consequences in flax fibre reinforced composites." *Polymer Testing* no. 20 (8):885–893. doi: https://doi.org/10.1016/S0142-9418(01)00017-4.

Verma, Deepak. 2012. "Bagasse fiber composites: A Review." *Journal of Materials and Environmental Science* no. 3:1079–1092.

Vieira, Melissa Gurgel Adeodato, Mariana Altenhofen da Silva, Lucielen Oliveira dos Santos, and Marisa Masumi Beppu. 2011. "Natural-based plasticizers and biopolymer films: A review." *European Polymer Journal* no. 47 (3):254–263. doi: https://doi.org/10.1016/j.eurpolymj.2010.12.011.

Visakh, P. M., Sabu Thomas, Kristiina Oksman, and Aji P. Mathew. 2012. "Crosslinked natural rubber nanocomposites reinforced with cellulose whiskers isolated from bamboo waste: Processing and mechanical/thermal properties." *Composites Part A: Applied Science and Manufacturing* no. 43 (4):735–741. doi: https://doi.org/10.1016/j.compositesa.2011.12.015.

Vu, Ngo, Hang Tran, and Toan Nguyen. 2018. "Characterization of polypropylene green composites reinforced by cellulose fibers extracted from rice straw." *International Journal of Polymer Science* no. 2018:1–10. doi: 10.1155/2018/1813847.

Wang, Cai, and Shaojun Chen. 2019. "Biodegradable and water-responsive shape memory PHA-based polyurethane for tissue engineering." *Materials Today: Proceedings* no. 16:1475–1479. doi: https://doi.org/10.1016/j.matpr.2019.05.326.

Wang, Shurong, Gongxin Dai, Haiping Yang, and Zhongyang Luo. 2017a. "Lignocellulosic biomass pyrolysis mechanism: A state-of-the-art review." *Progress in Energy and Combustion Science* no. 62:33–86. doi: https://doi.org/10.1016/j.pecs.2017.05.004.

Wang, X., L. Sheng, and X. Yang. 2017b. "Pyrolysis characteristics and pathways of protein, lipid and carbohydrate isolated from microalgae Nannochloropsis sp." *Bioresour Technol* no. 229:119–125. doi: 10.1016/j.biortech.2017.01.018.

Xu, J., and B. H. Guo. 2010. "Poly(butylene succinate) and its copolymers: research, development and industrialization." *Biotechnol J* no. 5 (11):1149–1163. doi: 10.1002/biot.201000136.

Xu, Xuezhu, Fei Liu, Long Jiang, J. Y. Zhu, Darrin Haagenson, and Dennis P. Wiesenborn. 2013. "Cellulose nanocrystals vs. cellulose nanofibrils: A comparative study on their microstructures and effects as polymer reinforcing agents." *ACS Applied Materials & Interfaces* no. 5 (8):2999–3009. doi: 10.1021/am302624t.

Yu, Min, Runzhou Huang, Chunxia He, Qinglin Wu, and Xueni Zhao. 2016. "Hybrid composites from wheat straw, inorganic filler, and recycled polypropylene: Morphology and mechanical and thermal expansion performance." *International Journal of Polymer Science* no. 2016:1–12. doi: 10.1155/2016/2520670.

Yu, Xiaochen, Yubin Zheng, Kathleen M. Dorgan, and Shulin Chen. 2011. "Oil production by oleaginous yeasts using the hydrolysate from pretreatment of wheat straw with dilute sulfuric acid." *Bioresource Technology* no. 102 (10):6134–6140. doi: https://doi.org/10.1016/j.biortech.2011.02.081.

Zdanowicz, Magdalena, Katarzyna Wilpiszewska, and Tadeusz Spychaj. 2018. "Deep eutectic solvents for polysaccharides processing. A review." *Carbohydrate Polymers* no. 200:361–380. doi: https://doi.org/10.1016/j.carbpol.2018.07.078.

Zeikus, J. G., M. K. Jain, and P. Elankovan. 1999. "Biotechnology of succinic acid production and markets for derived industrial products." *Applied Microbiology and Biotechnology* no. 51 (5):545–552. doi: 10.1007/s002530051431.

Zhang, Junyu, Ekaterina I. Shishatskaya, Tatiana G. Volova, Luiziana Ferreira da Silva, and Guo-Qiang Chen. 2018. "Polyhydroxyalkanoates (PHA) for therapeutic applications." *Materials Science and Engineering: C* no. 86:144–150. doi: https://doi.org/10.1016/j.msec.2017.12.035.

Zou, Y., S. Huda, and Y. Yang. 2010. "Lightweight composites from long wheat straw and polypropylene web." *Bioresour Technol* no. 101 (6):2026–2033. doi: 10.1016/j.biortech.2009.10.042.

3 Properties of Composite Materials

Arvind Kumar Chauhan, Amarjeet Singh,
Deepak Kumar and Kuldeep Mishra

CONTENTS

3.1 INTRODUCTION

Composite materials have a macroscopic structure containing two or more non-soluble materials. One old and well-known example of a composite material is mud brick, which is prepared by fire-drying mud. It has good compressive strength but poor tensile strength. Strong fibrous straw can be a good reinforcing material to be added to mud to make excellent building blocks. The straw is used to bind clay and concrete to form an admirable building material called cob. The most appropriate properties of composite materials are:

- High stiffness and strength across a wide temperature range
- High Young's modulus
- Highly resistive to corrosion/oxidation
- Low density and light weight
- High thermal and electrical conductivity
- High wear resistance.

Concrete, a mixture of small stones, cement, and sand, has good compressive strength. Its tensile strength is enhanced by adding metal rods or wires, when it is called reinforced concrete or reinforced cement concrete (Figure 3.1a). One constituent in composite materials, the reinforcing phase in the form of fibers, particles, or flakes, is embedded in the other continuous constituent, which is called the matrix. The properties of the reinforcing and matrix phases are complementary to provide the best performance of the composite material for a particular application. The physical, mechanical, and chemical properties of the matrix, the reinforcement size, morphology and distribution, and the interface between constituents are important factors in this regard. On the basis of the well-known solid matrices (polymers, ceramic, metals, and carbon), composite materials are, in general, classified in to these three major categories:

1. Polymer-matrix composites
2. Ceramic-matrix composites
3. Metal-matrix composites.

Polymer matrix with carbon-fiber reinforcement is used to prepare some advanced composites. These carbon-fiber-reinforced polymer composite materials are, generally, lighter and stronger than monolithic systems. They have good tensile strength and provide excellent resistance to compression, which makes them appropriate for manufacturing fuselages of aircraft. The fibrous nature of the composite material inculcates good tensile strength by aligning the fibers in the direction of the applied force. The polymer matrix provides adhesiveness and stiffness properties to the composite to keep the fibers in straight columns and prevent them from buckling. The polymer matrix may be composed of thermosetting polymers such as epoxy, polyester, phenolic and polyimide resins, and thermoplastic polymers such as polypropylene, nylon 6.6, polymethylmethacrylate, polyphenylene sulfide, and polyetheretherketone [1]. The fuselages of the Boeing 787 Dreamliner and the Airbus A350 XWB have ~80% and ~83%, respectively, of their volume made from composite materials, which accounts for ~50% of their weight [2]. The potential use of composite materials in the Boeing 787 dreamliner and the Airbus A350 XWB is shown in Figure 3.1b.

Fiberglass is another example of a composite material, using glass reinforcement in a plastic matrix. It is widely used in sports equipment, decorative items, building panels, car bodies, protective sheets, etc. In this composite, the strong but brittle glass is supported by a plastic matrix that holds the glass fibers together and also protects them from damage by dispersing the forces acting on them.

Ceramic materials are well known for their useful physical and mechanical properties, characterized by high stiffness and hardness, their resistance to corrosion, and their high-temperature operational capabilities. However, their applications are restricted because of their brittle nature, which makes them very sensitive to the presence of defects and creates poor resistance to thermal and mechanical shock. In such materials, the addition of reinforcing fibers, whiskers, and particles creates a composite with improved fracture toughness and improved thermal and mechanical shock resistance. The thermal and mechanical properties of ceramic-matrix composites mean they are suitable for high-temperature applications; they are widely used in aeronautic and astronautic fields.

Metal-matrix composites provide advantages over monolithic metals such as aluminum, iron, magnesium, titanium, etc. [5], as follows:

- Higher specific strength and modulus results from reinforcing low-density metals such as aluminum and titanium
- Lower coefficients of thermal expansion when reinforcing with carbon fibers with low coefficients of thermal expansion
- Maintenance of mechanical properties at elevated temperatures
- Higher elastic properties
- Insensitivity to moisture
- Higher electric and thermal conductivities
- Better wear, fatigue, and flaw resistance
- Higher fire resistance.

Owing to these excellent properties, metal-matrix composites have a wide range of applications in automotive, aerospace, and electronics industries, and in other consumer products. These composites are widely used in ground transportation industries, including in drive shafts, engine components, and brake components [6]. The enhanced stiffness and strength of the metal-matrix composites make these materials a great choice for fabricating aircrafts.

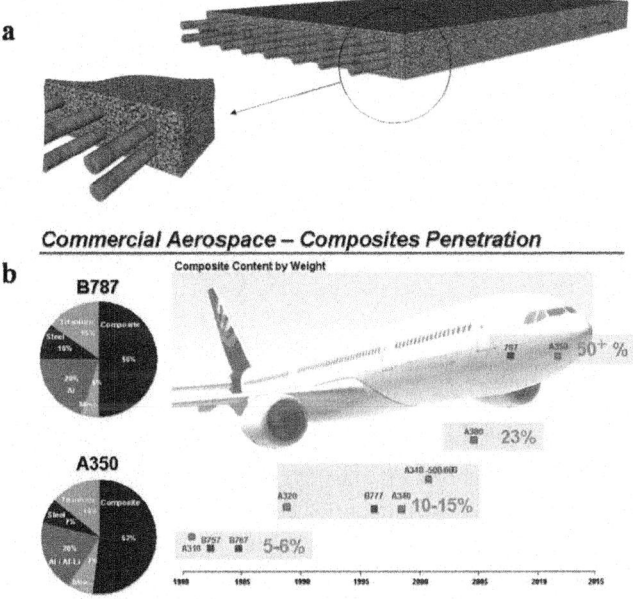

FIGURE 3.1 (a) Typical structure of reinforced concrete (Reprinted from Ref. 3); (b) the breakdown of materials used in the Boeing 787 dreamliner and the Airbus A350 XWB by weight. (Reprinted from Ref. 4)

These composite materials are different not only in terms of their composition but in their properties designed for specific applications. Recently, polymer-matrix composites have been given extraordinary attention; however, there are important applications of the other types also, revealing their potential in various social and technical applications. In this chapter, the properties of the above-mentioned composite materials will be discussed. The use of composite materials is not limited to constructions, sports, automobile, or such type of applications, but also has great importance in the energy sector. Many composite materials are used as electrodes and electrolytes in electrochemical devices like batteries, fuel cells, etc. The electrochemical properties of some composite electrode and electrolyte materials will also be discussed in the last section of this chapter.

3.2 PROPERTIES OF POLYMER-MATRIX COMPOSITES

As discussed above, polymer-matrix composites are well-exploited materials. Polymers are relatively weak materials with low stiffness and high viscoelasticity compared with the other two types of matrix. However, these matrices possess good deformability and shape versatility, and their electrical and mechanical properties can be greatly altered by using specific reinforcements. This makes them interesting for the construction of a range of structures, biomedical applications, electromagnetic shielding materials, and other devices.

Polymer matrices are of two types: thermoplastics and thermosetting. A thermoplastic is a resin that is solid at room temperature but becomes plastic and soft upon heating due to crystal melting or by virtue of crossing the glass transition temperature. Reversibility and shape versatility are some interesting properties of thermoplastics. Polyethylene, polycarbonate, polyvinyl chloride, polyamide imide, polyphenylene sulfide, polyarylsulfone, and polyetheretherketone are typical examples of thermoplastics.

Thermosetting polymers possess a crosslinked network structure and they do not soften on heating. These are, in general, liquid at room temperature and become irreversibly hard upon heating or chemical addition. The polymers form crosslinks when heated, and receive a specified shape. Some common thermosets include epoxy, polyester, polyimide, and phenolic. Both these types of polymer matrices and their composites have been explored well for various applications.

3.2.1 Electrical Properties of Polymer Composites

Classical polymers such as rubber, plastics, etc. show high resistance to electricity and come into the category of insulators. Since Heeger, MacDiarmid, and Shirakawa jointly received the Nobel Prize in Chemistry in 2000 for their work on conducting polymers, research on conducting polymers has accelerated around the world for their wide applications in electrochromic devices and sensors [7–9]. Conventional polymers, except for conjugate polymers, are electrical insulators with conductivity values of $\sim 10^{-14}$–10^{-17} S cm^{-1}. In the monomer of a typical polyethylene polymer, each carbon atom is sp^3 hybridized and, therefore, bonded to four neighboring atoms through sigma (σ) bonding. There is a large band gap between the σ-band

and the σ^*-band, which causes electrical conductivity in such polymers to be significantly low.

In contrast, graphite, a common form of carbon, possesses a layered structure with sp^2 hybridized carbon atoms and one free electron capable of forming co-planar and inter-planar bonds. Owing to this characteristic, graphite displays electrical conductivity, depending on its planar structure. The through-plane electrical conductivity of graphite is appreciably high (~50 S cm^{-1}) whereas in-plane conductivity is relatively low. Depending on the graphite loading level, the electrical conductivity of graphite-reinforced polymer composite can be as high as ~110 S cm^{-1}. The increase in the conductivity of composite is explained on the basis of the development of an intense conductive network in the matrix [10]. The processing technique is also an important factor influencing the conductivity of the composite. Polymer/graphite composite has an extremely high conductivity of up to 300 S cm^{-1} [11].

Therefore, it has been attempted to modify the electrical and physical characteristics of polymer composites by dispersing conducting reinforcements. Owing to their superior electrical properties, facile processing, and low cost, the polymer-matrix conducting composites have been largely studied. Carbon is one of the most interesting reinforcements for polymer-matrix composites. Various carbon reinforcements such as graphite, modified graphite, and carbon fiber have been used to fabricate conducting composites for several applications such as electromagnetic shielding, aerospace structural parts, electronic and biomedical functions, etc. [12]. The polymer/carbon composites offer low density, significant electrical conductivity, and other chemical and physical properties that are useful for device applications.

The electrical conductivity of these composites depends on the shape, size, concentration, fine dispersion and conductivity properties of the reinforcement. At an optimum loading of carbon, the percolation threshold, the resistivity of composite may significantly decrease and conductivity increase by several orders of magnitude [13]. Small spherical particles of up to nano-scale dimensions may result in a lower percolation threshold. Additionally, fibrous fillers, such as carbon fiber, that have an aspect ratio greater than unity possess a low percolation threshold. Percolation is also affected by the surface properties of the reinforcement and the polymer matrix. This percolation phenomenon is also observed in nano-structured carbons and metal-reinforced polymer-matrix composites. The processing technique also plays an important role in enhancing the conductivity of the composites.

Carbon black (CB) reinforcement is most commonly used for making conducting polymer matrices. It enhances the mechanical strength and, due to its high specific surface area, its presence in a polymer matrix, even in a small amount, forms a facile conductive network that greatly increases the conductivity of the polymers. Owing to its interesting electrical and mechanical properties, the research into CB-reinforced polymer composites crosses diverse fields. The concept of percolation threshold is also valid in such systems, i.e., beyond a certain concentration of CB in a polymer matrix, the variation is indistinct. The variation in electrical conductivity of some CB-reinforced polymer composites is shown in Figure 3.2. This variation of conductivity can be better understood on the basis of Figure 3.3, which shows the formation of a conductive path at the optimum concentration of CB reinforcement.

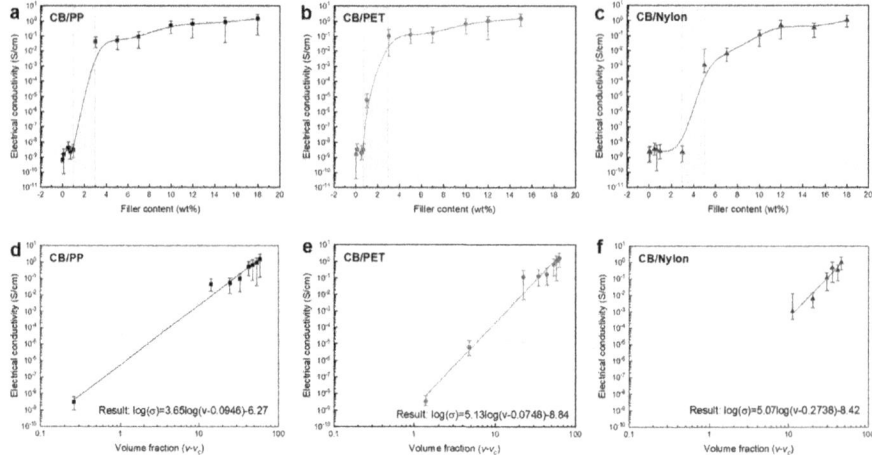

FIGURE 3.2 Electrical conductivity of (a) carbon black (CB)/polypropylene, (b) CB/polyethylene terephthalate, and (c) CB/nylon composite as a function of CB content. Shaded box: percolation threshold. Best-fit curves of electrical conductivity of (d) CB/polypropylene, (e) CB/polyethylene terephthalate, and (f) CB/nylon composites vs $(v-v_c)$. Reprinted from Ref. 14

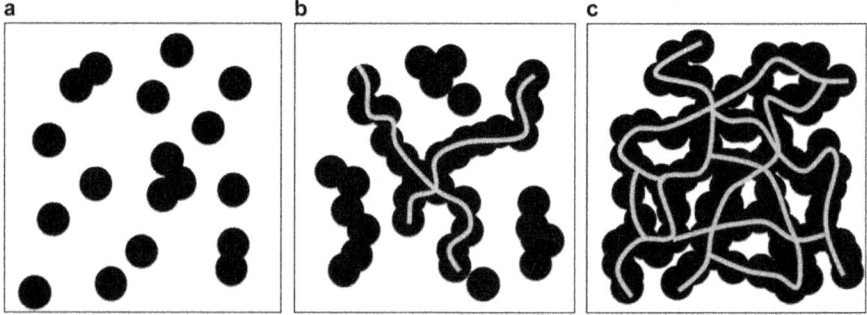

FIGURE 3.3 Schematics of percolation thresholds of carbon black (CB)/polymer composites. Orientation of CB in polymer matrix: (a) disconnected, (b) partially connected, (c) fully connected. Reprinted from Ref. 14

In the case of carbon nanotubes (CNTs), the interface between reinforcement and matrix becomes more significant, as interfacial layers around CNTs can accelerate the development of conductive networks [15]. Both the CNTs and the surrounding interface layers of CNTs can affect the size of the conductive networks and support an increase in the conductivity of polymer-CNT composite (Figure 3.4). The percolation threshold is not only associated with the concentration of CNTs in the polymer matrix but the CNT/polymer interface and the tunneling spaces also play significant roles. Liu et al. explained the conductivity of such composites on the basis of the formula:

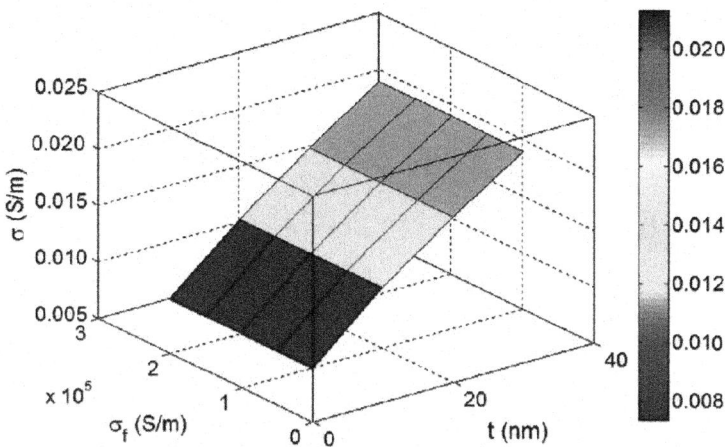

FIGURE 3.4 Variation of the conductivity of a polymer–carbon nanotube composite as a function of thickness of the interface (t) and conductivity of CNT (σ_f). Reprinted from Ref. 15, Copyright (2018), with permission from The Royal Society of Chemistry

$$\sigma = \frac{l\left(\phi_N + \phi_{iN} + \phi_{tN}\right)^{2x+1}}{2\pi\left(R + t + d\right)^2\left(R_f + R_t + R_i\right)} \tag{3.1}$$

where where l and R are the length and radius of CNTs, t is the thickness of the interface layer, d is the tunneling distance between adjacent CNTs, R_i is the intrinsic resistance of interface regions, R_t is tunneling resistance related to the resistance due to the CNTs and polymer matrix in the tunneling spaces, R_f is the intrinsic resistance of wavy CNTs, and ϕ_N, ϕ_{iN}, and ϕ_{tN} are percolation thresholds associated with the volume fraction of CNT in the networks, interface regions in the networks, and tunneling spaces in the networks, respectively.

Carbon-fiber-reinforced polymer composites display mechanical properties comparable to those of structural metals but with much lighter weight. This property makes these materials the frontline structural systems, especially in aircraft. In these composites, the carbon-fiber reinforcement is implanted in a polymer matrix (Figure 3.5). In order to obtain engineering components, some composite layers are stacked, with different orientations according to the required thickness, to form a laminate.

Although carbon fibers are good conductors, they are randomly distributed within an insulating matrix resulting in poor electrical conductivity in reinforced polymer composites [16], which restricts their applications. Interestingly, the directionality of fiber in the composite is an important characteristic. The conductivity in the direction of the fiber is much higher than in the transverse direction [17]. The electrical conductivity of these composites may be enhanced by adding conductive fillers such as CNTs, conducting graphene, CB, metal fibers, etc.

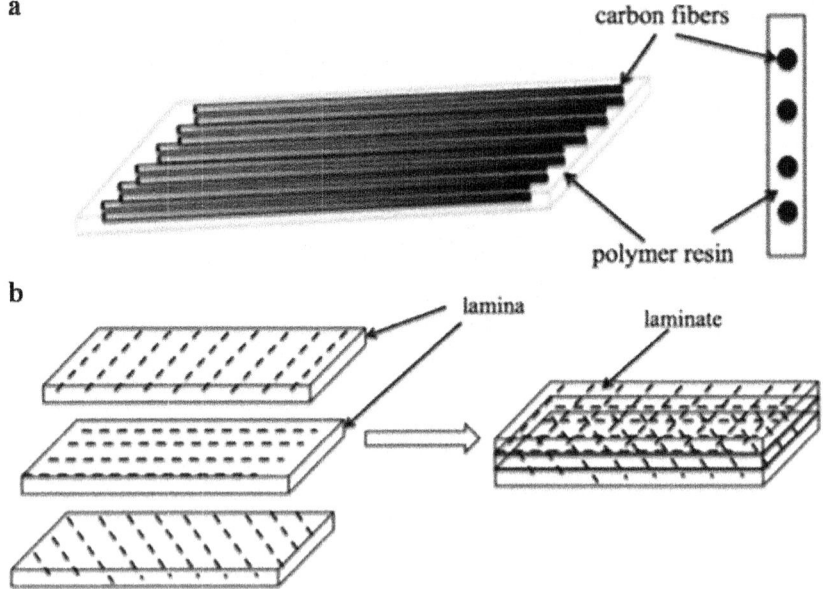

FIGURE 3.5 (a) A typical structure of carbon-fiber-reinforced polymer; (b) a laminate formed by stacking single lamina. Reprinted from Ref. 16

3.2.2 MECHANICAL PROPERTIES OF POLYMER COMPOSITES

Fiber-reinforced polymer composites possess a non-elastic structure that can absorb and dissipate vibration. This property makes these materials suitable for constructions like bridges and aircraft that are frequently exposed to vibration from the wind. The properties of these composites are dependent on the form, volume/weight fraction, and geometry of the reinforcement. Table 3.1 illustrates some constants for epoxy polymer and its composites. As can be seen from the table, unidirectional reinforcements give more mechanical strength to the polymer matrix than cross-ply. The fibers may be of two types, as discontinuous fibers (randomly oriented, wicker shaped) or as straight, parallel continuous fibers. For mechanical strength, continuous fibers are more efficient than discontinuous. In the polymer-matrix composites, the randomly oriented fiber reinforcements are likely to have lower volume fractions than those made with aligned fibers or fabrics.

Nanoparticles with high surface-to-volume ratio possess significant interface in the polymer matrix which strongly affect the mechanical property of the composites in terms of the strength and the toughness. Excellent mechanical properties can be achieved by incorporation of the proper type, size/shape, and loading of nanoparticles into the epoxy. Domun et al. [19] rigorously reviewed the state of epoxy-matrix-based composites with different reinforcements. Figure 3.6 demonstrates the enhanced mechanical strength of the epoxy-matrix-based composites. This behavior is the same for most of the other polymer-matrix composites. In general, the concept of threshold in terms of reinforcement loading is also valid for mechanical strength.

TABLE 3.1
Mechanical Properties of Epoxy and Its Composites [18]

Material	Class	Specific Gravity	Young's Modulus (GPa)	Ultimate Strength (MPa)
Epoxy	Polymer	~1.14	~3.44	~68.92
Unidirectional graphite/epoxy	Composite	1.6	180.90	1499.65
Unidirectional glass/epoxy	Composite	1.8	38.60	1061.34
Cross-ply graphite/epoxy	Composite	1.6	95.98	372.85
Cross-ply glass/epoxy	Composite	1.8	23.58	88.21
Quasi-isotropic graphite/epoxy	Composite	1.6	69.64	276.48
Quasi-isotropic glass/epoxy	Composite	1.8	18.96	73.08

The mechanical properties of polymer-matrix composites are strongly dependent on temperature. The tensile strength of polymer composites decreases as the surrounding temperature increases. The loss in tensile strength is mainly associated with softening of the matrix in the laminates. Depending on the glass transition temperature of the polymer matrix, brittleness may also be introduced.

3.3   PROPERTIES OF CERAMIC-MATRIX COMPOSITES

Ceramics are hard and brittle materials with low tensile strength and low density, comprising strong covalent and ionic bonds. In general they comprise one or more metals combined with non-metals such as oxygen, nitrogen, and carbon. These matrix materials possess low failure strains and low toughness or fracture energies. They also have low thermal and mechanical shock resistance but display good stability at very high temperatures, which makes them interesting as structural materials for elevated temperature applications, as aircraft components and energy storage/conversion devices such as sodium–sulfur batteries and fuel cells.

However, thermal stress, as a result of a non-uniform distribution of temperature in different parts of the ceramic structure and some constraint on thermal expansion or contraction imposed by adjacent layers of the ceramic, is an important issue to resolve. Thermal stress may result in malfunctioning of aircraft components, buildings, and other structures. Fiber reinforcement is the most commonly used approach to make ceramic matrices useful for applications. Fiber-reinforced ceramic-matrix composites are widely used as thermal structural materials in aeronautic and astronautic fields. These composites demonstrate remarkable thermal and mechanical properties at high temperatures.

3.3.1   ELECTRICAL PROPERTIES OF CERAMIC-MATRIX COMPOSITES

Ceramic materials, in general, are dielectric in nature and therefore do not conduct electricity. However, they are capable of conducting specific ions, which makes them

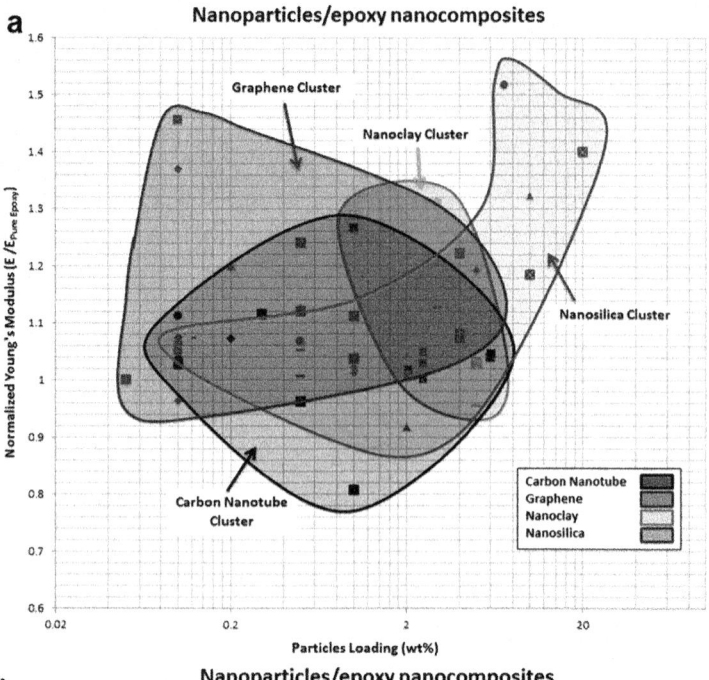

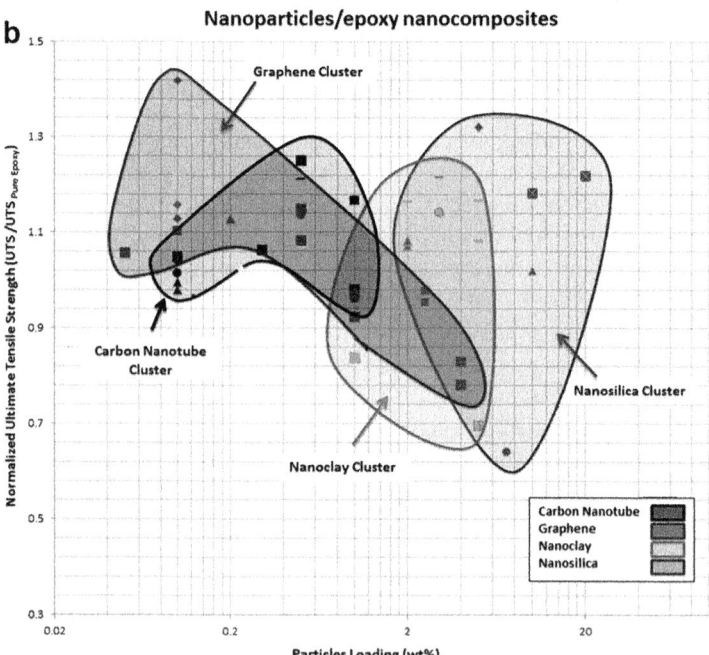

FIGURE 3.6 Variation of (a) normalized Young's modulus and (b) ultimate tensile strength of nanoparticles/epoxy composites with respect to particle loading. Reprinted from Ref. 19, Copyright (2015), with permission from The Royal Society of Chemistry

interesting materials to use between the two electrodes of energy devices such as fuel cells and capacitors. The solid oxide fuel cell is an example in which a solid ceramic electrolyte is a main component. This electrolyte is capable of conducting oxygen ions and this makes it suitable for a fuel cell electrolyte.

The electrical conductivity of ceramic matrix can be greatly enhanced by dispersing conducting carbon or metal reinforcements, which can be useful for electromagnetic interference (EMI) shielding applications at elevated temperature, which is superior over low melting polymer composites and highly dense metal composites. The electrical conductivity behavior of such composites, in general, resembles that of polymer-matrix composites discussed in an earlier section of this chapter.

3.3.2 MECHANICAL PROPERTIES OF CERAMIC-MATRIX COMPOSITES

Ceramic matrix, in general, is characterized by high stiffness and hardness. In ceramic-matrix composites, reinforcements such as fibers, whiskers, and particles are dispersed to improve fracture toughness, which enhances the materials' capability to resist thermal and mechanical shocks. Figure 3.7 demonstrates the better mechanical strength of ceramic-matrix composites as compared to monolithic ceramics. In the case of monolithic ceramics, brittle fracture without any plastic deformation is observed. However, fiber-reinforced ceramic-matrix composites exhibit an elastic region like monolithic ceramics in the beginning, and then reach an ultimate tensile strength followed by so-called plastic deformation and finally fracture [20]. The above behavior reveals the applicability of these composites in high-temperature

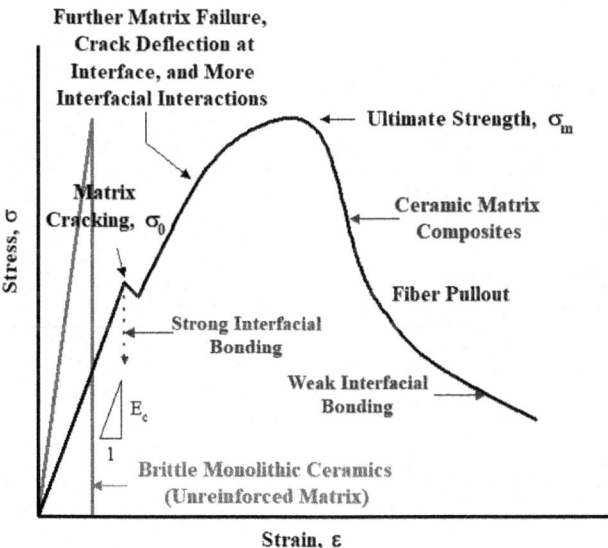

FIGURE 3.7 The stress–strain behavior in ceramic-matrix composites as compared with monolithic ceramics. Reprinted from Ref. 20

applications with some immunity to the issues associated with low fractural toughness.

Not only fibers but interface nanoparticle reinforcements also enhance the mechanical viability of ceramic-matrix composite materials. Owing to their high surface-to-volume ratio, these reinforcements cover a large interface. Therefore, even small loads of such reinforcements in a ceramic matrix may provide significantly enhanced mechanical properties such as fracture toughness, wear resistance, and flexural strength. Kim et al. [21] demonstrated that the fracture toughness and strength of a graphene/Al_2O_3 composite increased by ~75% and ~25%, respectively, compared with pristine Al_2O_3, with remarkable increase in wear resistance (Figure 3.8). They observed that the hard, lubricating graphene layer on the alumina grains protects the specimen surface from wear and hinders the cracking or falling out of alumina grains.

3.4 PROPERTIES OF METAL-MATRIX COMPOSITES

Metals are very versatile materials. The high electrical conductivity of metals makes them suitable for electronic applications including EMI shielding. The electrical

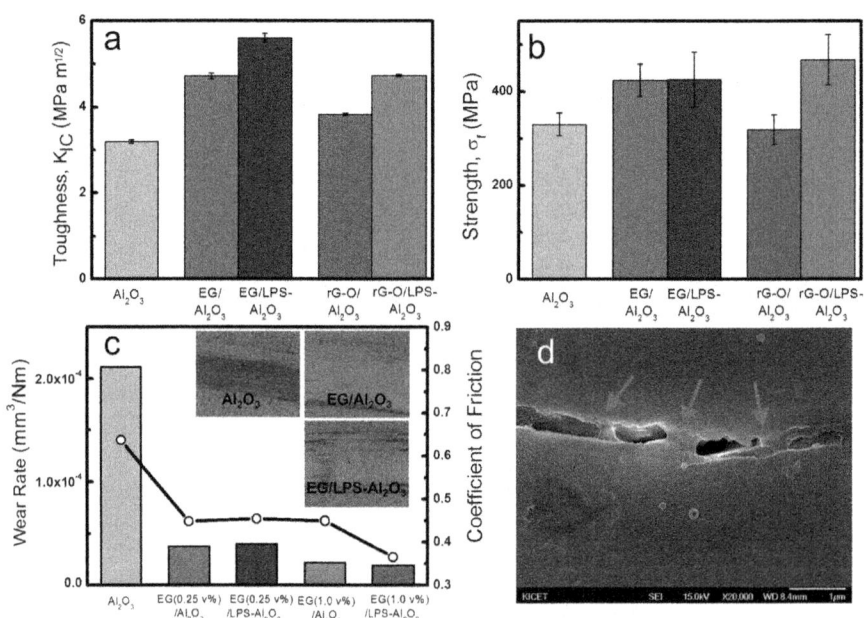

FIGURE 3.8 (a) Fracture toughness and (b) flexural strength for Al_2O_3 and composites. (c) Effect of unoxidized graphene (EG) on the wear rate of the EG/Al_2O_3 and EG/ liquid-phase-sintered alumina (LPS-Al_2O_3) composites in a severe wear environment (load: 25 N, sliding speed: 10 cm s^{-1}) and long sliding regime (27,700 cycle; 2,000 m) (inset shows optic images of wear tracks on sample surface) and (d) a crack on the polished surface of 1.0 vol% EG/Al_2O_3 composite by indentation. The arrows in (d) suggest crack bridging by embedded EG nanoplatelets. Reprinted from Ref. 21, Copyright (2014), with permission from Springer Nature

property of metal-matrix composites is significantly influenced by graphite reinforcement. These composites show excellent conductivity and anti-seizing properties, which make them good materials for industrial applications like sliding electrical contacts.

They are strong, tough, and can be plastically deformed. These properties of metals have established them as a frontline material for construction. Stiffness is an issue in such matrices; however, it can be cured by incorporating high-modulus fibers or particles in a metal matrix. Additionally, by choosing proper reinforcements, fibers, whiskers, and particles, higher strength and better wear resistance with better chemical stability can be achieved. There are four factors affecting the mechanical strength of these composites:

1. Load transfer effect, associated with the transfer of load from the soft and compliant matrix to the stiff and hard reinforcement under an applied external load
2. Hall–Petch strengthening, associated with the grain size and grain orientations, as grain boundaries can obstruct the movement of dislocations
3. Orowan strengthening, associated with the interaction of nanoparticles with dislocations in the matrix that attract crossing dislocations around the particles under an external load
4. Mismatch in the coefficient of thermal expansion and the elastic modulus of the reinforcements and the matrix during cooling and straining [22].

The continuous fiber-reinforced metal-matrix composites have much greater transverse strengths, which ensure their utility in a unidirectional configuration [23]. The axial moduli of these composites are much greater than those of the monolithic metals. On the other hand, the addition of discontinuous fibers to the metal matrix improves the wear resistance and elevated-temperature strength and fatigue properties of aluminum, which makes them a choice for internal engine components.

Particle-reinforced metal-matrix composites have also been studied and used in engineering applications. Titanium carbide particle-reinforced steel and silicon carbide particle-reinforced aluminum are two examples of such materials. Titanium carbide particle-reinforced steel composite has a higher modulus (304 GPa) than monolithic metal (193 GPa) with significantly lower specific gravity ($\sim 80\%$), which results in higher specific stiffness of the composite [23]. The composite possesses higher wear resistance than the metal. In silicon carbide reinforced aluminum, the modulus and yield strength increase whereas fracture toughness and tensile ultimate strain decrease with an increase in particle volume fraction. The silicon carbide particle also improves the short-term elevated-temperature strength properties.

In addition to these factors in CNT reinforcements, the length of the CNTs plays an important role in mechanical strength of the composite. Chen et al. demonstrated the effect of CNT length on the strengthening mechanism of an aluminum metal-matrix composite [24]. They showed that the CNTs reinforced aluminum metal-matrix composites. Different CNT aspect ratios exhibited distinct mechanical properties with an ultimate tensile strength range from 172 MPa to 368 MPa compared with 133 MPa for the pure metal (Figure 3.9).

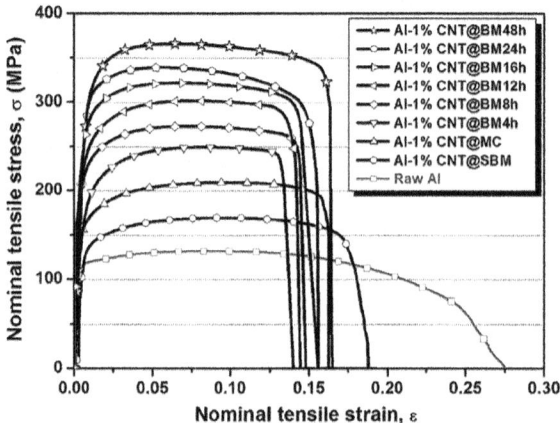

FIGURE 3.9   Nominal tensile stress–strain curves of composites processed by ball milling (BM), solution ball milling (SBM), and mechanical coating (MC). Reprinted from Ref. 24, Copyright (2017), with permission from Elsevier

3.5   PROPERTIES OF COMPOSITE MATERIALS USED IN ENERGY STORAGE/CONVERSION DEVICES

Composite materials have been extensively proposed and used as electrodes (anode and cathode) and electrolytes for well-known secondary lithium-ion batteries and newly proposed sodium-ion batteries. These materials have also shown their potential in fuel cell applications.

Silicon has recently been proposed as one of the most promising anode materials for lithium-ion batteries due to its high theoretical lithium storage capacity of ~3579 mA h g^{-1} for $Li_{15}Si_4$ (better than the commercial graphite anodes, which show ~372 mA h g^{-1} for LiC_6), high volumetric capacity, a relatively low discharge voltage, and low cost because of its high abundance [25]. However, the commercialization of silicon as anode is not that feasible due to the major drawback of its volume expansion upon lithium insertion. Additionally, the silicon forms Li_xSi alloys, which inculcate cracking and pulverization of silicon particles causing adhesion problems in the silicon anode. These problems lead to the loss of electrical contact and a shorter cycle life. In this regard, use of silicon and carbon composite instead of pure silicon is one of the most appreciated strategies. The carbon channels, with good electronic conductivity and superior mechanical properties, act as stress buffering matrices into silicon to accommodate the strain during intercalation/de-intercalation of lithium ions. In this composite system, silicon materials act as active components contributing a high lithium storage capacity while the carbon matrix can significantly buffer volume expansion of silicon, improve electronic conductivity, and stabilize the solid-electrolyte interface layers of the silicon-based anodes. Hence, coupling of nano-sized silicon with carbon proves to be an effective method of improving capacity and cycling stability of the anode system. This is not only true for silicon: composites of carbon with tin, phosphorous, and various metal oxides such as MnO_2, TiO_2, lithium-titanate (LTO), and various transition metal oxides have been proposed as potential anode materials for lithium batteries.

As cathode materials of the lithium-ion batteries, layered transition metal oxides LiM_xO_y (M: transition metals such as Fe, Co, Mn, Ti, Cr, etc.) and their derivatives, and some metal fluorides (metal: Fe, V, Ti, Co, and Mn) or phosphates have been proposed, but the poor electronic conductivity of these materials hinders their application. In order to enhance their electronic conductivity, conductive carbonaceous materials such as graphene, hard carbon, carbon fibers, etc., are used in the composite form with these active materials. Carbonaceous materials not only enhance electronic conductivity but also provide enough space for volume expansion during intercalation/de-intercalation of lithium ions.

Metal/metal oxide modified graphene electrodes have been well studied as composite electrodes for microbial fuel cells. In these composite electrodes, nano-structured metal oxide semiconductors such as SnO_2, TiO_2, MnO_2, and Co_3O_4, are used due to their high surface area, non-toxicity, good biocompatibility, and chemical stability, while graphene displays properties like high chemical stability, high conductivity, high specific surface area, and low cost [26].

For the electrolyte, the dispersion of micro/nano-sized metal oxides/ceramic particles in conventionally used liquid electrolytes results in the formation of a highly ionic conducting space-charge layer on the boundaries of the dispersed grains [27, 28]. On increasing the filler contents in these liquid/ceramic composite electrolytes, the layers overlap and lead to the formation of high conducting pathways that help in better mobility of ions and hence better ionic conductivity. The concept of composite electrolytes has been well accepted, as demonstrated by the currently trending polymer-based electrolytes [29, 30]. The composite polymer electrolytes are basically polymer-matrix composites with reinforcing micro/nano-sized metal oxides/ceramic particles with specific salts and solvents to introduce ionic conductivity. Various inert reinforcing filler particles such as Al_2O_3, SiO_2, TiO_2, $BaTiO_3$, etc. have been used in the preparation of composite polymer electrolytes. It has been observed that dispersion of these ceramic particles not only enhances the ionic conductivity of polymer electrolyte systems but also improves other physical, mechanical, and electrochemical properties.

Another approach uses active filler particles such as $Li_{6.75}La_3Zr_{1.75}Ta_{0.25}O_{12}$, $Li_{0.33}La_{0.557}TiO_3$, Na-β-alumina, Na-NASICON, and MgO in specific polymer electrolyte systems [30]. The dispersion of such fillers supports transport of cations (Li^+, Na^+, Mg^{2+}, etc.) through the electrolyte, which, in turn, enhances the electrolytic performance in the secondary batteries. In proton (H^+) conducting polymer electrolyte systems for fuel cell applications, the inorganic additives such as SiO_2 particles improve proton conductivity at high temperatures and low relative humidity [31]. The mechanism of proton transport is based on the surface and chemical properties of the interface between the inorganic and organic phases in the hybrid system (Figure 3.10). The hydrophilic nanoparticles, consisting of ionic clusters of the sulfonic acid moiety, are interconnected by a network of channels in the hydrophobic domain of the polymeric membrane [32, 33].

a Polymer Membrane (e.g. Nafion)

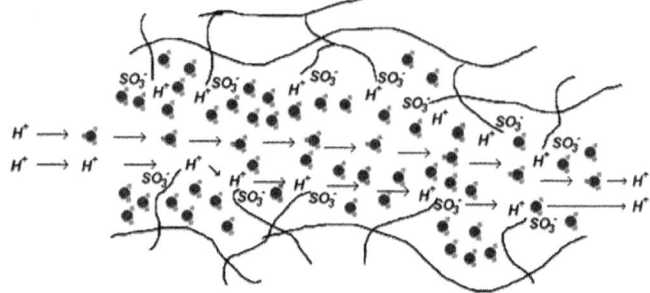

b Hydroscopic Nanocomposite Polymer Membrane

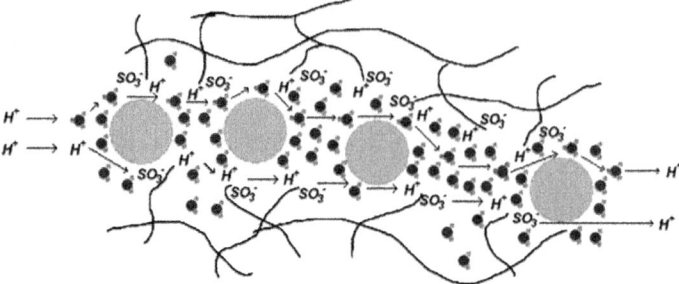

c Layered Metal Phosphate with functionalised surface

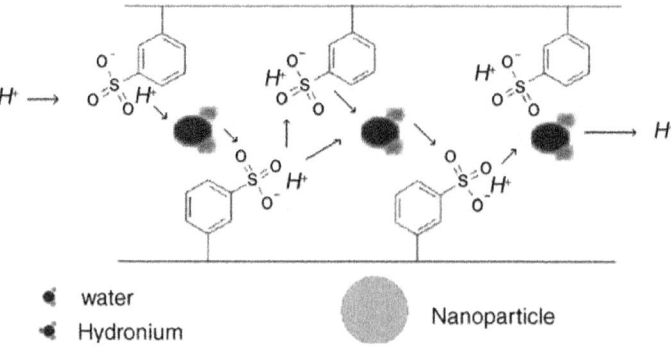

FIGURE 3.10 Proton transport in (a) Nafion (polymer) membranes, (b) polymer/nanoparticle composite membranes, and (c) surface functionalized solid acid membranes. Reprinted from Ref. 32, Copyright (2005), with permission from Elsevier

3.6 CONCLUSIONS

Composite materials, a mixture of two or more non-soluble systems, are interesting in terms of their unique characteristics compared with monolithic systems. The composite materials are classified, on the basis of their matrix, into three main categories: polymer-matrix composites, ceramic-matrix composites, and metal-matrix composites. The electrical and mechanical properties of the monolithic systems can be greatly improved upon dispersion of specific reinforcements. The overall properties of composite materials are mainly decided by the individual constituents and the synthesis/preparation technique; therefore, reinforcement and the matrix are chosen to acquire specific properties meeting the requirement of the application.

REFERENCES

1. Mangalgiri, P. D. "Composite materials for aerospace applications." *Bulletin of Materials Science* 22, no. 3 (1999): 657–664.
2. Kinsley-Jones, M. 2006 *"Airbus's A350 vision takes shape—Flight takes an in-depth look at the new twinjet"*, Flight International 12 Dec. 2006 http://www.flightglobal.com/news/articles/airbus39s-a350-vision-takesshape-flight-takes-an-in-depth-look-at-the-new-211028/
3. http://www.technologystudent.com/joints/reinforc1.html
4. Anon. *"Composites Penetration—Step Change Underway with Intermediate Modulus Carbon Fiber as the Standard"*, Hexcel Corporationx, http://www.sec.gov/Archives/edgar/data/717605/000110465908021748/g97851bci012.jpg
5. Haghshenas, M. "Metal–matrix composites." *Reference Module in Materials Science and Materials Engineering* (2016): 03950–03953. DOI:10.1016/B978-0-12-803581-8.03950-3
6. Business Communications Company, *RGB-108N Metal–matrix composites in the 21st Century: Markets and Opportunities*, 2006. http://www.bccresearch.com/market-research/advanced-materials/metal-matrix-compositesmarket-avm012d.html
7. Thakur, Vijay Kumar, Guoqiang Ding, Jan Ma, Pooi See Lee, and Xuehong Lu. "Hybrid materials and polymer electrolytes for electrochromic device applications." *Advanced materials* 24, no. 30 (2012): 4071–4096.
8. Park, Chul Soon, Changsoo Lee, and Oh Seok Kwon. "Conducting polymer based nano-biosensors." *Polymers* 8, no. 7 (2016): 249.
9. Omar, FatinSaiha, Navaneethan Duraisamy, K. Ramesh, and S. Ramesh. "Conducting polymer and its composite materials based electrochemical sensor for Nicotinamide Adenine Dinucleotide (NADH)." *Biosensors and Bioelectronics* 79 (2016): 763–775.
10. Taherian, Reza. "Experimental and analytical model for the electrical conductivity of polymer–based nanocomposites." *Composites Science and Technology* 123 (2016): 17–31.
11. Taherian, Reza, Ahmad NozadGolikand, and Mohammad JaffarHadianfard. "The effect of mold pressing pressure and composition on properties of nanocomposite bipolar plate for proton exchange membrane fuel cell." *Materials & Design* 32, no. 7 (2011): 3883–3892.
12. Waltman, R. J., and J. Bargon. "Electrically conducting polymers: a review of the electropolymerization reaction, of the effects of chemical structure on polymer film properties, and of applications towards technology." *Canadian Journal of Chemistry* 64, no. 1 (1986): 76–95.
13. Ayesha Kausar, Reza Taherian, *Book Chapter: Electrical Conductivity in Polymer Composite Filled With Carbon Microfillers, Electrical Conductivity in Polymer-Based CompositesExperiments, Modelling, and Applications*, Plastics Design Library, Elsevier, UK, 2019, Pages 19–40
14. Choi, Hyun-Jung, Moo Sung Kim, DamiroAhn, Sang Young Yeo, and Sohee Lee. "Electrical percolation threshold of carbon black in a polymer matrix and its application to antistatic fibre." *Scientific reports* 9, no. 1 (2019): 1–12.
15. Liu, Zhenling, Wanxi Peng, Yasser Zare, David Hui, and Kyong Yop Rhee. "Predicting the electrical conductivity in polymer carbon nanotube nanocomposites based on the

volume fractions and resistances of the nanoparticle, interphase, and tunneling regions in conductive networks." *RSC advances* 8, no. 34 (2018): 19001–19010.

16. Zhao, Qian, Kai Zhang, Shuang Zhu, Hanyang Xu, Dianguo Cao, Lina Zhao, Ronghua Zhang, and Wuliang Yin. "Review on the Electrical Resistance/Conductivity of Carbon Fiber Reinforced Polymer." *Applied Sciences* 9, no. 11 (2019): 2390.

17. Li, Xin. *"Eddy current techniques for non-destructive testing of carbon fibre reinforced plastic (CFRP)."* PhD diss., The University of Manchester (United Kingdom), 2012.

18. Kaw, Autar K. *Mechanics of composite materials.* CRC Press, Boca Raton 2005.

19. Domun, Nadiim, H. Hadavinia, T. Zhang, T. Sainsbury, G. H. Liaghat, and S. Vahid. "Improving the fracture toughness and the strength of epoxy using nanomaterials–a review of the current status." *Nanoscale* 7, no. 23 (2015): 10294–10329.

20. Kim, Jeongguk. "Tensile Fracture Behavior and Characterization of Ceramic Matrix Composites." *Materials* 12, no. 18 (2019): 2997.

21. Kim, Hyo Jin, Sung-Min Lee, Yoon-Suk Oh, Young-Hwan Yang, Young Soo Lim, Dae Ho Yoon, Changgu Lee, Jong-Young Kim, and Rodney S. Ruoff. "Unoxidized graphene/ alumina nanocomposite: fracture-and wear-resistance effects of graphene on alumina matrix." *Scientific reports* 4, no. 1 (2014): 1–10.

22. Casati, Riccardo, and Maurizio Vedani. "Metal matrix composites reinforced by nano-particles—a review." *Metals* 4, no. 1 (2014): 65–83.

23. Carl Zweben, Composite Materials, *Mechanical Engineers' Handbook*, Fourth Edition, edited by Myer Kutz, Copyright © 2015 John Wiley & Sons, Inc]

24. B Chen, B., J. Shen, X. Ye, L. Jia, S. Li, J. Umeda, M. Takahashi, and K. Kondoh. "Length effect of carbon nanotubes on the strengthening mechanisms in metal matrix composites." *ActaMaterialia* 140 (2017): 317–325.

25. Shen, Xiaohui, Zhanyuan Tian, Ruijuan Fan, Le Shao, Dapeng Zhang, Guolin Cao, Liang Kou, and Yangzhi Bai. "Research progress on silicon/carbon composite anode materials for lithium-ion battery." *Journal of Energy Chemistry* 27, no. 4 (2018): 1067–1090.

26. Zhang, Yizhen, Lifen Liu, Bart Van der Bruggen, and Fenglin Yang. Nanocarbon based composite electrodes and their application in microbial fuel cells." *Journal of Materials Chemistry A* 5, no. 25 (2017): 12673–12698.

27. Kumar, Binod. "From colloidal to composite electrolytes: properties, peculiarities, and possibilities." *Journal of Power Sources* 135, no. 1–2 (2004): 215–231.

28. Bhattacharyya, Aninda J., Joachim Maier, Ryan Bock, and Frederick F. Lange. "New class of soft matter electrolytes obtained via heterogeneous doping: Percolation effects in "soggy sand" electrolytes." *Solid State Ionics* 177, no. 26–32 (2006): 2565–2568.

29. Stephan, A. Manuel, and K. S. Nahm. "Review on composite polymer electrolytes for lithium batteries." *Polymer* 47, no. 16 (2006): 5952–5964.

30. Boaretto, Nicola, LeireMeabe, Maria Martinez-Ibañez, Michel Armand, and Heng Zhang. "Polymer Electrolytes for Rechargeable Batteries: From Nanocomposite to Nanohybrid." *Journal of The Electrochemical Society* 167, no. 7 (2020): 070524.

31. Kim, DeukJu, Min Jae Jo, and Sang Yong Nam. "A review of polymer–nanocomposite electrolyte membranes for fuel cell application." *Journal of Industrial and Engineering Chemistry* 21 (2015): 36–52.

32. Hogarth, Warren HJ, JC Diniz Da Costa, and GQ Max Lu. "Solid acid membranes for high temperature (¿ 140° C) proton exchange membrane fuel cells." *Journal of Power Sources* 142, no. 1–2 (2005): 223–237.

33. Peighambardoust, S. Jamai, Soosan Rowshanzamir, and Mehdi Amjadi. "Review of the proton exchange membranes for fuel cell applications." *International Journal of Hydrogen Energy* 35, no. 17 (2010): 9349–9384.

4 Synthesis of a Hybrid Self-Cleaning Coating System for Glass

A. Syafiq, B. Vengadaesvaran,
Nasrudin Abd. Rahim, A. K. Pandey,
A. R. Bushroa, K. Ramesh and S. Ramesh

CONTENTS

4.1 INTRODUCTION

Extreme weather and high humidity cause the accumulation of water spots and dusty surfaces on glass. This is because the fog droplets leave behind mineral-laden precipitate after evaporating from glass surfaces. Depending on environment and weather, skyscrapers may need to be washed twice a year. In general, it takes 4–10 weeks to clean all the window glass of a sky-scraper; cleaning the Burj Khalifa sky-scraper in Dubai requires even longer, about 3 months. The cost of cleaning windows is approximately $55 per hour, including the window cleaners' salary, detergents, tools, etc. In addition, cleaning the skyscrapers is risky because of strong winds: according to data from the International Window Cleaning Association, approximately one high-rise window cleaner is killed each year. Under low relative humidity, fog vapor forms a water droplet suspension in the air that deposits on the glass surface [1]. Small water droplets then collide and merge during a collision–coalescence process. The accumulation of fog droplets scatters the incident light, reducing the distance visibility of glass windows. This is because the partial light scattering causes curvature of the air/fog drop interface. Spherical fog drops deflect the incident rays; consequently the glass surface experiences total internal reflection [2].

Self-cleaning is the ability of a coated surface to remove dust particles through the action of water droplets [3]. Studies into self-cleaning mechanisms found in nature have engineered a great revolution in self-cleaning glass. The natural self-cleaning mechanism is commonly known as the "lotus effect". Lotus leaf was the first biological surface to inspire a hydrophobic phenomenon. The synergy of rough micro-papillae and nano-wax tubules creates roughness at the solid–liquid interphase [4]. At minimum solid–water interphase energy, the water droplet does not adhere to the solid surface, leading to formation of spherical water droplets. The hierarchical papillae (size 100 nm) have a smaller surface area compared with dirt particles, which prevents the adhesion of dust [5]. The spherical droplets then roll off or slide away dust particles on the hydrophobic surface. In recent years, researchers have developed a self-cleaning hydrophobic coating using polydimethylsiloxane (PDMS) resin. The PDMS has great chemical stability, low toxicity, high transparency [6], and competitive price [7]. Su et al. [8] have developed a self-cleaning hydrophobic coating using porous silica/PDMS microsphere film. The coated surface has a high water contact angle (WCA), which causes the water droplet to roll graphite powder off the surfaces. Yu et al. [9] have synthesized a nano-TiO_2 (P25)/attapulgite/PDMS coating that exhibits a WCA as high as $150°$. They have reported that applied methyl blue can roll ultrafine carbon powder off the coated surface.

Anti-fog coating is an effective approach to prevent the accumulation of fog droplets; the anti-fog performance depends on the geometrical nanostructure and chemical composition of surface [10, 11]. As an example, a hydrophilic surface spreads the fog droplets into a thin film of water. This is known as filmwise mode. A hydrophobic surface causes tiny mist droplets to form on its surface, known as dropwise mode. According to Snell's law, a hydrophilic coating with a WCA less than $48°$ is sufficient to prevent total internal reflection and light scattering. On a hydrophilic surface, the light rays are completely transmitted through the fog layer, without any light scattering; therefore the transparent surface remains optically clear under very humid conditions. However, a hydrophobic surface also can prevent total internal reflection. Hydrophobic surfaces have a water-repellent property that restricts the coalescence of fog on the surface. Nanostructures, especially a modified metal oxide sol, can create a protective layer on a glass or plastic surface to repel the fog vapor [12, 13]. The problem with metal oxide nanoparticles such as ZnO_2 and TiO_2 is their high opacity, which can obstruct the optical transparency of glass and allow minimum light absorption [14–18]. A drawback of SiO_2 nanoparticles is that the nano-fillers are not mechanically strong against certain abrasion tests [19–21]. Nano-calcium carbonate ($CaCO_3$) particles have potential for anti-fog coating because nano-$CaCO_3$ has a great effect in binder resin [22], is compatible interface with inorganic and organic matrices [23], and is cost-effective [24]. In addition, a nano-$CaCO_3$ coating could be highly transparent because its refractive index is lower than those of nano-TiO_2 and nano-ZnO_2 [25].

Several previous studies have reported the great impact of self-cleaning on dust and dirt particles indoors; however, the self-cleaning performance has not been validated in an outdoor environment. Outdoor exposure is a real threat, especially for window glass, because the outdoor atmosphere includes heavy rainfall, prolonged UV exposure, scratching, and organic contaminants that shorten the lifespan of glass windows. For example, organic contaminants adhere strongly to the glass surface and cannot be removed by simple self-cleaning [26]. These organic contaminants cause the formation of hydrophilic chains that degrade the hydrophobicity of coating. The durability of the self-cleaning hydrophobic coating should be given much attention because present hydrophobic coatings are capable of retaining their hydrophobic property for only 1.5 years. In comparison, the more durable and strong adhesion coating lasts for up to 3 years [27].

In this study, we attempted to develop a hybrid self-cleaning coating with good performance in both laboratory tests and a real outdoor environment. The anti-dust surface of coated glass should maintain its high transparency of above 80% after 3 months of outdoor exposure. The proposed self-cleaning coating exhibits strong adhesion onto glass substrate as the silane coupling binder 3-aminopropyltriethoxysilane (APTES) generates strong interaction between assembled nano-$CaCO_3$ in a polydimethylsiloxane (PDMS) matrix [28, 29]. As a result, the coating can withstand 5 N of scratch and maintains its hydrophobic property after prolonged outdoor exposure. The focus of our research is to develop a large-scale coating for the construction window glass industry; therefore, the novel invention requires a simple synthetic process, easy spray-fabrication, high transparency (above 87%), and a short tack-free time at ambient conditions.

4.2 MATERIALS AND EXPERIMENTAL PROCEDURE

4.2.1 RAW MATERIALS

Hydroxyl-terminated PDMS was supplied by Sigma Aldrich, Malaysia. The viscosity of PDMS is 0.000025 Pa.s; it was used as a surface modifier. Absolute ethanol (analytical grade; C_2H_6O) was supplied by Friendemann Schmidt, USA, with viscosity (20°C) 1.2 mPa s and density 0.79 g/cm³. The ethanol was used as solvent. The APTES resin used as binding agent was supplied by Shin-Etsu Singapore with a purity of above 99%. The technical data for the solvent and resin are presented in Table 3.1. $CaCO_3$ nano-powder has been used as nano-fillers and has an average size of 50 nm. The nano-$CaCO_3$, treated by 98% stearic acid, was supplied by US Research Nanomaterials, Inc., USA.

4.2.2 SYNTHESIS OF SELF-CLEANING COATING

The hybrid self-cleaning coating was synthesized by blending 2 g APTES with 50 wt% PDMS resin. The hydrophobic PDMS polymer acts as surface modifier to hinder the hydrophobic properties of nano-$CaCO_3$. The nano-$CaCO_3$ particles were subjected to a dispersion process via ultrasonication in 250 mL of ethanol at 50°C. Dispersed nano-$CaCO_3$ (0.80 wt%) was then blended with the APTES/PDMS resin mixture by magnetic stirrer for 30 min at 27°C. Glass slides were coated with the prepared hydrophobic resins by a dip-coating technique whereas glass panels were fabricated using spray-coating. Finally, the prepared systems were cured at room temperature with a short tack-free time of around 5–10 mins. Testing and characterization of the coating system included WCA, surface morphology, a self-cleaning test, an anti-fog test, a scratch hardness test, and an outdoor test.

4.2.3 CHARACTERIZATION AND TESTING(S)

The hydrophobic property of coated glass was measured using a 15EC Optical Contact Angle machine. The droplet volume was 5 µL, which was dispensed by an automatic dispensing system at a velocity of 2 µL/s. The image of the water droplet was captured by 15EC Optical Contact Angle software. The surface morphology of the coating was observed by field emission scanning electron microscope (FESEM) FEI Quanta 450 FEG. The magnification of FESEM was adjusted at 1 µm in low vacuum mode. The ASTM 2546 micro-scratch test was adjusted in order to study the mechanical hardness of the coating. Scratches on the coated surface were observed by optical microscope. The adhesion strength of the coating has been tested with indenter load forces of 1500 mN (1.5 N), 2000 mN (2.0 N), 2500 mN (2.5 N), 4000 mN (4.0 N), and 5000 mN (5.0 N).

The BS EN 168 anti-fog test was conducted with hot-steam vapor. The glass slides were exposed to a water bath at 130°C for 5 min. The glass panels were exposed for 10 min. After fogging exposure, both glass panels were placed above printed paper so the adhered fog droplets could be observed clearly.

The indoor self-cleaning test was conducted using silicon powder (35 µm) as dust, at different weight ratios of 10 wt% and 30 wt %. The dust solutions were injected on the inclined glass substrate ($\theta = 60°$) using a 150 mL syringe. The glass substrates were also tested with a 50 wt% ketchup solution and 100 wt% concentrated sucrose syrup, which is strongly attracted to the glass surface.

The EN 1096-5 2016 outdoor test was conducted on the rooftop of Wisma R&D for 3 months. Based on the European standard, the outdoor test is necessary to establish the self-cleaning performance of a coated substrate that utilizes rain and sun to clean the glass surface. The optical transmission of glass substrates has been measured in order to compare the degradation in glass transparency due to dust accumulation. A good self-cleaning coating that possesses anti-dust and anti-dirt properties would exhibit a low degradation rate . The WCA of glass substrates after prolonged outdoor exposure have been measured using the 15EC Optical Contact Angle machine. The objective of this measurement is to investigate the durability of coated glass against rainfall impact, organic contaminants, and prolonged UV exposure.

4.3 RESULTS AND DISCUSSION

4.3.1 WATER CONTACT ANGLE OF COATING

Figure 4.1 shows the WCA of bare glass and nano-CaCO$_3$ coated glass. It is observed clearly that the incorporation of nano-CaCO$_3$ filler improves the hydrophobic property of the glass. The bare glass substrate possesses hydrophilic properties with a WCA of

57.90°. This hydrophilicity is attributed to Si-OH species on the glass substrate. In comparison, the nano-CaCO$_3$ coated glass exhibits a WCA of 132.80° because the nano-fillers create a number of air pockets between nanoparticles, which reduces contact of water molecules to glass substrate at the liquid-air-solid interface [30].

Figure 4.2 shows the chemical reactions between polar nano-CaCO$_3$ and hydrophobic PDMS polymer. The chemical reactions of the coating system are rendered by covalent bond between the Si-O bond of glass substrate and the Si-OH hydroxyl group of PDMS and have been classified into three steps. First, the silane-coupling agent APTES is crosslinked to the organic PDMS resin via the reactive amine group, NH$_2$. Second, the C-N attachment of the PDMS chain creates a longer alkyl (C-H)

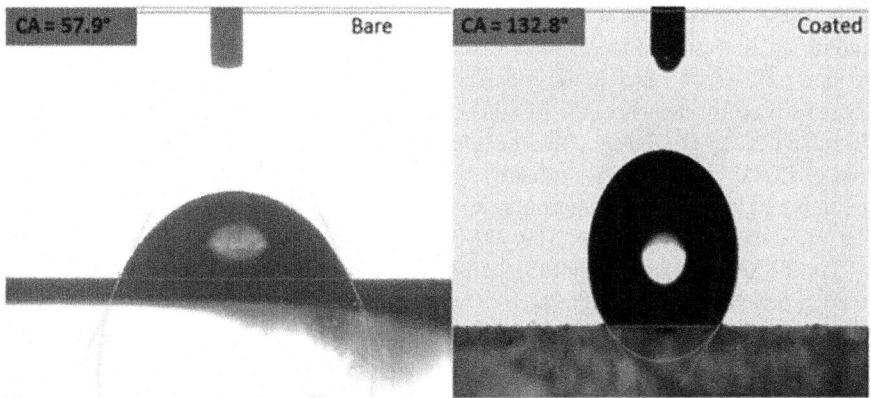

FIGURE 4.1 The water contact angle of bare and coated glass

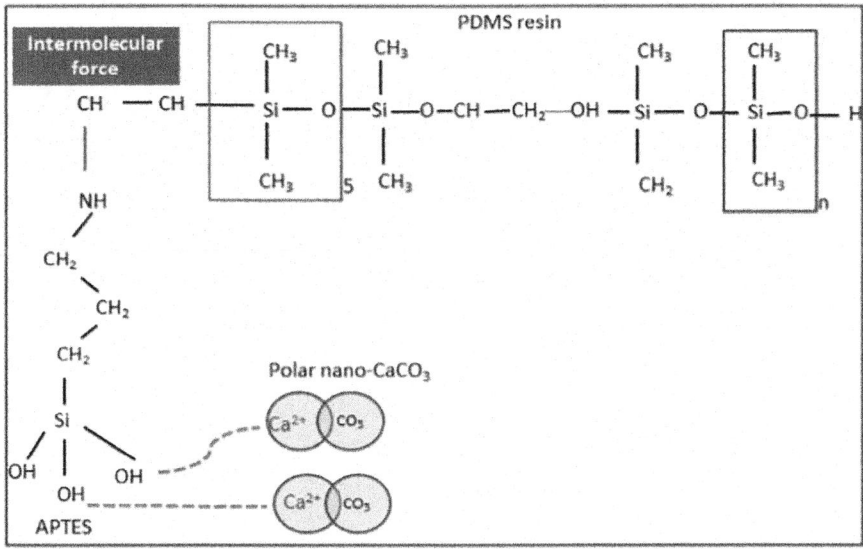

FIGURE 4.2 The chemical bonds of PDMS/nano-CaCO$_3$ coating

chain, which creates the non-wettability property and compatibility of coating. APTES consists of the silanol species Si-OH, which is highly reactive for intermediate crosslinking with the nano-CaCO$_3$ particles [31]. The polar Si-O linkage is strongly attracted to Ca^{2+}[CO$_3$]$^{2-}$ molecules under the action of van der Waals forces [32]. At the final stage, nano-CaCO$_3$ assembles inside the PDMS matrix and forms hydrophobic nanostructures.

4.3.2 SURFACE MORPHOLOGY

Figure 4.3 presents FESEM images of glass coated with PDMS and PDMS/nano-CaCO$_3$. The FESEM images illustrate the smooth surface of PDMS, indicating the good compatibility of PDMS with the glass substrate. The homogeneous surface has achieved its hydrophobicity by creating a low surface energy at the glass interface, mainly as the result of the flexible primary bond, Si-O-Si and the compact methyl groups, CH$_3$ which attach at the end of PDMS molecules. These covalent bonds have been verified by the presence of silicon, oxygen, and carbon elements in energy-dispersive X-ray (EDX) spectra. The nitrogen spectra revealed the blending formulation of PDMS and APTES, probably from an amine group, NH. Amine groups can assemble agglomerate nanoparticles and improve the interfacial adhesion of nano-CaCO$_3$ in the polymer matrix. [29, 33].

The FESEM image of PDMS/nano-CaCO$_3$ coating exhibits the aggregation of nanoparticles at the solid interface. It was observed that the coated substrate possesses multiple interphase layers after incorporating nano-CaCO$_3$. The great dispersion of nano-CaCO$_3$ in the PDMS matrix is inferred to lead to the capability of APTES to form attractive intermolecular forces such as hydrogen bonds and van der Waals bonds, and long-range forces between nanoparticles. The embedded

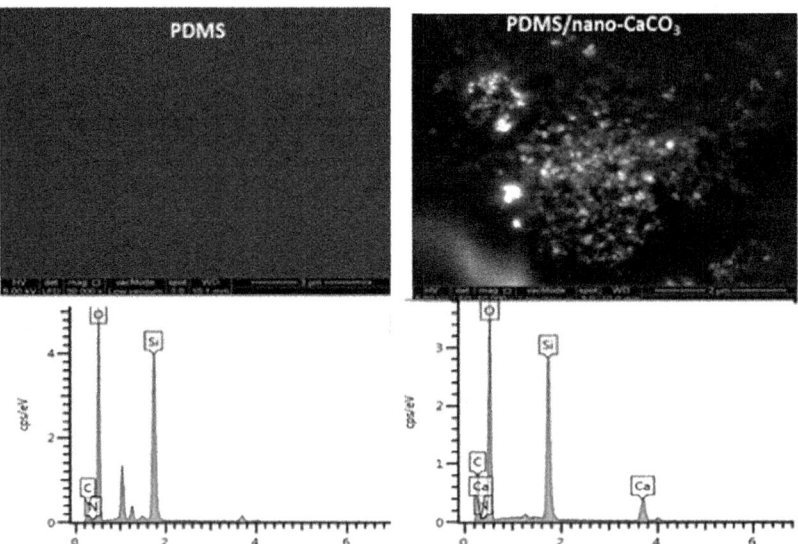

FIGURE 4.3 FESEM images and EDX spectra of coated glass substrate

nano-CaCO$_3$ creates a rougher surface and denser nanostructures, which form a number of air pockets as a protective layer against water droplet and dust particles. The presence of calcium element observed in the EDX spectra has validated the presence of embedded nano-CaCO$_3$ in the coating system.

4.3.3 Anti-Fog Properties

Figure 4.4 shows the performance of bare glass and nano-coated glass substrate. It was observed that the bare glass was impacted by large fog droplets due to spontaneous initial fogging within 30 s. After 2 min, the bare glass was completely covered by fog droplets and vision is blurry. The nano-CaCO$_3$ coated glass substrate produces a slightly hazy image due to the adhesion of tiny mist droplets only, because the trapped air pockets delay the penetration of fog droplets within the nano-CaCO$_3$ [34]. Based on the anti-fog industry standard BS EN 168, initial fogging on the coated substrate must exceed 60 s [35]: here, the initial fogging on the coated glass took about 6.16 min. This is because the trapped air pockets create a capillary adhesion force that reduces the contact of fog droplets to glass substrate. If the critical weight of the fog droplets (during the coalescence process) exceeds the capillary adhesion force, the fog droplets detach from the coated glass as the drops are attracted by gravitational force. If the critical weight of the fog droplets is below the capillary adhesion force, the fog droplets remain adhered to the coated glass [36].

The hydrophilic and silanol species on the bare glass surface have a strong attraction to the fog droplet molecules during the convection process. As a result, the bare glass is covered by large fog droplets that obscure visibility. The large fog droplets completely evaporate from bare substrate after 30 min but evaporation time is reduced by the applied coating system: mist droplets on the coated glass had dried completely within 4 min at ambient temperature. The anti-fog performance of both glass substrates is detailed in Table 4.1.

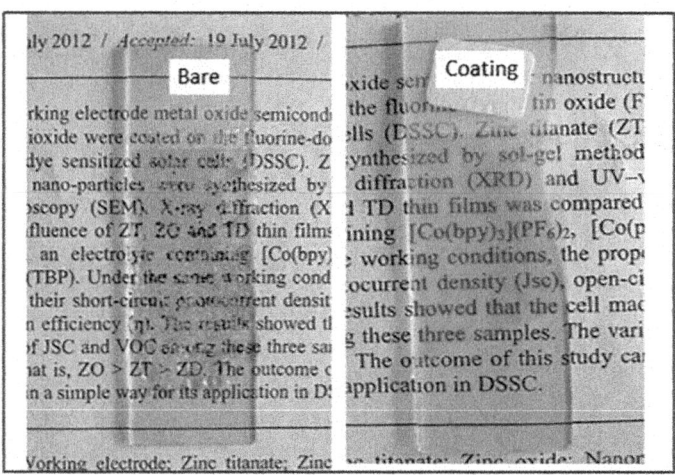

FIGURE 4.4 Anti-fog performance of bare and coated glass

TABLE 4.1
Details of Initial Fog Time and Evaporation Time

Type of Glass	Type of Testing	Result	Initial Fog Time (min)	Evaporation Time (min)
Bare	Initial fogging above 60 s	Fail	0.30	30.00
	Fog's screening on the glass's surface	Positive		
Coated	Initial fogging above 60 s	Pass	6.16	4.00
	Fog's screening on the glass's surface	Positive		

The fogging test has been extended on both glass panels according to industry standard BS EN 168. Figure 4.5 shows the anti-fog performance of bare and coated glass. The bare glass panel was adhered by numerous fog droplets. The hot fog vapor condenses spontaneously on the bare glass, causing a shorter period of initial fogging. The condensed fog droplets scatter the incident light, which causes the blurry image [37]. The bare glass possesses a hydrophilic surface, which has strong attraction to large fog droplets. Large fog droplets then coalesce due to slow fog removal. Besides that, the large fog droplets are formed because the smooth homogenous bare glass surface is unable to prevent the adhesion of hot fog vapor. As a result, the large fog droplets had completely evaporated after about 20 min, which seems impractical for real-life glass application.

The coated glass panel was covered by tiny mist droplets only. This great anti-fog performance is inferred to result from the hydrophobic surface delaying the formation of fog droplets during the convection process. Long alkyl chains are oriented outwards at the glass interface, which disrupts the van der Waals interaction with fog molecules. In addition, the aggregated nano-$CaCO_3$ restricts fog penetration, which prevents the coalescence of fog droplets. It was observed that the initial fogging is faster for the coated glass than the bare glass: the tiny mist droplets started to appear within 6 min. After the fogging test, the coated glass was exposed at ambient

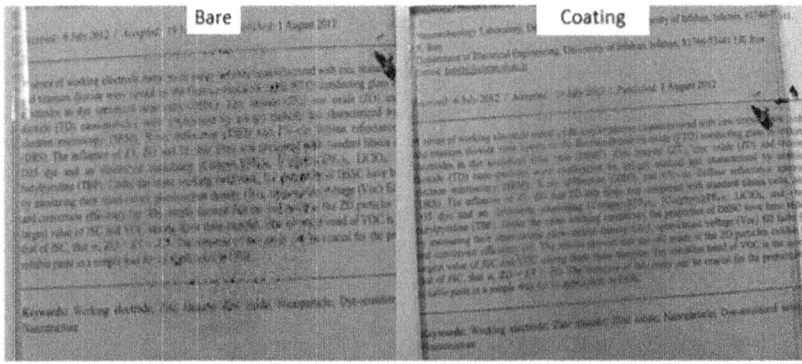

FIGURE 4.5   Anti-fog performance of bare and coated glass panels

temperature. The tiny mist droplets completely evaporated within a short period since nano-coating on the glass gives more air pockets and a rougher surface, consequently providing a large surface area (combination of smooth surface area and rough surface area) [38]. The initial fogging and evaporation times are recorded in Table 4.2.

4.3.4 SELF-CLEANING ANALYSIS

Figure 4.6 presents the effect of 10 wt% of silicon powder solution on the coated glass. The PDMS coating was impacted by dust layers and dirt streaks because of strong van der Waals forces between dust particles and the glass surface [39]. However, it was observed that all the nano-$CaCO_3$ coated glass panels can expel dust solution from their surface. The incorporation of nano-$CaCO_3$ particles improved the self-cleaning performance by lowering the contact adhesion of dust to glass surface, and the rough surface of nano-$CaCO_3$ modified with low-surface-energy PDMS further reduced the van der Waals forces. Under ambient conditions, the adhesion of dust particles primarily depends on van der Waals forces, which are short-range forces that occur at distances below 10 nm. When the dust particles are further than 10 nm from the glass substrate, van der Waals attraction decreases with separation distance [40, 41].

The self-cleaning performance of nano-$CaCO_3$ coated glasses has been tested against 30 wt% of dust solution as demonstrated in Figure 4.7, which shows the different self-cleaning efficiencies at various weight compositions of nano-$CaCO_3$ in PDMS matrix. Nano-$CaCO_3$ particles at 0.60 wt% and 1.00 wt% fail to expel the dust layer and dust streaks. The agglomeration of nano-$CaCO_3$ particles at these weight compositions is unable to create sufficient air pockets to restrict the penetration of the concentrated dust solution [42]. A 0.80 wt% composition of nano-$CaCO_3$ provides a cleaning glass surface, indicating the coated glass successfully repels 30 wt% dust solution. The absence of dirt streaks suggests that this 0.80 wt% of nano-$CaCO_3$ is the best composition. This is probably because 0.80 wt% nano-$CaCO_3$ could create sufficient air pockets. When the distance between nanoparticles is smaller than the size of the dust particle, the contact area of the dust particle to the coated surface is smaller and consequently the dust particle experiences low particle-surface adhesion forces.

TABLE 4.2
Initial Fogging and Evaporation Times of Glass Panels

Type of Glass	Type of Testing	Result	Initial Fog Time (min)	Evaporation Time (min)
Bare	Initial fogging above 60 s	Fail	0.52	20.00
	Fog's screening on the glass's surface	Positive		
Coated	Initial fogging above 60 s	Pass	6.00	6.00
	Fog's screening on the glass's surface	Positive		

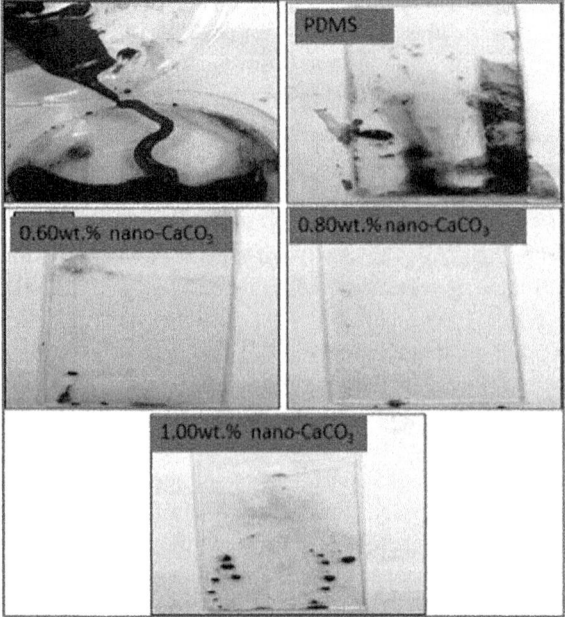

FIGURE 4.6 Self-cleaning performance against 10 wt% of dust solution

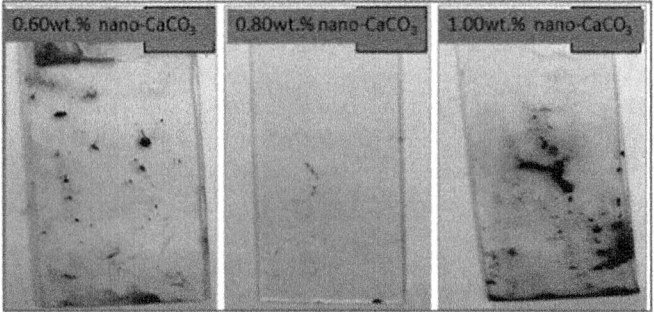

FIGURE 4.7 Self-cleaning performance against 30 wt% of dust solution

Figure 4.8 presents the self-cleaning performance of bare and coated glass panels against 100 wt% of syrup and 50 wt% of ketchup solution. The bare panel failed to expel the applied contaminants: the atoms of ketchup solution were diffused across the glass interface when they achieved molecular contact by wetting [1]. The coated glass panel demonstrated self-cleaning and completely expelled these dirt solutions. The rougher nano-$CaCO_3$ structures reduced the van der Waals attraction and contact adhesion of dirt contaminants [39].

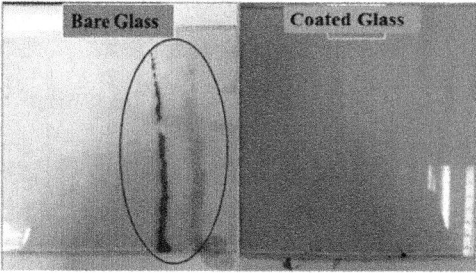

FIGURE 4.8 Self-cleaning performances of bare and coated glass against concentrated dirt solutions

4.3.5 ADHESION PROPERTIES

Figure 4.9 presents various types of crack failures on the glass substrate. The cracking points on the scratch tracks include (i) transverse cracking, (ii) partial ring cracks, (iii) crack/tensile cracking, (iv) delamination, (v) chipping, and (vi) total failure [44]. During transverse cracking, the coating system fails to maintain its elasticity. This is because transverse cracking causes splitting of the polymer matrix and delaminates the coating on the glass substrate [45]. As transverse cracks grow larger, they create large stress concentration on the coating at an earlier stage [46]. The coated glass then experiences a tensile crack due to the disruption of covalent bonds between coating and glass substrate. The tensile effects can be prevented if the coated glass possesses a high degree of Si-O covalent bonds, which undergo slow crack

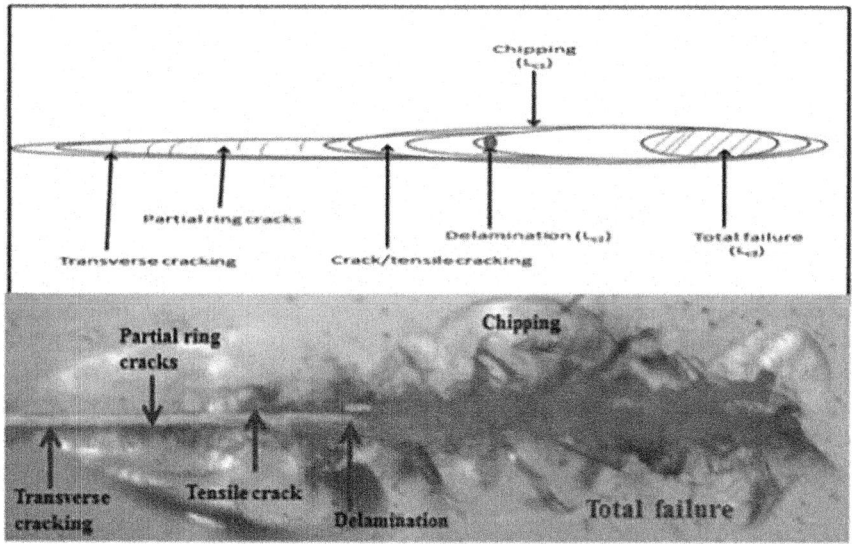

FIGURE 4.9 Illustration of crack failures on the scratch tracks

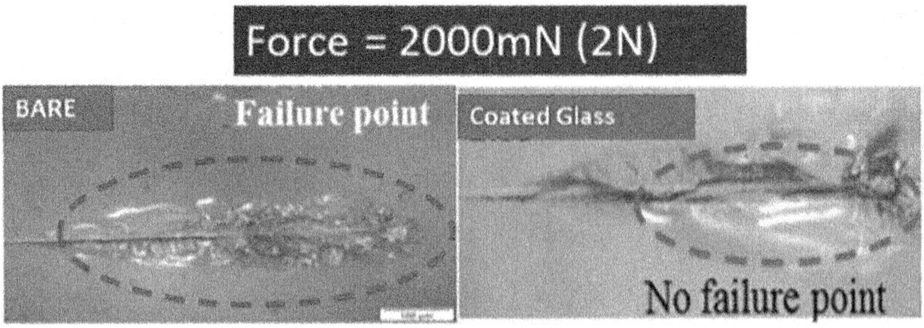

FIGURE 4.10   The anti-scratch performance of bare and coated glass

propagation at constant tensile stress. Meanwhile, the weak covalent Si-O bonds undergo unstable crack propagation; later the covalent bonds break at higher tensile stress [2] . At this stage, the coated glass experiences destructive fracture, resulting from certain failure points namely cohesive failure (L_{c1}) due to parallel cracking or chipping of coating, adhesive failure (L_{c2}) due to delamination of coating, and total failure (L_{c3}) due to the complete exposure of surface material [48].

The scratch test has been conducted based on the ASTM 2546 industry standard in order to develop a coating that is durable under real outdoor conditions. Figure 4.10 shows an optical micrograph of scratch track profiles on the bare and coated glass substrate after it was impacted with a 2 N load. Bare glass experiences larger initial cracking and failure point due to its brittleness. The glass fracture was surrounded by glass debris, which is attributed to propagation damage during glass breakage, which occurs as the cracks concentrated at the origin damage the glass surface [3]. The relationship between crack size and failure stress is derived from Equation 4.1.

$$\sigma = \frac{1}{Y}\frac{K_{lc}}{\sqrt{c}} \tag{4.1}$$

where Y is the constant dependent on sample shape and crack, K_{lc} is fracture toughness, and the crack size is symbolized by c. Fractured glass would exhibit a large crack size, c, at lower failure stress during glass breakage [49]. Unlike the bare glass, the coated glass showed no failure point at the same indention force, indicating its better mechanical hardness, strength, and anti-scratch performance. The best anti-scratch performance results from a 10:1 ratio of base to crosslinker in PDMS resin, which develops a stiff surface [50]. Another factor is the ethanol, which shows great compatibility with PDMS and the glass substrate. Because of its low solubility for PDMS, the ethanol does not detach the PDMS coating from the glass panel under certain applied stress [51].

Figure 4.11 provides an optical micrograph of scratch profiles on the coated substrate. The coating system exhibits transverse cracks only at an indenter load force of 1.5 N. Tensile cracks do not appear on the coated surface because the coating strongly adheres to the glass substrate due to the abundant hydroxyl groups of PDMS [52].

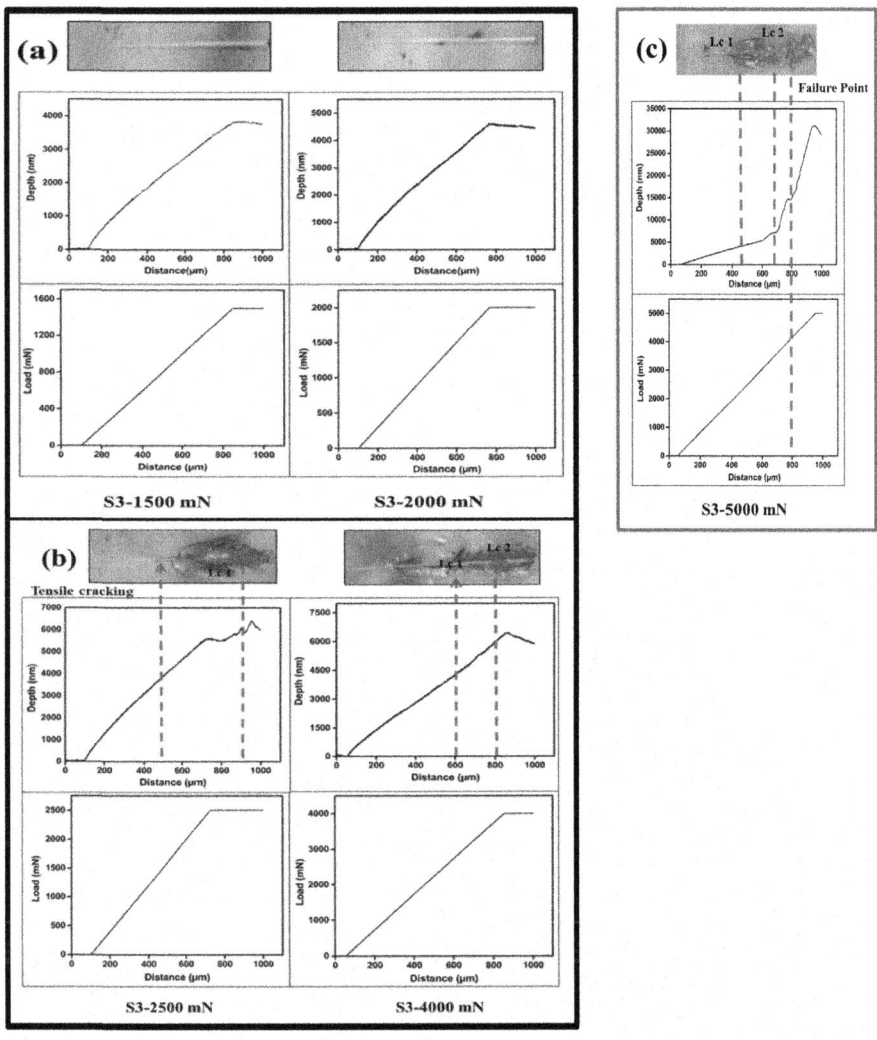

FIGURE 4.11 Optical micrograph of scratch failure on the coated surface at indenter loads of 1.5 N, 2.0 N, 2.5 N, 4.0 N, and 5.0 N

When the indenter load increases to 2 N, the coated glass exhibited tensile cracking at a scratch distance of 524.28 µm. The coated glass shows no cohesive failure (L_{c1}) at this indenter force, indicating strong adhesion of coating. PDMS at 50 wt% creates a high degree of crosslinking in the APTES matrix, which prevents the polymer chains from uncoiling and improves the elasticity of polymer resin [50, 53]. In addition, saturated PDMS-APTES crosslinking generates more identical terminating methyl groups, which the methyl groups are adhered strongly to the glass substrate [54]. As the indenter force increases to 2.5 N, the coated surface is impacted by initial cohesive failure (L_{c1}) at a scratch distance of at 900.18 µm.

TABLE 4.3
Crack Failures

Coating System	Failure Point	Scratch Distance at 1.5 N (μm)	Scratch Distance at 2 N (μm)	Scratch Distance at 2.5 N (μm)	Scratch Distance at 4 N (μm)	Scratch Distance at 5 N (μm)
S3	–	–	–	900.18	604.75	449.13
	L_{c2}	–	–	–	796.69	667.62
	L_{c3}	–	–	–	–	782.58

Coating delamination occurred when the indenter load increased to 4 N, when the indenter penetrates deeper into the coating's surface. Delamination indicates the breaking point of the coating system-substrate bond and is tagged as adhesive failure (L_{c2}). At a lower loading force, the weak coating-substrate bond would be delaminated at a shorter distance. From the scratch micrograph, the adhesive failure (L_{c2}) of coated glass was measured at a distance of 796.69 μm. The coated glass showed no total failure point or glass breakage, suggesting that the strong adhesion coating undergoes slow crack propagation, which delays the initial glass fracture [55]. The coated substrate experienced total coating failure (L_{c3}) as the indenter load was increased to 5 N. However, total coating failure (L_{c3}) occurred at the long scratch tracks about 782.58 μm. The details of crack failure on coated glass are presented in Table 4.3.

4.3.6 SELF-CLEANING OUTDOORS

Figure 4.12 presents the optical transmittances of bare and coated glass before and after prolonged outdoor exposure. It was observed that bare glass exhibits an average optical transmission of about 95% in the UV-Vis region; it is employed as reference. Average optical transmission is classified into two UV-Vis regions: the high visible region (600–800 nm) and the low visible region (400–550 nm). In the high visible region, the coated glass possesses an average optical transmission of about 87.09±0.38%, which is considered highly transparent. In the low visible region, the coated glass demonstrates an average optical transmission of about 84.93±0.87%. The high transparency is attributed to the thin (<60 nm) coating. Besides that, PDMS has a low refractive index of around 1.4 [56].

After 3 months of outdoor exposure, the bare glass displays significantly decreased transparency due to the accumulation of dust due to strong van der Waals forces with the hydrophilic bare surface: a high force such as mechanical cleaning is necessary to remove the dust. The average transmission of bare glass was reduced to 47.31% in the 600–800 nm range and to 49.76% in the 400–550 nm range. In comparison, the coated glass exhibits low degradation in optical transparency because of its self-cleaning ability. The degradation rate in optical transmission is about 6.21% in the 600–800 nm range and 6.71% in the 400–550 nm range. The details of optical transmission of both glass substrates before and after outdoor exposure are presented in Tables 4.4 and 4.5.

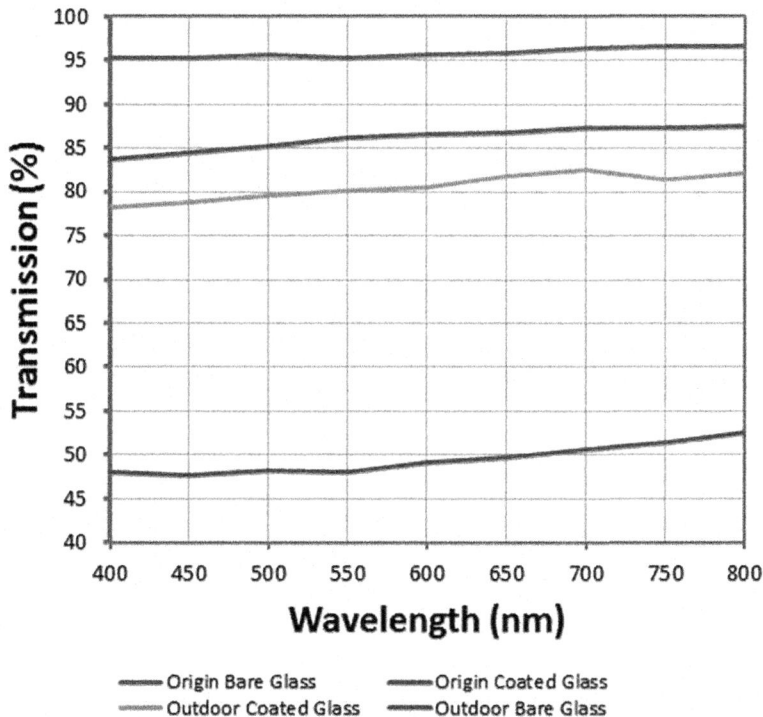

FIGURE 4.12   Average optical transmittance of bare and coated glass, (a, b) before and (c, d) after outdoor exposure

---

TABLE 4.4
Optical Transmittance of Glass Substrates at 600–800 nm

Glass	Average Transmittance (%) at 600–800 nm (Before Outdoor Exposure)	Average Transmittance (%) at 600–800 nm (After Outdoor Exposure)	Degradation (%)
Bare	96.18 ± 0.42	50.67 ± 0.52	47.31
Coated	87.09 ± 0.38	81.68 ± 0.71	6.21

TABLE 4.5
Optical Transmittance at 400–550 nm

Glass	Average Transmittance (%) at 400–550 nm (Before Outdoor Exposure)	Average Transmittance (%) at 400–550 nm (After Outdoor Exposure)	Degradation (%)
Bare	95.37 ± 0.18	47.91 ± 0.91	49.76
Coated	84.93 ± 0.87	79.23 ± 0.73	6.71

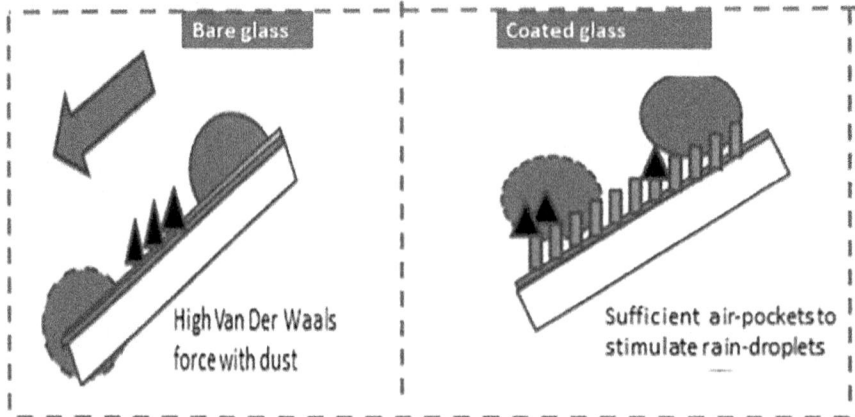

FIGURE 4.13   The self-cleaning mechanism

TABLE 4.6
The Degradation in WCA of Glass Substrates After Prolonged Outdoor Exposure

Type of Glass	Water Contact Angle ($\Theta°$) (Before Outdoor Exposure)	Water Contact Angle ($\Theta°$) (After Outdoor Exposure)	Degradation Rate (%)
Bare	57.90	70.00	+20.89
Coated	132.80	131.10	−1.28

Figure 4.13 shows the self-cleaning mechanism on the coated glass surface. The smooth surface of the bare glass is unable to expel dust particles due to strong van der Waals attraction forces. As mentioned previously, embedded nano-$CaCO_3$ creates a number of air pockets at the solid-liquid-air interphase. The self-cleaning mechanism of nano-$CaCO_3$ can be explained by two simple steps. First, the trapped air pockets reduce the contact area b etween dust particles and glass substrate; the air pockets create an anti-dust surface [57]. Second, the trapped air pockets reduce the van der Waals attraction force for rain droplets; consequently, the air pockets stimulate the rain droplets to wash away dust particles [24].

The outdoor environment includes heavy rainfall, prolonged UV exposure, organic contaminants, and moisture. During heavy rainfall, the rain droplets generate water hammer pressure, P_{WH}, on the nanostructures. The coated surface then generates several micro-cracks and holes that contribute to weak adhesion of coating and low hydrophobicity [58]. Organic contaminants and moisture derive new hydroxyl, hydroperoxides, and carbonyl groups, which degrade the hydrophobicity of the coating. Figure 4.14 presents the WCA of bare and coated glass after prolonged outdoor exposure. From the figure, the WCA of bare glass increased to 70.00° due to the

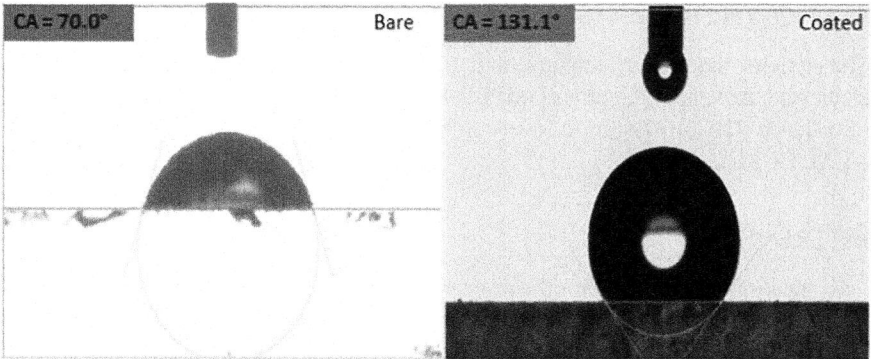

FIGURE 4.14   The average WCA of bare and coated glass after outdoor exposure

buildup of contaminants. It was noticed that the hydrophobic properties of the coated glass substrates have been maintained during prolonged outdoor exposure. The coating system exhibited the reduction in WCA by 1.28% only, suggesting that the coating possesses great durability against organic contaminants and rainfall impact. However, the WCA of coated glass substrate was above 125°, which is adequate for effective self-cleaning [59]. The details of degradation in WCA of both glasses are presented in Table 4.6.

4.4   CONCLUSION(S)

The hybrid self-cleaning coating is engineered with user-friendly properties such as a highly transparent surface, simple spray-fabrication, low-cost resin materials, and a short tack-free time at ambient temperature. The transparency of coated glass was above 87% in the UV-Vis region, which is comparable with non-coated glass. The incorporation of nano-$CaCO_3$ creates a rough surface, which consists of air pockets at the solid-liquid-air interface. The number of air pockets ameliorates the self-cleaning effect, anti-fogging, and durability of the coating under real outdoor conditions. The embedded nano-$CaCO_3$ nanoparticles exhibit great anti-fog performance, delaying the formation of initial fog for 6 mins. The adhered tiny mist droplets completely evaporated within a shorter time compared with the bare glass substrate. The hybrid coating provides a dirt-repellent, anti-dust surface that has been tested in laboratory and outdoor conditions.

This excellent self-cleaning effect is attributed to strong adhesion of coating, which can withstand a scratch-load as high as 5 N. After 3 months of outdoor exposure, the transparency of coated glass was degraded by only around 6%. As the coated glass possesses an anti-dust surface and great adhesion strength, the WCA of coating showed a slight drop by only 1.28% under real environment. In conclusion, the proposed hybrid self-cleaning glass has showed its multi-functionality as a great novel invention for building glass. The coated glass is expected to extend the lifespan and durability of windows.

ACKNOWLEDGMENTS

The authors thank the technical and financial assistance of UM Power Energy Dedicated Advanced Centre (UMPEDAC) and the Higher Institution Centre of Excellence (HICoE) Program Research Grant, UMPEDAC – 2019 (MOHE HICoE – UMPEDAC).

REFERENCES

1. Martikainen, A.L. *Comparative evaluation of fogging phenomenon in the ramp of three mines in Finland.* in *Proceedings of the 8th International Mine Ventilation Congress.* 2005.
2. Duran, I.R. and G. Laroche, Current trends, challenges, and perspectives of anti-fogging technology: Surface and material design, fabrication strategies, and beyond. *Progress in Materials Science*, 2018.
3. Nosonovsky, M. and B. Bhushan, Superhydrophobic surfaces and emerging applications: non-adhesion, energy, green engineering. *Current Opinion in Colloid & Interface Science*, 2009. 14(4): p. 270–280.
4. Ensikat, H.J., et al., Superhydrophobicity in perfection: the outstanding properties of the lotus leaf. *Beilstein Journal of Nanotechnology*, 2011. 2(1): p. 152–161.
5. Cao, M., et al., Water-repellent properties of superhydrophobic and lubricant-infused "slippery" surfaces: A brief study on the functions and applications. *ACS Applied Materials & Interfaces*, 2015. 8(6): p. 3615–3623.
6. Gao, S., et al., Rational construction of highly transparent superhydrophobic coatings based on a non-particle, fluorine-free and water-rich system for versatile oil-water separation. *Chemical Engineering Journal*, 2018. 333: p. 621–629.
7. Liu, X., et al., Transparent, durable and thermally stable PDMS-derived superhydrophobic surfaces. *Applied Surface Science*, 2015. 339: p. 94–101.
8. Su, Q., et al., Abrasion resistant semitransparent self-cleaning coatings based on porous silica microspheres and polydimethylsiloxane. *Ceramics International*, 2019. 45(1): p. 401–408.
9. Yu, N., et al., Facile preparation of durable superhydrophobic coating with self-cleaning property. *Surface and Coatings Technology*, 2018. 347: p. 199–208.
10. Han, Z., et al., Flourishing bioinspired antifogging materials with superwettability: Progresses and challenges. *Advanced Materials*, 2018. 30(13): p. 1704652.
11. Utech, S., et al., Tailoring re-entrant geometry in inverse colloidal monolayers to control surface wettability. *Journal of Materials Chemistry A*, 2016. 4(18): p. 6853–6859.
12. Satoh, M. and S. Kitajima, *Anti-Fogging Agent.* 2018, Google Patents.
13. Varshney, P., S. Mohapatra, and A. Kumar, Fabrication of mechanically stable superhydrophobic aluminium surface with excellent self-cleaning and anti-fogging properties. *Biomimetics*, 2017. 2(1): p. 2.
14. Lai, Y., et al., Transparent superhydrophobic/superhydrophilic TiO 2-based coatings for self-cleaning and anti–fogging. *Journal of Materials Chemistry*, 2012. 22(15): p. 7420–7426.
15. Yang, F. and Z. Guo, Bio-inspired design of a transparent TiO2/SiO2 composite gel coating with adjustable wettability. *Journal of Materials Science*, 2016. 51(16): p. 7545–7553.
16. Faraz, M., M.Z. Ansari, and N. Khare, Synthesis of nanostructure manganese doped zinc oxide/polystyrene thin films with excellent stability, transparency and super-hydrophobicity. *Materials Chemistry and Physics*, 2018. 211: p. 137–143.
17. Guo, S., et al., Controllable preparation of transparent superhydrophobic TiO$_2$ nanoarrays. *Materials Technology*, 2015. 30(1): p. 43–49.

18. Li, B.-j., et al., Superhydrophobic and anti-reflective ZnO nanorod-coated FTO transparent conductive thin films prepared by a three-step method. *Journal of Alloys and Compounds*, 2016. 674: p. 368–375.

19. Xu, L., et al., Transparent, superhydrophobic surfaces from one-step spin coating of hydrophobic nanoparticles. *ACS Applied Materials & Interfaces*, 2012. 4(2): p. 1118–1125.

20. Luo, G., et al., Preparation and performance enhancements of wear-resistant, transparent PU/SiO$_2$ superhydrophobic coating. *Surface Engineering*, 2018. 34(2): p. 139–145.

21. Schaeffer, D.A., et al., Optically transparent and environmentally durable superhydrophobic coating based on functionalized SiO$_2$ nanoparticles. *Nanotechnology*, 2015. 26(5): p. 055602.

22. Kezuka, Y., et al., Calcium carbonate chain-like nanoparticles: Synthesis, structural characterization, and dewaterability. *Powder Technology*, 2018. 335: p. 195–203.

23. Sun, S., et al., Effects of organic modifiers on the properties of TiO$_2$-coated CaCO 3 composite pigments prepared by the hydrophobic aggregation of particles. *Applied Surface Science*, 2018.

24. Zhang, H., et al., A facile method to prepare superhydrophobic coatings by calcium carbonate. *Industrial & Engineering Chemistry Research*, 2011. 50(6): p. 3089–3094.

25. Lide, D.R., Physical constant of inorganic compound. *Handbook of Chemistry and Physics*, 2005. 474.

26. Zimmermann, J., et al., Long term environmental durability of a superhydrophobic silicone nanofilament coating. *Colloids and Surfaces A: Physicochemical and Engineering Aspects*, 2007. 302(1–3): p. 234–240.

27. Shea, T.M., *Durable hydrophobic surface coatings using silicone resins.* 2008, Google Patents.

28. Rahman, I., M. Jafarzadeh, and C.S. Sipaut, Synthesis of organo-functionalized nanosilica via a co-condensation modification using γ-aminopropyltriethoxysilane (APTES). *Ceramics International*, 2009. 35(5): p. 1883–1888.

29. Rowley, J. and N.H. Abu-Zahra, Synthesis and characterization of polyethersulfone membranes impregnated with (3-aminopropyltriethoxysilane) APTES-Fe3O4 nanoparticles for As (V) removal from water. *Journal of Environmental Chemical Engineering*, 2019. 7(1): p. 102875.

30. Ammar, S., et al., Amelioration of anticorrosion and hydrophobic properties of epoxy/PDMS composite coatings containing nano ZnO particles. *Progress in Organic Coatings*, 2016. 92: p. 54–65.

31. Arkles, B., et al., Factors contributing to the stability of alkoxysilanes in aqueous solution. *Journal of Adhesion Science and Technology*, 1992. 6(1): p. 193–206.

32. Lork, A., I. König-Lumer, and H. Mayer, Silicone resin networks: The structure determines the effect. *European Coatings Journal*, 2003(4): p. 132–137.

33. Zheng, Y., et al., Preparation of superhydrophobic coating using modified CaCO$_3$. *Applied Surface Science*, 2013. 265: p. 532–536.

34. Guo, Z., W. Liu, and B.-L. Su, Superhydrophobic surfaces: from natural to biomimetic to functional. *Journal of Colloid and Interface Science*, 2011. 353(2): p. 335–355.

35. Dain, S.J., et al., Assessment of fogging resistance of anti-fog personal eye protection. *Ophthalmic and Physiological Optics*, 1999. 19(4): p. 357–361.

36. Seo, D., et al., The effects of surface wettability on the fog and dew moisture harvesting performance on tubular surfaces. *Scientific Reports*, 2016. 6: p. 24276.

37. Bai, S., et al., Enhancing antifogging/frost-resisting performances of amphiphilic coatings via cationic, zwitterionic or anionic polyelectrolytes. *Chemical Engineering Journal*, 2019. 357: p. 667–677.

38. Chen, Y., et al., Transparent superhydrophobic/superhydrophilic coatings for self-cleaning and anti-fogging. *Applied Physics Letters*, 2012. 101(3): p. 033701.

39. Quan, Y.-Y. and L.-Z. Zhang, Experimental investigation of the anti-dust effect of transparent hydrophobic coatings applied for solar cell covering glass. *Solar Energy Materials and Solar Cells*, 2017. 160: p. 382–389.

40. Thio, B.J.R. and J.C. Meredith, Measurement of polyamide and polystyrene adhesion with coated-tip atomic force microscopy. *Journal of Colloid and Interface Science*, 2007. 314(1): p. 52–62.

41. Israelachvili, J.N., *Intermolecular and surface forces*. 2011: Academic Press.

42. Tan, C.L.C., et al., Adhesion of dust particles to common indoor surfaces in an air-conditioned environment. *Aerosol Science and Technology*, 2014. 48(5): p. 541–551.

43. Griesser, H.J., *Thin film coatings for biomaterials and biomedical applications*. 2016, Duxford: Woodhead Publishing.

44. Alireza, R., *Mechanical and biological evaluations of smart antibacterial nanostructured TI-6AL-7NB implant/Alireza Rafieerad*. 2017, California: University of California.

45. Hoover, J., D. Kujawski, and F. Ellyin, Transverse cracking of symmetric and unsymmetric glass-fibre/epoxy-resin laminates. *Composites Science and Technology*, 1997. 57(11): p. 1513–1526.

46. Nikforooz, M., et al., Assessment of failure toughening mechanisms in continuous glass fiber thermoplastic laminates subjected to cyclic loading. *Composites Part B: Engineering*, 2019. 161: p. 344–356.

47. Wang, D., et al., Study on the long-term behaviour of glass fibre in the tensile stress field. *Ceramics International*, 2019. 45(9): p. 11578–11583.

48. Vengadaesvaran, B., et al., Scratch resistance enhancement of 3-glycidyloxypropyltrimethoxysilane coating incorporated with silver nanoparticles. *Surface Engineering*, 2014. 30(3): p. 177–182.

49. Ono, T. and R. Allaire, Fracture analysis, a basic tool to solve breakage issues. *Taiwan FPD Expo*, 2000. 201: p. 1–9.

50. Carrillo, F., et al., Nanoindentation of polydimethylsiloxane elastomers: Effect of crosslinking, work of adhesion, and fluid environment on elastic modulus. *Journal of Materials Research*, 2005. 20(10): p. 2820–2830.

51. Whitesides, M., Solvent compatibility of PDMS-based microfluidic devices *Analytical Chemistry*, 2003. 75: p. 6544–6554.

52. Zhao, X., et al., Fabrication of a scratch & heat resistant superhydrophobic SiO 2 surface with self-cleaning and semi-transparent performance. *RSC Advances*, 2018. 8(44): p. 25008–25013.

53. Brazel, C.S. and S.L. Rosen, *Fundamental principles of polymeric materials*. 2012. Hoboken, NJ: John Wiley & Sons.

54. Liu, H., et al., Robust translucent superhydrophobic PDMS/PMMA film by facile one-step spray for self-cleaning and efficient emulsion separation. *Chemical Engineering Journal*, 2017. 330: p. 26–35.

55. Zhao, D., W. Wang, and Z. Hou, Tensile initial damage and final failure behaviors of glass plain-weave fabric composites in on-and off-axis directions. *Fibers and Polymers*, 2019. 20(1): p. 147–157.

56. Ko, Y.H., et al., High transparency and triboelectric charge generation properties of nano-patterned PDMS. *RSC Advances*, 2014. 4(20): p. 10216–10220.

57. Isaifan, R.J., et al., Evaluation of the adhesion forces between dust particles and photovoltaic module surfaces. *Solar Energy Materials and Solar Cells*, 2019. 191: p. 413–421.

58. Syafiq, A., et al., Advances in approaches and methods for self-cleaning of solar photovoltaic panels. *Solar Energy*, 2018. 162: p. 597–619.

59. Li, X.-M., D. Reinhoudt, and M. Crego-Calama, What do we need for a superhydrophobic surface? A review on the recent progress in the preparation of superhydrophobic surfaces. *Chemical Society Reviews*, 2007. 36(8): p. 1350–1368.

5 Experimental and Characterization Techniques

Rahul, Rakesh K. Sonker, P. K. Shukla,
Pramod K. Singh and Zishan H. Khan

CONTENTS

5.1   SAMPLES STUDIED

We have synthesized different dye-/perovskite-sensitizer complex systems and their applications in dye-sensitized solar cells (DSSC) and perovskite-sensitized solar cells (PSSCs) have been analyzed. Table 5.1 shows the sensitizer systems with the electrolytes used in solar cell synthesis.

PEO, polyethylene oxide

All experimental details regarding the synthesis of solar cells using the above-mentioned sensitizers and electrolytes are discussed in detail below.

In this work, synthesized materials and devices were characterized using different techniques such as X-ray diffraction (XRD), scanning electron microscopy (SEM), transmission electron microscopy (TEM), current density-voltage (J-V) characteristics, and UV-visible spectroscopy.

5.2   MATERIALS USED

For the synthesis of different functional layers of DSSCs and PSSCs studied here, different chemicals and materials were used. Some of the functional materials and chemicals were used as purchased, whereas others were synthesized using basic chemicals.

Blocking layer solution was prepared from Ti(IV) bis(acetoacetato) diisopropoxide and gamma-butyrolactone, which were purchased from Thomas Baker, India. Lead iodide (PbI_2), stannous chloride ($SnCl_2$), polyethylene oxide (PEO), and hexachloroplatinic acid (H_2PtCl_6) were purchased from Sigma Aldrich, USA. Ethylamine (C_2H_7N), hydroiodic acid (HI, 55 wt.% in water), methylamine (CH_5N), ethanol, petroleum ether, and acetonitrile (C_2H_3N) were purchased from Thomas Baker Company, India. Titanium dioxide (TiO_2) paste and Ruthenizer 535 N3 ($C_{26}H_{16}O_8N_6S_2Ru$) were purchased from Solaronix, Switzerland. Potassium iodide (KI) and iodine (I_2) were purchased from Himedia, India. Methanol (CH_3OH),

TABLE 5.1   Sensitizer Systems with the Electrolytes Used in Solar Cell Synthesis

Perovskite-/dye-sensitizer System	Electrolyte
Natural Dye	$PEO + KI + I_2$
$CH_3CH_2NH_3PbI_3$	$PEO + KI + I_2$
$CH_3NH_3PbI_3$ (Powder)	$PEO + KI + I_2$
$CH_3NH_3PbI_3$ (Crystal)	$PEO + KI + I_2$
$CH_3NH_3SnCl_3$ (Powder)	$PEO + KI + I_2$
$CH_3NH_3SnCl_3$ (Crystal)	$PEO + KI + I_2$

acetone (C_3H_6O), *N,N*-dimethyl formamide (DMF; C_3H_7NO), and hydrochloric acid (HCl) were purchased from Thomas Baker, India.

5.3 SAMPLE PREPARATION

Perovskite sensitizers have been synthesized using various methods.

5.3.1 Thin Film Deposition Techniques

As we discussed earlier, DSSCs and PSSCs are made up of different functional layers. In most cases, these layers are synthesized using solution processing. In our study, we have deposited the functional layers using a spin-coating method and the doctor's blade method.

5.3.1.1 Spin coating

Spin coating is one of the simplest techniques used to deposit a thin film on a glass substrate using solution processing. The main advantage of spin coating is its simplicity and ease of use (Figure 5.1). With this method, a highly uniform film can be deposited with a thickness ranging from few nanometers to several hundred microns. In DSSC technology, different functional layers may be deposited using this technique.

In this method, a small amount of the solution of the material in a volatile solvent is put onto the substrate fixed on a rotating chunk. The chunk is rotated at high speed so that the solution spreads evenly on the substrate. In order to optimize the thickness of the film, the viscosity of the solution, spinner speed, and amount of solution are varied. The whole assembly specially manufactured for this process is called a spin coater.

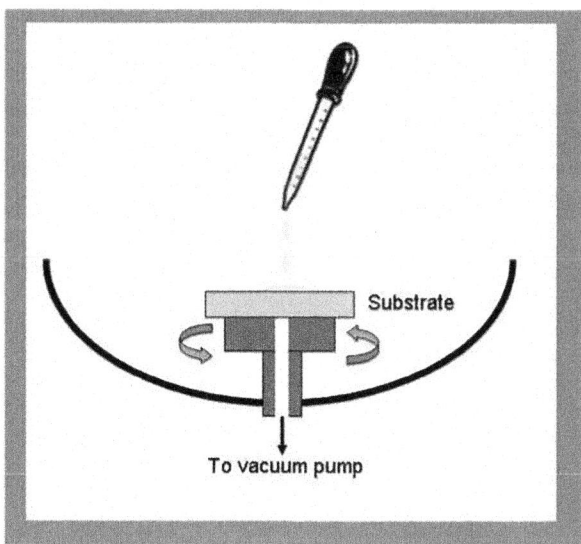

FIGURE 5.1 Schematic of thin-film deposition using spin coating

Spin coaters are widely used to prepare solution-processable thin films. It is a cost-effective deposition technique, but not ideal for mass production. For mass production other techniques are used [1].

5.3.1.2   Doctor's blade

In this method, a slurry of the desired material is placed on the substrate and spread, using a sharp-edged blade, to the desired length and width, which is commonly restricted by masks. Figure 5.2 shows the procedure. For small- and laboratory-scale fabrication of solar cells, the doctor's blade is one of the best methods, whereas for large-scale fabrication, screen printing is used [2–6].

5.3.2   Synthesis of Photoactive Layers Based on Perovskite Materials

5.3.2.1   Solution-Processed Two-Step Method

Proposed by Mitzi et al. [7], in the solution-processed two-step method, PbI_2 is deposited on the substrate prior to methyl ammonium iodide (MAI) treatment by spin coating. A saturated methanol solution of PbI_2 is used as the precursor solution for the spin-coating process, then the PbI_2-coated substrate immersed in a 2-propanol solution containing MAI before rinsing with 2-propanol. Dipping time is crucial to the final product. This two-step method has been applied for the synthesis of perovskite solar cells by some other workers also [8–10].

Owing to some drawbacks associated with dip-coating of MAI, it has been replaced by spin coating in later works. Figure 5.3 schematically explains the solution-processed two-step deposition method.

5.3.2.2   Solution-Processed One-Step Method

In the one-step method, the precursor of MAI and PbI_2 is dissolved in a mixed solution of N,N-dimethyl sulfoxide and gamma-butyrolactone, which is then spin-coated on the substrate. The one-step method is better in terms of minimizing the processing steps, but the photovoltaic performance of devices made in this way is inferior to those produced by a two-step process because of significant differences in morphology of the $MAPbI_3$ [11–16].

Anti-solvent engineering was proposed to control the crystal growth kinetics [17]. An anti-solvent such as toluene is dripped while spinning the precursor solution, which leads to a homogeneous, flat perovskite film with well-developed large grains.

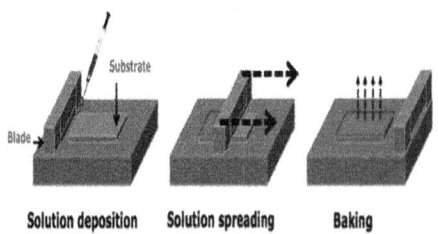

Solution deposition   Solution spreading   Baking

FIGURE 5.2   Doctor's blading

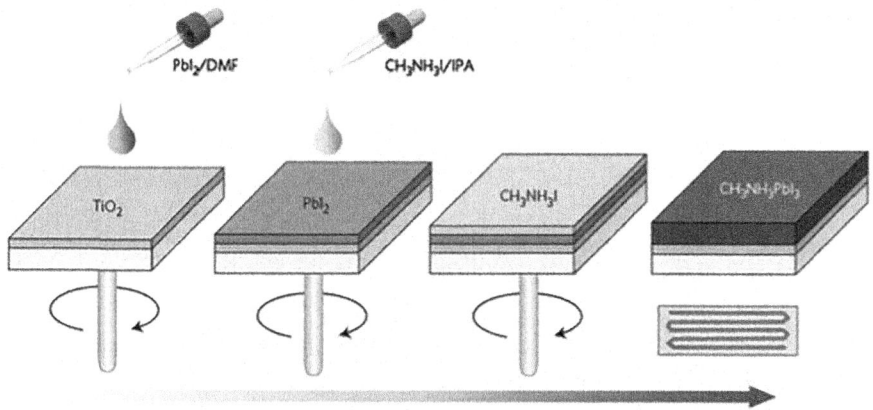

FIGURE 5.3 Solution-processed two-step deposition method

By utilizing the solvent engineering technique, a high photo conversion efficiency of 18.4% has been achieved from the solid solution of $FAPbI_3$ and $MAPbBr_3$ with a ratio of 85:15 [18–21].

5.3.3 SYNTHESIS OF DYE AND PEROVSKITE-BASED SENSITIZERS AND ELECTROLYTES

5.3.3.1 Extraction of Pigments from Natural Dyes and Preparation of Dye-Sensitizer Solutions

Solutions of various fruit and vegetable pigments were obtained from fresh fruit and vegetables. All fresh fruits and vegetables were crushed and added to ethanol and acetone. The mixture was then centrifuged to get the required dye. Figures 5.4–5.8 shows some of the fruit and vegetables used in this study.

a. Pomegranate (Anthocyanin) Dye

To extract anthocyanin pigment from pomegranate, fresh fruits were peeled. After drying at room temperature, the peel was ground into a powder then 100 g was weighed and placed in a round-bottomed flask. Solvent (40:60 ethanol:water; 500 ml) was added. The flask was heated in a water bath at 60°C for 60 mins. The solution was then filtered to obtain the required dye.

FIGURE 5.4 Some fruits and vegetables used in the study: (a) pomegranate, (b) raspberry, (c) spinach, (d) orange peel, and (e) tomato

FIGURE 5.5 Images of extracted natural dye solution: (a) pomegranate, (b) raspberry, (c) spinach, (d) orange peel, and (e) tomato

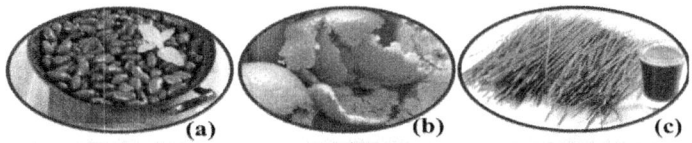

FIGURE 5.6 The used part of fruits and vegetables: (a) red beans, (b) orange peel, (c) barley grass

FIGURE 5.7 Extracted natural dye solutions: (a) red beans, (b) orange peel, (c) barley grass

b. Raspberry (Anthocyanin) Dye

Raspberry contains a red/purple anthocyanin pigment that can be attached to a TiO_2 layer and works as a sensitizer. To extract the pigment, ripe raspberries were washed multiple times with water to remove impurities and contaminations. The cleaned fruits were ground to a paste using a pestle and mortar. Ethanol (20 ml) was added and the mixture refluxed for 90–200 minutes at 60°C. The resulting liquid was filtered using filter paper to extract the dye.

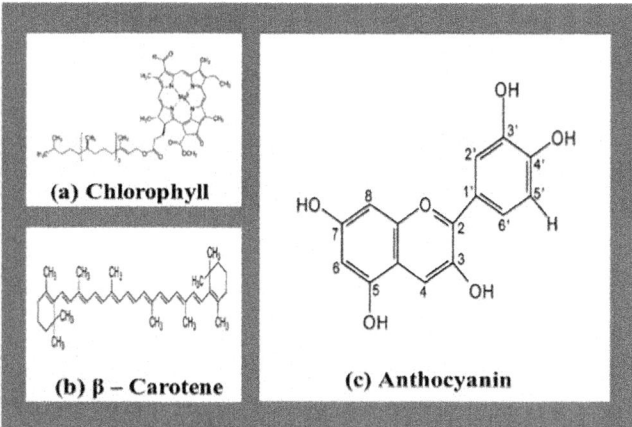

FIGURE 5.8 Structures of extracted natural dye pigments: (a) chlorophyll, (b) β-carotene, (c) anthocyanin

c. Spinach (Chlorophyll) Dye

Spinach leaves were first scrubbed with acetone then transformed into a paste by grinding. Acetone was added and the paste was ground again. The prepared solution was dispensed over filter paper in a funnel and acetone was poured over to extract the dye.

d. Orange Peel (β-carotene) Dye

Fresh oranges were purchased from a nearby market. After washing with de-ionized water, the oranges were precisely peeled. The peel was dried in a vacuum heater for around 10 h at 50°C then ground into powder in a mortar. Around 50 g of the powdered sample was put into a measuring cylinder and 150 ml supreme ethanol added. Subsequently, the blend was shaken for 5 h, shielded from daylight and kept at 50°C. This concentrate solution was utilized as a sensitizer in the synthesis of DSSC.

e. Tomato (Lycopene) Dye

Lycopene extract is produced from a tomato variety that has a high lycopene content, within the range of 150 to 250 mg/kg. This variety is not generally sold in market for edible use but produced specially for lycopene extract. The lycopene extract was prepared by crushing tomatoes into crude tomato juice. Ethanol (2 ml) was added to the tomato juice and filtered with filter paper.

f. Red Bean (Anthocyanin) Dye

To extract the anthocyanin pigment from red beans, we purchased fresh red bean seeds, crushed them and eliminated their white part. The resulting material was ground using a pestle and mortar to get anthocyanin dye.

g. Barley Grass (Chlorophyll) Dye

Barley grass leaves were scrubbed with acetone and transformed into a paste by grinding. Acetone was added and the paste ground again. A filter paper was attached

FIGURE 5.9   Stepwise synthesis of ethyalammonium tri-lead iodide: (a) ice bath process; (b) precipitate collection; (c) collected precipitate

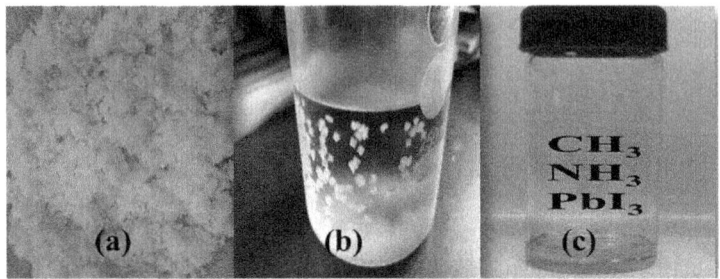

FIGURE 5.10   The formation of $CH_3NH_3PbI_3$. (a) Synthesized precipitate of methyl ammonium iodide; (b) recrystallized perovskite; (c) solution of methyl ammonium tri-lead iodide

to a funnel and the paste dispensed over the paper. Acetone (30 ml) was poured drop by drop on the paste. The barley grass chlorophyll was collected from the funnel, allowed to settle in a beaker, and collected in a clean petri dish [Figure 5.6 (a-c) and Figure 5.7].

5.3.3.2   Synthesis of Perovskite-Based Sensitizers

a. Synthesis of ethylammonium tri-lead iodide ($CH_3CH_2NH_3PbI_3$)

The perovskite sensitizer $CH_3CH_2NH_3PbI_3$ was synthesized as follows. Ethylamine (10 ml) and HI (10 ml) were stirred in a conical flask at 0°C for 2 h then heated at 60 °C for 2 h. The precipitate was collected then washed twice with petroleum ether and dried at 100°C in vacuum oven for 24 h. The as-synthesized, dried $CH_3CH_2NH_3I$ powder (2.234 g) was mixed with PbI_2 (6.016 g) at a 1:1 molar ratio in 10 ml DMF solvent. The solution was stirred at 60°C for 2 h. The resulting $CH_3CH_2NH_3PbI_3$ solution was used as the sensitizer.

b. Synthesis of Crystals and Powder of Methyl Ammonium Tri-Lead Iodide ($CH_3NH_3PbI_3$)

Methyl ammonium iodide (CH_3NH_3I) was prepared by mixing methylamine (27.86 ml) in HI (30 ml) followed by ice-bath treatment at ~0°C for 2 h. The resultant product was put into the oven (60°C) overnight for complete evaporation. The precipitate of CH_3NH_3I was collected in a round-bottomed flask and washed thoroughly using diethyl ether. The final white precipitate was dried at

FIGURE 5.11 Solid polymer electrolyte preparation (a) PEO film poured in a petri dish, and (b) electrolyte solution containing PEO+KI+I_2

100°C in a vacuum oven for 24 h (Figure 5.10a) then dissolved in ethanol and kept in a refrigerator for crystallization. The resultant CH_3NH_3I was put in the vacuum oven at 60°C for 24 hours prior to further processing.

$CH_3NH_3PbI_3$ solution was prepared by taking an equimolar ratio (1:1) of CH_3NH_3I and PbI_2. The as-synthesized CH_3NH_3I powder (0.395 g) was mixed with PbI_2 (1.157 g) in gamma-butyrolactone (2 ml). The overall solution was continuously stirred for ~4 h at 60°C, resulting in a solution of $CH_3NH_3PbI_3$ (Figure 5.10c). This perovskite solution was used as a light sensitizer by spin coating it on a working electrode for making PSSC.

c. Synthesis of Crystals and Powder of $CH_3NH_3SnCl_3$: A New Class of Lead-Free Perovskite Material for Low-Cost Solar Cell Applications

Methyl ammonium chloride (CH_3NH_3Cl) was prepared by mixing methyl-amine (30 ml) and HCl (32.3 ml) in a round-bottomed flask. The product was put into a vacuum oven (60°C) overnight to dry. The CH_3NH_3Cl precipitate was washed thoroughly using diethyl ether and the resultant white precipitates dried at 100°C in a vacuum oven for 24 h, to get a powdered form of methyl ammonium chloride.

To produce CH_3NH_3Cl crystals, CH_3NH_3Cl powder was dissolved in ethanol and the solution heated at 60° C for 4 h then kept in the refrigerator for 3 days until recrystallization occurred. The product was filtered out and kept in a vacuum oven at 80°C for 48 h.

$CH_3NH_3SnCl_3$ powder and crystal solutions were synthesized by dissolving CH_3NH_3Cl powder and crystals, respectively, (0.395 g) with $SnCl_2$ (1.157 g) in DMF (2 ml), followed by stirring at 60°C.

5.4 PREPARATION OF ELECTROLYTE SOLUTION

The solid polymer electrolyte solution containing PEO, 20% KI, and I_2 (10% of KI) in acetonitrile solvent was prepared. KI and I_2 were mixed in acetonitrile and stirred at 50°C for 3–4 hours using a magnetic stirrer. The KI:I_2 composition was fixed at 10:1 (w/w). This polymer electrolyte solution was poured into a petri dish to form a film. Figure 5.11(a) shows the PEO film in the petri dish after drying in vacuum at room temperature for ~7 days to eliminate traces of solvent. On characterization, the ionic conductivity was found to be 2.02×10^{-5} S/cm.

5.5 FABRICATION OF DYE-SENSITIZED SOLAR CELLS

As mentioned elsewhere, a conventional DSSC/PSSC has the following components:

1. Glass substrate coated in fluorine-doped tin oxide (FTO)
2. Nano-/micro-porous, nanocrystalline, semiconducting TiO_2 for the working electrode
3. A dye or perovskite sensitizer
4. Electrolyte composed of solid polymer electrolyte film having maximum ionic conductivity + I_2 (for I^-/ I_3^- redox couple)
5. Platinum-coated counter electrode

In the subsequent discussions, the techniques adopted for fabrication of DSSC/PSSC and its components are described. Selection of material for a component plays an important role in maximizing cell efficiency [22–25]. In our laboratory, we have successfully developed DSSC/PSSC assemblies using FTO glass as the substrate.

For laboratory-scale DSSC/PSSC fabrication, we cut FTO (Sigma Aldrich, USA) glass plate (1×1 cm^2) and cleaned it thoroughly in an ultrasonic bath with acetone, chloroform, and isopropyl alcohol. A blocking layer of TiO_2 solution (Sigma Aldrich, USA) was spin-coated on the FTO glass plate. The TiO_2-coated FTO substrate was annealed at 500°C for 30 min in a programmable muffle furnace [26–28]. To prepare the porous TiO_2 working electrode, TiO_2 paste was coated on as-annealed

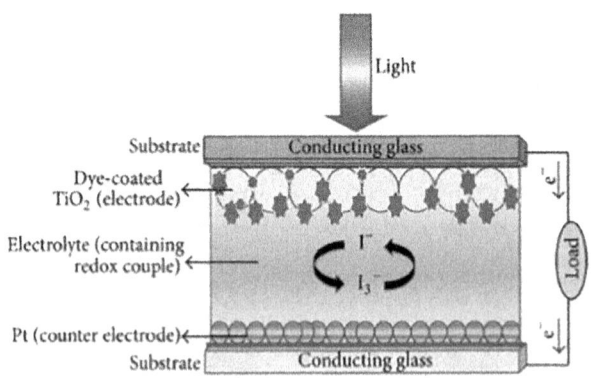

FIGURE 5.12   Assembly of DSSC

titania-coated FTO substrate using the doctor's blade method, followed by heating at 500°C for 30 min. This provided a nice porous (10–15 nm pore diameter) TiO_2 film of approximately 10 mm thickness. The counter electrode was prepared by coating a layer of platinum (spin-coating chloroplatinic acid) on another piece of FTO-coated glass followed by calcination at 400°C for 30 min [29]. The working electrode (FTO/TiO_2) was dipped into the (N3) dye solution overnight. The solid polymer electrolyte with redox couple was sandwiched between the working and the counter electrode. The assembly of a typical DSSC/PSSC is shown in Figure 5.12.

5.6 CHARACTERIZATIONS

5.6.1 X-Ray Diffraction

The phenomenon of diffraction of X-rays from a crystal was first discovered in 1913 by W. L. Bragg and W. H. Bragg using ZnS crystal. There are four components in a typical XRD instrument (Figure 5.13):

1. X-ray source
2. Sample stage
3. Receiving optics
4. X-ray detector.

The optics and X-ray detector are situated on the focusing circle while the sample stage lies at the center of the circle. The angle between the sample and X-ray source is θ (Bragg's angle) and the angle between the X-ray detector and the projection of X-rays 2θ. In a typical XRD investigation, the sample is mounted on the sample holder and illuminated by the X-ray beam. The X-rays are scattered by each atom in the sample [30–35]. These scattered beams interfere constructively or destructively depending whether the beams superimpose in phase or out of phase with each other.

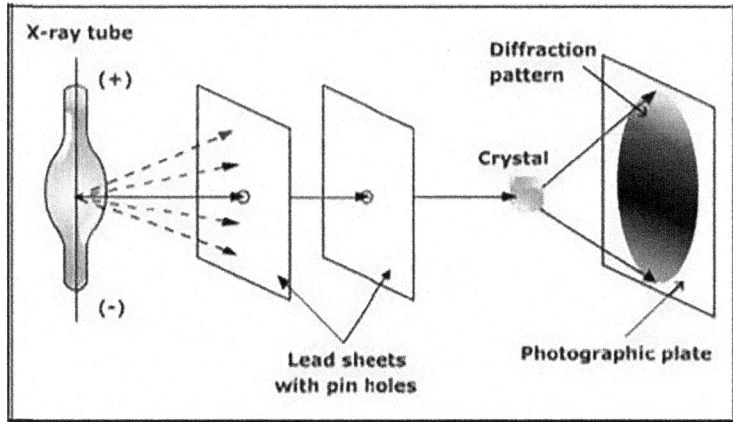

FIGURE 5.13 Schematic of X-ray diffraction by a crystal

The planes of the sample at which the X-rays scatter are called reflecting planes. The processing system records the position of the interference maxima, corresponding to constructive interference. Each peak position corresponds to different planes present in the sample.

5.6.1.1 X-Ray Diffraction Data Analysis

After collecting the XRD patterns, the first task is to assign its peaks to the different planes present in the sample. Each plane is represented by a unique set of Miller indices. Therefore, every peak is assigned a Miller index corresponding to the plane it represents. There are three methods of assigning Miller indices to the diffraction peaks:

1. Comparing the measured XRD pattern with the standard database (JCPDS cards)
2. Analytical methods
3. Graphical methods.

XRD is an important characterization technique that not only provides information regarding the crystallinity of the sample but it is also used to measure the particle size present in the sample indirectly. All the parameters regarding the crystalline arrangement of the system can be evaluated by XRD analysis in Figures 5.14 and 5.15. Analysis of XRD data gives the unit cell parameters of the system. In our study, we used XRD technique to analyze the course of the reaction as well as to study the effect of heat treatment over time.

In our study, XRD played an important role in analyzing the crystallite size. Actually, at the grain boundaries, recombination sites exist. Large crystallites enhances the efficiency of photovoltaic devices. Therefore, an investigation of crystallite size is critical in order to optimize the performance of the devices [36].

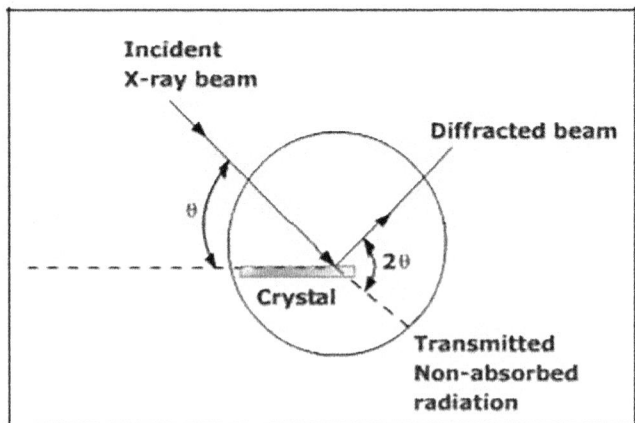

FIGURE 5.14   A schematic representation of X-ray diffraction

FIGURE 5.15   Typical XRD instrument

5.6.2   RAMAN SPECTROSCOPY

Raman spectroscopy is a powerful and non-destructive technique, named after Sir C. V. Raman, for the evaluation of vibrational, rotational, and other low-frequency modes in materials. It provides molecular vibrational–rotational spectra, which are complementary to infrared spectroscopy. The process is based on inelastic scattering of radiation. Raman spectroscopy can detect a change in vibrational, rotational, or electronic energy of a molecule. For a vibrational mode in a molecule to be Raman active, it must be associated with changes in polarization. It is very useful for the analysis of vibrational modes in single crystals, due to their anisotropic nature. The molecular vibrations and phonons of the system interact inelastically with incident radiation, which leads to a shift in its energy known as Raman (Stokes or anti-Stokes) scattering. This shift in the energy is equal to the energy of vibration of the scattering molecule. The electric field E of the incident electromagnetic radiation distorts the electron cloud of the molecule and induces dipole moment P via the relation

$$P = \alpha\,E$$

Molecular vibration changes the polarizability (α); thereby the energy of the scattered electromagnetic radiation gets changed.

The plot of intensity of scattered light vs. energy difference/Raman shift in cm^{-1} gives the Raman spectrum. Features such as peak position, intensity, width, etc. are very sensitive to the chemical composition, functional groups, structural phase, defects, and impurities.

The Raman spectrum of a material has specific characteristic features, variations in which depict changes in the molecular/structural symmetry of materials. Renishaw In. Via Micro-Raman equipment was employed in the present study. The schematic for optics and a photograph of the spectrometer are shown in Figure 5.16.

Raman spectroscopy plays a significant role in analyzing the orientational disorder of a material. A Raman spectroscopic study of organo-halides provides the basis for the optimization of photovoltaic devices. Raman spectroscopy is very sensitive to environmental effects such as moisture-induced degradation. At very low excitation energies, we were able to obtain the Raman spectra of the pristine, environmentally exposed perovskite films. Careful analysis of the results obtained during this investigation provides lots of important information.

5.6.3   UV-Visible Spectroscopy

UV-visible spectroscopy is the atomic spectroscopic technique used to investigate electronic transitions. In our study, all the UV-visible spectra were recorded in the wavelength range of 900–1100 nm. A typical UV-visible spectrometer has the provision to analyze liquid as well as solid samples. Figure 5.17 shows a picture of a typical UV-visible spectrometer. Films of pristine and irradiated samples were mounted

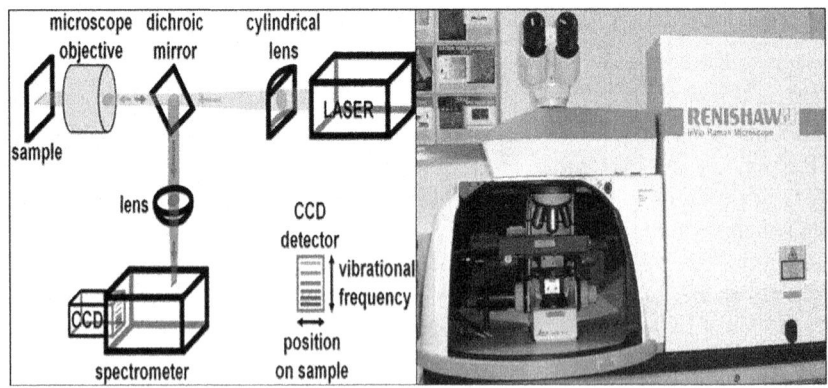

FIGURE 5.16   Renishaw In. Via Raman spectrometer

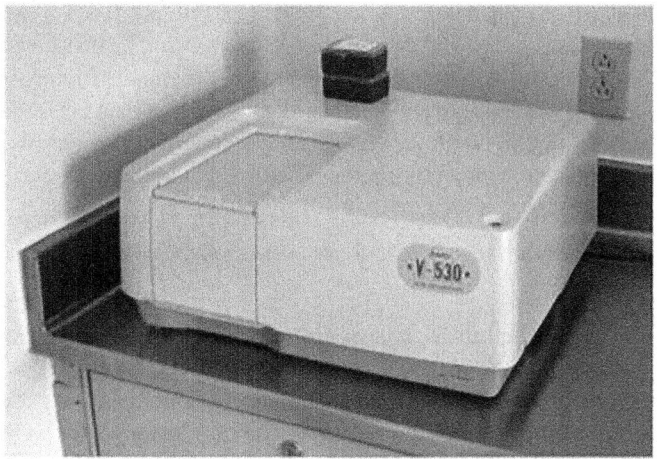

FIGURE 5.17   UV-visible spectrophotometer

on a black sheet, having a window of area 0.5 cm^2. All the measurements took air as reference.

The UV-visible absorption spectroscopy is the measurement of the attenuation of a beam of light with wavelength after it passes through a sample or after reflection from a sample surface. The short-wavelength limit for simple UV-visible spectrometers is 180 nm, due to absorption of ultraviolet wavelength below 180 nm by atmospheric gases. The absorbance A is related to the input and output intensities according to the Beer-Lambert Law [37], given by the equation

$$\frac{1}{lo} = e^{-A}$$

The absorbance A is divided by the path length l to yield the absorption coefficient, which qualifies the absorbance per meter. Thus, taking the film thickness into account one can write

$$A\alpha(\lambda) = 2.303$$

On illumination with UV-visible radiation, the atoms or molecules of the material get excited, and the atoms present in the molecules vibrate and rotate with respect to each other. The absorption of UV or visible radiation corresponds to the excitation of outer electrons. When an atom or molecule absorbs energy, electrons move from the ground state to an excited state. Only certain functional groups in a molecule are Raman active. These functional groups are called chronophers. The vibrational and rotational modes of the molecules superimpose on the absorption spectrum. Therefore, the UV-visible absorption spectra of the chromophores become complex and their careful investigation provides a lot of information regarding the sample [38].

$CH_3NH_3PbI_3$ perovskite is known to have direct band-gap properties, which can be confirmed by UV-visible spectroscopic data, as shown with detailed explanation in Chapter 4. The band-gap determination of $CH_3NH_3PbI_3$ perovskite is near 1.55 eV, which corresponds to an onset of light absorption at 800 nm. It can also be seen from these data that the absorption maximum of perovskite increases from 550 to 720 nm upon increases in PbI_2 concentration. When the absorption data are compared with the SEM data of the films with increasing PbI_2 concentrations, increases in absorbance can be correlated with a morphology change in the films.

5.6.4   SCANNING ELECTRON MICROSCOPE AND ENERGY-DISPERSIVE X-RAY SPECTROSCOPY

The SEM is a type of electron microscope capable of producing high-resolution images of a sample surface. Because of the way the image is created, SEM images have a characteristic three-dimensional appearance and are useful for observing the surface structure of the sample. The microstructures of hexaferrites were examined using an SEM.

5.6.4.1   Working of an SEM Instrument

Electrons from a thermionic or field-emission cathode are accelerated by a voltage of 1–50 kV between cathode and anode. The smallest beam cross-section at the gun (the crossover), with a diameter of the order of 10–50 μm for thermionic and 100–100nm for field-emission guns, is de-magnified by a two- or three-stage electromagnetic lens system, so that an electron probe of diameter 1–10 nm carrying a current of 10^{-10} to 10^{-12} A is formed at the specimen's surface.

A deflection coil system in front of the last lens scans the electron probe in a raster across the specimen, synchronized with the electron beam of a separate cathode ray tube (CRT). The intensity of the CRT is modulated by one of the signals recorded to

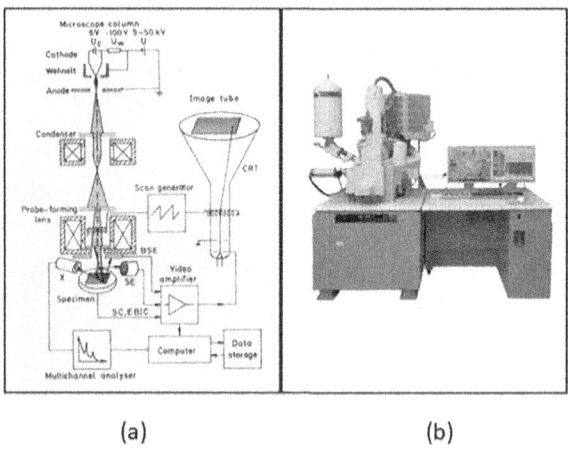

(a)   (b)

FIGURE 5.18   (a) Schematic diagram of scanning electron microscope BSE = back-scattered electrons, SE = secondary electrons, SC = specimen current, EBIC = electron-beam-induced current, X = X-rays, CRT = cathode ray tube, and (b) typical SEM instrument

form an image. The magnification can be increased simply by decreasing the scan-coil current and keeping the image size on the CRT constant.

An advantage of SEM is the wide variety of electron–specimen interactions that can be used to form an image to give qualitative and quantitative information. The large depth of focus, excellent contrast, and the straightforward preparation of solid specimen are advantages of SEM. Here an electron beam scans the object (the specimen) and because of synchronized scans of electron beam and the CRT screen (nowadays, monitor), there is a one-to-one correspondence between the spot on the specimen and the spot on the screen. Unlike optical microscopy, SEM requires a vacuum environment and for the specimen surface to be electrically conductive. Figure 5.18(a) shows a general schematic diagram of SEM.

The electron beam is produced by a hairpin-shaped tungsten filament by thermionic emission. An acceleration voltage of 5 to 50 kV can be applied and hence we get an electron beam of such energy. This diverged electron beam passes through a pair of electromagnetic lenses (coils) and through a probe-forming lens, which gives the beam the form of a fine probe (~10 nm diameter). This fine electron probe scans the specimen area in a linear manner. Another electron beam in synchronization with this beam scans on the CRT (or monitor) with the help of same scan generator. In SEM, the formation of image takes place because of electron beam–specimen interaction.

The importance of SEM analysis as a diagnostic tool for analyzing the degradation of a polycrystalline photovoltaic cell has been studied. The main aim of this study is to characterize the surface morphology of degraded and undegraded PSSCs.

5.6.4.2 Energy-Dispersive X-ray Spectroscopy

Energy-dispersive X-Ray analysis (EDX), also referred to as EDS or EDAX, is an X-ray technique used to identify the elemental composition of materials. Applications include materials and product research, troubleshooting, and de-formulation. EDX systems are attached to electron microscopy (SEM or TEM) instruments where the imaging capability of the microscope identifies the specimen of interest. The data

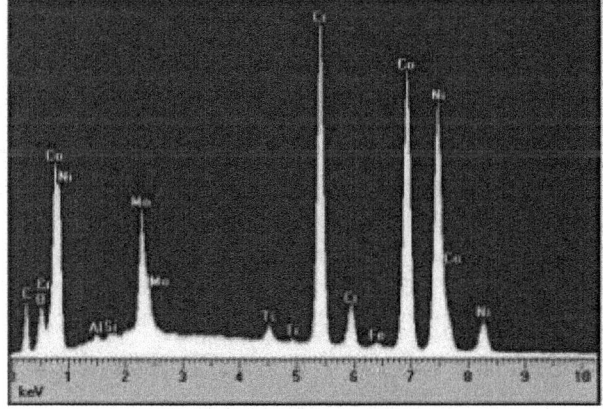

EDS Spectrum for Alloy MP35N

FIGURE 5.19 EDS spectrum for alloy MP35N

generated by EDX analysis consist of spectra showing peaks corresponding to the elements making up the sample being analyzed. Elemental mapping of a sample and image analysis are also possible.

Energy Dispersive X-Ray Spectroscopy (EDS or EDX) is a chemical microanalysis technique used in conjunction with SEM. The EDS technique detects X-rays emitted from the sample during bombardment by an electron beam to characterize the elemental composition of the sample shown in Figure 5.19. Features or phases as small as 1 µm or less can be analyzed. When the sample is bombarded by the SEM's electron beam, electrons are ejected from the atoms at the sample's surface. The resulting electron vacancies are filled by electrons from a higher state, and an X-ray is emitted to balance the energy difference between the two states. The X-ray energy is characteristic of the element from which it was emitted.

The EDS X-ray detector measures the relative abundance of emitted X-rays versus their energy. The detector is typically a lithium-drifted silicon, solid-state device. When an incident X-ray strikes the detector, it creates a charge pulse that is proportional to the energy of the X-ray. The charge pulse is converted to a voltage pulse (which remains proportional to the X-ray energy) by a charge-sensitive preamplifier. The signal is then sent to a multichannel analyzer where the pulses are sorted by voltage. The energy, as determined from the voltage measurement, for each incident X-ray is sent to a computer for display and further data evaluation. The spectrum of X-ray energy versus counts is evaluated to determine the elemental composition of the sample.

5.6.4.3   Benefits of EDX Analysis

- Improved quality control and process optimization
- Rapid identification of contaminants and source
- Full control of environmental factors, emissions, etc.
- Greater on-site confidence, higher production yield
- Identifying the source of a problem in a process chain.

The EDS technique was used to analyze the elemental composition and element ratio of the $MAPbI_3$ powder and the $MAPbI_3$ film prepared as described earlier. There are feature peaks recorded that are assigned to the Pb and I elements, respectively. The EDS spectral line pattern of the perovskite film is like that of the source perovskite powder, clearly indicating their homogeneity. It is a fantastic tool to analyze our synthetic procedure.

In a multi-technique approach EDX becomes very powerful, particularly in contamination analysis and industrial forensic science investigations. The technique can be qualitative, semi-quantitative, or quantitative, and provide spatial distribution of elements through mapping. The EDX technique is non-destructive and specimens of interest can be examined in situ with little or no sample preparation [39–45].

5.6.5   Transmission Electron Microscope

TEM has been used to study the size, shape, and distribution of nanoscale materials [46]. The schematic diagram of a typical TEM is represented in Figure 5.20 [47]. A

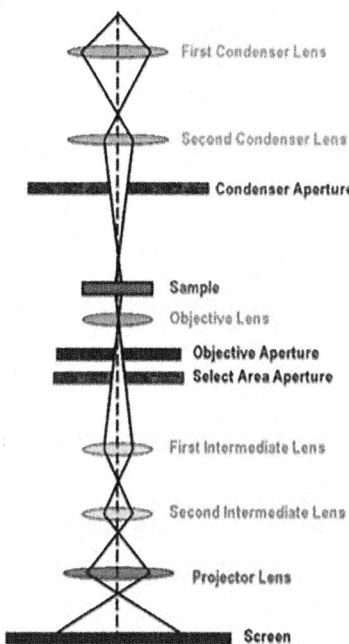

First Condenser Lens

Second Condenser Lens

Condenser Aperture

Sample

Objective Lens

Objective Aperture

Select Area Aperture

First Intermediate Lens

Second Intermediate Lens

Projector Lens

Screen

FIGURE 5.20   Schematic diagram of a typical transmission electron microscope

thin, solid specimen (< 200 nm) is bombarded in vacuum with a highly focused, monoenergetic beam of electrons. The smaller de Broglie wavelength associated with the high-energy electron beam is responsible for the high resolution and the ability to focus the electron beam. As an example, electrons with an energy of 100 keV correspond to a de Broglie wavelength of 3.7×10^{-3} nm. In general, TEM is expected to use electron beams having energy in the range of 20 to 200 keV. In this energy range, the beam has enough energy to propagate through the specimen. A series of electromagnetic lenses are used to magnify the transmitted electron signals [48].

There are two modes of TEM employed: image and diffraction. In image mode, contrast must be induced. There are many contrast-forming mechanisms, but the interpretation of images is complicated due to the interplay of the different mechanisms. The most common imaging techniques in TEM are mass thickness imaging, diffraction imaging, and phase contrast imaging.

In diffraction imaging mode, a pattern of diffracted electrons is obtained from the electron-illuminated sample. When the electron beam is incident on the sample, scattering events occur because all the illuminated parts of the sample act as scattering sources. Interference between scattering beam causes coherently scattered beams when Bragg's law is fulfilled. The scattered beams are recorded in the form of a "spot". The spot pattern of diffracted electron beams from the selected sample area is called the electron-diffraction pattern and provides information about the crystalline and crystal orientation. The dependence of electron transpiring lies in the sample thickness, as a thick sample would cause too many interactions leaving no intensity in the transmitted beam.

A thick sample also increases the risk that an electron is scattered on multiple occasions, making the resulting image difficult to interpret. In the work presented here, samples for TEM imaging have been prepared by dispersing powder in isopropanol using sonication; a small drop of that solution was cast onto the carbon-coated copper grid. Thus, TEM has been used to analyze the shape, size, and particle size distribution of perovskite light sensitizers.

High-resolution electron microscopy can provide information that is difficult to obtain in any other way, but it needs skill and experience. Lattice fringe images have been obtained from a wide range of polymers. Here, the particle size and morphology of perovskite thin films based on Pb and Sn have been examined using a high-resolution transmission electron microscope (Tecnai G2 F30 S-Twin) operating at an accelerating voltage of 300 kV and having a point resolution of 0.2 nm and a lattice resolution of 0.14 nm.

TEM observations have been performed for the perovskite materials. Although TEM is a powerful tool for nanostructured materials, sample damage by electron beam irradiation should be avoided, because $CH_3NH_3PbI_3$ is known to be unstable during annealing at elevated temperatures. Several TEM results have been reported for $CH_3NH_3PbI_3$ and $CH_3CH_2NH_3PbI_3$, and the structures revealed by electron diffraction and high-resolution images discussed.

5.6.6 J-V CHARACTERISTICS

To calculate photocurrent conversion efficiencies of fabricated DSSCs/PSSCs the fabricated cells were illuminated at 1 sun condition (100 mW/cm^2) (Figure 5.21) and the J-V curves were recorded on a Keithley 2400 Source meter. Using the obtained short circuit current density (J_{sc}) data, open circuit voltage (V_{oc}) and photocurrent conversion efficiencies were calculated.

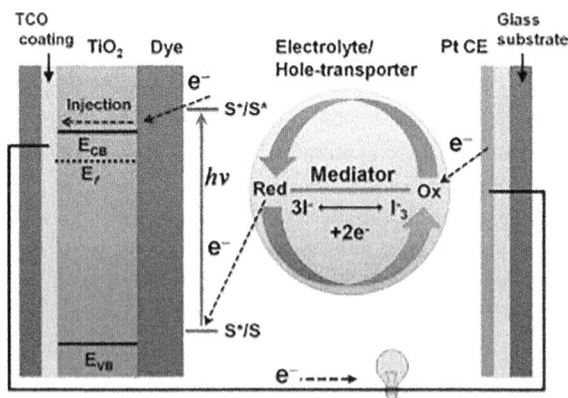

FIGURE 5.21   Schematic diagram showing process in DSSC

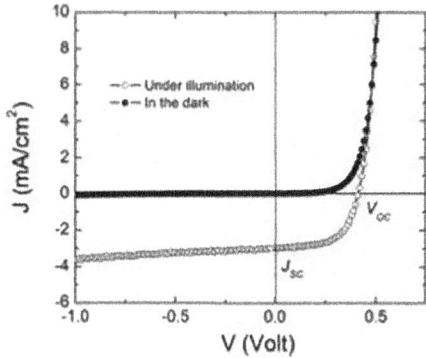

FIGURE 5.22   J-V characteristics of DSSC in dark and light

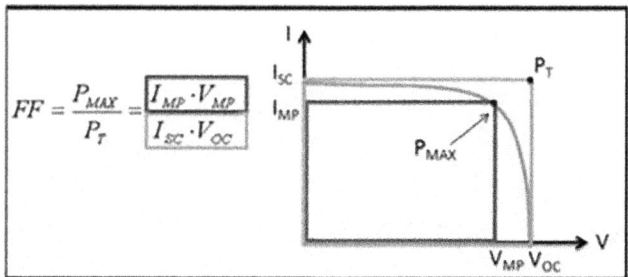

FIGURE 5.23   The I-V curve of DSSC exposed to sunlight

DSSCs/PSSCs can be correlated with a diode: in the absence of light, the electrons do not get excited and no current is generated, so in the dark the solar cell behaves like a diode and shows diode characteristic. As high-energy light falls on the solar cells, electrons are excited and current generated in the cell (Figure 5.22).

Figure 5.23 shows the current–voltage (I-V) curve obtained by exposing the DSSC to sunlight. Voltage is measured from zero to V_{oc} against load. Using data obtained from this curve, other DSSC/PSSC parameters can be calculated.

5.6.6.1   Short Circuit Current (I_{sc})

When the impedance of DSSC is very low, at zero voltage, the condition is called the short circuit condition. The current at this point is the short circuit current and denoted as I_{sc}.

In a power quadrant, the I_{sc} is the maximum current value. Under ideal conditions the total current produced by a cell is equal to the maximum current value. I_{sc} is generally given per unit area, or short circuit current density (J_{sc}).

5.6.6.2   Open Circuit Voltage (V_{oc})

The voltage at zero current passing to the cell, V_{oc}, is the maximum voltage difference in the power quadrant of the cell when it is connected in forward-bias.

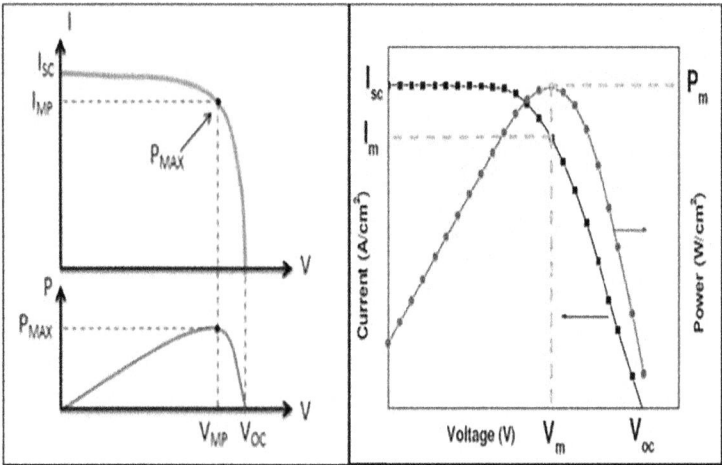

FIGURE 5.24 P_{max} in an I-V curve

5.6.6.3 Maximum Power of Solar Cell (P_{max})

There is a point at which the maximum power in watts can be obtained at a power quadrant during an I-V sweep in a solar cell, as shown in Figure 5.24. As maximum power of this product of current and voltage, it is always zero at I_{sc} as well as V_{oc} points; in between these two there will be maximum power. At maximum power, the corresponding voltage is known as V_{max} and the current I_{max}.

5.6.6.4 Fill Factor (FF)

The quality of a fabricated cell is indicated by its fill factor, which is calculated by comparing the obtained maximum power with the theoretical maximum power. In simple words we can say that fill factor represents the deviation from ideal of solar cell characteristics.

5.6.6.5 Efficiency (η)

Efficiency is the ratio of maximum power obtained from any solar cell to the input power provided to that solar cell. It is calculated using the following formula:

$$\eta = \frac{P_m}{P_{in}} = \frac{V_{oc} \cdot I_{sc} \cdot FF}{P_{in} \cdot A_c}$$

where P_m is the maximum power and P_{in} is the input solar energy.

REFERENCE

1. Yuan, Meng, Xiaoman Zhang, Jun Kong, Wenhui Zhou, Zhengji Zhou, Qingwen Tian, Yuena Meng, Sixin Wu, and Dongxing Kou. "Controlling the Band Gap to Improve Open-Circuit Voltage in Metal Chalcogenide based Perovskite Solar Cells." *Electrochimica Acta* 215 (2016): 374–379.
2. Wang, Jiang, Yu Chen, Fusheng Li, Xueping Zong, Jinlin Guo, Zhe Sun, and Song Xue.

"A new carbazole-based hole-transporting material with low dopant content for perovskite solar cells." *Electrochimica Acta* 210 (2016): 673–680.

3. Perumallapelli, Goutham Raj, Sandhya Rani Vasa, and Jin Jang. "Improved morphology and enhanced stability via solvent engineering for planar heterojunction perovskite solar cells." *Organic Electronics* 31 (2016): 142–148.

4. Tang, Jian-Fu, Zong-Liang Tseng, Lung-Chien Chen, and Sheng-Yuan Chu. "ZnO nanowalls grown at low-temperature for electron collection in high-efficiency perovskite solar cells." *Solar Energy Materials and Solar Cells* 154 (2016): 18–22.

5. Singh, Pramod K., Kang-Wook Kim, Ki-Il Kim, Nam-Gyu Park, and Hee-Woo Rhee. "Nanocrystalline porous TiO_2 electrode with ionic liquid impregnated solid polymer electrolyte for dye sensitized solar cells." *Journal of nanoscience and nanotechnology* 8, no. 10 (2008): 5271–5274.

6. Jung, Young-Sam, AR Sathiya Priya, Min Ki Lim, Seung Yong Lee, and Kang-Jin Kim. "Influence of amylopectin in dimethylsulfoxide on the improved performance of dye-sensitized solar cells." *Journal of Photochemistry and Photobiology A: Chemistry* 209, no. 2 (2010): 174–180.

7. Liang, Kangning, David B. Mitzi, and Michael T. Prikas. "Synthesis and characterization of organic– inorganic perovskite thin films prepared using a versatile two-step dipping technique." *Chemistry of materials* 10, no. 1 (1998): 403–411.

8. Finkenstadt, Victoria, and J. L. Willett. "Preparation and characterization of electroactive biopolymers." In *Macromolecular symposia*, vol. 227, no. 1, pp. 367–372. WILEY-VCH Verlag, 2005.

9. Singh, Pramod K., Kang-Wook Kim, Nam-Gyu Park, and Hee-Woo Rhee. "Mesoporous nanocrystalline TiO_2 electrode with ionic liquid-based solid polymer electrolyte for dye-sensitized solar cell application." *Synthetic Metals* 158, no. 14 (2008): 590–593.

10. French, David N. "X-Ray Stress Analysis of WC-Co Cermets: II, Temperature Stresses." *Journal of the American Ceramic Society* 52, no. 5 (1969): 271–275.

11. Grätzel, Michael. "Dye-sensitized solar cells." *Journal of Photochemistry and Photobiology C: Photochemistry Reviews* 4, no. 2 (2003): 145–153.

12. Jung, Young-Sam, AR Sathiya Priya, Min Ki Lim, Seung Yong Lee, and Kang-Jin Kim. "Influence of amylopectin in dimethyl sulfoxide on the improved performance of dye-sensitized solar cells." *Journal of Photochemistry and Photobiology A: Chemistry* 209, no. 2 (2010): 174–180.

13. Sequeira, César, and Diogo Santos, eds. *Polymer electrolytes: fundamentals and applications.* Elsevier, 2010.

14. Finkenstadt, Victoria, and J. L. Willett. "Preparation and characterization of electroactive biopolymers." In *Macromolecular symposia*, vol. 227, no. 1, pp. 367–372. WILEY-VCH Verlag, 2005.

15. Singh, Pramod K., Kang-Wook Kim, Nam-Gyu Park, and Hee-Woo Rhee. "Mesoporous nanocrystalline TiO 2 electrode with ionic liquid-based solid polymer electrolyte for dye-sensitized solar cell application." *Synthetic Metals* 158, no. 14 (2008): 590–593.

16. Sutter-Fella, Carolin M., Yanbo Li, Matin Amani, Joel W. Ager III, Francesca M. Toma, Eli Yablonovitch, Ian D. Sharp, and Ali Javey. "High photoluminescence quantum yield in band gap tunable bromide containing mixed halide perovskites." *Nano Lett* 16, no. 1 (2016): 800–806.

17. Burschka, Julian, Norman Pellet, Soo-Jin Moon, Robin Humphry-Baker, Peng Gao, Mohammad K. Nazeeruddin, and Michael Grätzel. "Sequential deposition as a route to high-performance perovskite-sensitized solar cells." *Nature* 499, no. 7458 (2013): 316–319.

18. Im, Jeong-Hyeok, In-Hyuk Jang, Norman Pellet, Michael Grätzel, and Nam-Gyu Park. "Growth of $CH_3NH_3PbI_3$ cuboids with controlled size for high-efficiency perovskite solar cells." *Nature nanotechnology* 9, no. 11 (2014): 927–932.

19. Ahn, Namyoung, Seong Min Kang, Jin-Wook Lee, Mansoo Choi, and Nam-Gyu Park. "Thermodynamic regulation of $CH_3NH_3PbI_3$ crystal growth and its effect on photovoltaic performance of perovskite solar cells." *Journal of Materials Chemistry A* 3, no. 39 (2015): 19901–19906.

20. Mastroianni, S., F. D. Heinz, J-H. Im, W. Veurman, M. Padilla, M. C. Schubert, U. Würfel, M. Grätzel, N-G. Park, and A. Hinsch. "Analysing the effect of crystal size and structure in highly efficient $CH_3NH_3PbI_3$ perovskite solar cells by spatially resolved photo-and electroluminescence imaging." *Nanoscale* 7, no. 46 (2015): 19653–19662.

21. Song, Zhaoning, Antonio Abate, Suneth C. Watthage, Geethika K. Liyanage, Adam B. Phillips, Ullrich Steiner, Michael Graetzel, and Michael J. Heben. "Perovskite Solar Cell Stability in Humid Air: Partially Reversible Phase Transitions in the PbI_2-CH_3NH_3I-H_2O System." *Advanced Energy Materials* 6, no. 19 (2016).

22. Ko, Hyun-Seok, Jin-Wook Lee, and Nam-Gyu Park. "15.76% efficiency perovskite solar cells prepared under high relative humidity: importance of PbI 2 morphology in two-step deposition of CH_3 NH_3 PbI_3." *Journal of Materials Chemistry A* 3, no. 16 (2015): 8808–8815.

23. Xu, Yuzhuan, Lifeng Zhu, Jiangjian Shi, Xin Xu, Junyan Xiao, Juan Dong, Huijue Wu, Yanhong Luo, Dongmei Li, and Qingbo Meng. "The Effect of Humidity upon the Crystallization Process of Two-Step Spin-Coated Organic–Inorganic Perovskites." *ChemPhysChem* 17, no. 1 (2016): 112–118.

24. El-Henawey, M. I., Ryan S. Gebhardt, M. M. El-Tonsy, and Sumit Chaudhary. "Organic solvent vapor treatment of lead iodide layers in the two-step sequential deposition of $CH_3NH_3PbI_3$-based perovskite solar cells." *Journal of Materials Chemistry A* 4, no. 5 (2016): 1947–1952.

25. Yang Bin, Ondrej Dyck, Jonathan Poplawsky, Jong Keum, Sanjib Das, Alexander Puretzky, Tolga Aytug et al. "Controllable Growth of Perovskite Films by Room-Temperature Air Exposure for Efficient Planar Heterojunction Photovoltaic Cells." *Angewandte Chemie International Edition* 54, no. 49 (2015): 14862–14865.

26. Huang, Jin-hua, Ke-jian Jiang, Xue-ping Cui, Qian-qian Zhang, Meng Gao, Mei-ju Su, Lian-ming Yang, and Yanlin Song. "Direct conversion of $CH_3NH_3PbI_3$ from electrodeposited PbO for highly efficient planar perovskite solar cells." *Scientific reports* 5 (2015).

27. Yang, Woon Seok, Jun Hong Noh, Nam Joong Jeon, Young Chan Kim, Seungchan Ryu, Jangwon Seo, and Sang Il Seok. "High-performance photovoltaic perovskite layers fabricated through intramolecular exchange." *Science* 348, no. 6240 (2015): 1234–1237.

28. Uzu, Hisashi, Mitsuru Ichikawa, Masashi Hino, Kunihiro Nakano, Tomomi Meguro, José Luis Hernández, Hui-Seon Kim, Nam-Gyu Park, and Kenji Yamamoto. "High efficiency solar cells combining a perovskite and a silicon heterojunction solar cells via an optical splitting system." *Applied Physics Letters* 106, no. 1 (2015): 013506.

29. Jeon, Nam Joong, Jun Hong Noh, Young Chan Kim, Woon Seok Yang, Seungchan Ryu, and Sang Il Seok. "Solvent engineering for high-performance inorganic–organic hybrid perovskite solar cells." *Nature materials* 13, no. 9 (2014): 897–903.

30. Jeon, Nam Joong, Jun Hong Noh, Woon Seok Yang, Young Chan Kim, Seungchan Ryu, Jangwon Seo, and Sang Il Seok. "Compositional engineering of perovskite materials for high-performance solar cells." *Nature* 517, no. 7535 (2015): 476–480.

31. Xiao, Manda, Fuzhi Huang, Wenchao Huang, Yasmina Dkhissi, Ye Zhu, Joanne Etheridge, Angus Gray-Weale, Udo Bach, Yi-Bing Cheng, and Leone Spiccia. "A fast deposition-crystallization procedure for highly efficient lead iodide perovskite thin-film solar cells." *Angewandte Chemie* 126, no. 37 (2014): 10056–10061.

32. Ahn, Namyoung, Dae-Yong Son, In-Hyuk Jang, Seong Min Kang, Mansoo Choi, and Nam-Gyu Park. "Highly reproducible perovskite solar cells with average efficiency of 18.3% and best efficiency of 19.7% fabricated via Lewis base adduct of lead (II) iodide." *Journal of the American Chemical Society* 137, no. 27 (2015): 8696–8699.

33. Chen, Yani, Jiajun Peng, Diqing Su, Xiaoqing Chen, and Ziqi Liang. "Efficient and balanced charge transport revealed in planar perovskite solar cells." *ACS applied materials & interfaces* 7, no. 8 (2015): 4471–4475.

34. Rahul Bhattacharya, B. Pramod K. Singh, Roja Singh, and Zishan H. Khan. "Perovskite sensitized solar cell using solid polymer electrolyte." *International Journal of Hydrogen Energy* 41, no. 4 (2016): 2847–2852. Liu, Mingzhen, Michael B. Johnston, and Henry J. Snaith. "Efficient planar heterojunction perovskite solar cells by vapour deposition." *Nature* 501, no. 7467 (2013): 395–398.

35. Chen, Chang–Wen, Hao-Wei Kang, Sheng-Yi Hsiao, Po-Fan Yang, Kai-Ming Chiang, and Hao-Wu Lin. "Efficient and uniform planar-type perovskite solar cells by simple sequential vacuum deposition." *Advanced Materials* 26, no. 38 (2014): 6647–6652.

36. Park, Nam-Gyu. "Methodologies for high efficiency perovskite solar cells." *Nano convergence* 3, no. 1 (2016): 1–13.

37. Chen, Qi, Huanping Zhou, Ziruo Hong, Song Luo, Hsin-Sheng Duan, Hsin-Hua Wang, Yongsheng Liu, Gang Li, and Yang Yang. "Planar heterojunction perovskite solar cells via vapor-assisted solution process." *Journal of the American Chemical Society* 136, no. 2 (2013): 622–625.

38. De Angelis, Filippo. "Modeling materials and processes in hybrid/organic photovoltaics: from dye-sensitized to perovskite solar cells." *Accounts of chemical research* 47, no. 11 (2014): 3349–3360.

39. O'regan, Brian, and Michael Grätzel. "A low-cost, high-efficiency solar cell based on dye-sensitized colloidal TiO2 films." *Nature* 353, no. 6346 (1991): 737.

40. Rahul Bhattacharya, B., and Zishan H. Khan. "Effect of crystal and powder of CH3NH3I on the CH3NH3PbI3 based Perovskite sensitized solar cell." *Materials Research Bulletin* 89 (2017): 292–296.

41. Kojima, Akihiro, Kenjiro Teshima, Yasuo Shirai, and Tsutomu Miyasaka. "Organometal halide perovskites as visible-light sensitizers for photovoltaic cells." *Journal of the American Chemical Society* 131, no. 17 (2009): 6050–6051.

42. Lee, Hyo Joong, Jun-Ho Yum, Henry C. Leventis, Shaik M. Zakeeruddin, Saif A. Haque, Peter Chen, Sang Il Seok, Michael Grätzel, and Md K. Nazeeruddin. "CdSe quantum dot-sensitized solar cells exceeding efficiency 1% at full-sun intensity." *The Journal of Physical Chemistry C* 112, no. 30 (2008): 11600–11608.

43. Kojima, Akihiro, Kenjiro Teshima, Yasuo Shirai, and Tsutomu Miyasaka. "Organometal halide perovskites as visible-light sensitizers for photovoltaic cells." *Journal of the American Chemical Society* 131, no. 17 (2009): 6050–6051.

44. Jeon, Nam Joong, Jun Hong Noh, Young Chan Kim, Woon Seok Yang, Seungchan Ryu, and Sang Il Seok. "Solvent engineering for high-performance inorganic–organic hybrid perovskite solar cells." *Nature materials* 13, no. 9 (2014): 897.

45. Chiba Yasuo, Ashraful Islam, Yuki Watanabe, Ryoichi Komiya, Naoki Koide, and Liyuan Han. "Dye-sensitized solar cells with conversion efficiency of 11.1%." *Japanese Journal of Applied Physics* 45, no. 7L (2006): L638.

46. Bach, Udo, D. Lupo, P. Comte, J. E. Moser, F. Weissörtel, J. Salbeck, H. Spreitzer, and M. Grätzel. "Solid-state dye-sensitized mesoporous TiO$_2$ solar cells with high photon-to-electron conversion efficiencies." *Nature* 395, no. 6702 (1998): 583–585.

47. Rahul, Singh Pramod K., R. Singh, Vijay Singh, S. K. Tomar, B. Bhattacharya, and Zishan H. Khan. "Effect of crystal and powder of CH$_3$NH$_3$I on the CH$_3$NH$_3$PbI$_3$ based Perovskite sensitized solar cell." *Materials Research Bulletin* 89 (2017): 292–296.

48. Snaith, Henry J., Adam J. Moule, Cédric Klein, Klaus Meerholz, Richard H. Friend, and Michael Grätzel. "Efficiency enhancements in solid-state hybrid solar cells via reduced charge recombination and increased light capture." *Nano Letters* 7, no. 11 (2007): 3372–3376.

6 Electrical Characterization of Electro-Ceramics

Sheela Devi and Shilpi Jindal

CONTENTS

6.1   INTRODUCTION

Materials are milestones for monitoring the progress of technology. Civilizations have been classified by the materials they used: the Bronze Age, the Stone Age, the Steel Age, etc. These periods in history lasted for centuries but in recent times the rates of both material revolutions and their improvements have increased more rapidly. Over the 20th century, virtually every aspect of the conscious world, from clothing to construction, was replaced by novel materials. The century's most improbable accomplishments like spacecraft, microchips, airplanes, magnetic disks, lasers, and fiber-optics were made possible by special materials and behind them lies the ability of scientists and engineers to customize matter for particular applications by controlling its microstructure and composition. Generally, ferroelectric materials are categorized into metals, ceramics, and polymers. Together, various combinations of these materials have the capability to produce a new material, known as a composite, whose properties may not be attainable by conventional means.

Ceramics are considered as prime engineering materials. Ceramics are providing technological advancements in many industries due to their exclusive electrical, magnetic, and optical properties, their hardness, and their heat resistance.

The discovery in the early 1940s of the uncommon dielectric properties of various mixed oxides that crystallize with the perovskite structure led to a new era of ferroelectric ceramics, which are now commonly used in electronic applications [1]. Since then, the electro-ceramics field has continued to develop, promoting the growth of electro-ceramic industries.

6.2   FERROELECTRICITY

Various dielectric and non-conducting materials exhibit spontaneous electric polarization without an external applied field; this is called ferroelectricity. The name ferroelectricity is analogous to its magnetic counterpart, ferromagnetism [2]. Ferroelectricity [3–5] was initially discovered in Rochelle salt (double tartrate of sodium and potassium; sodium potassium tartrate tetrahydrate; $(KNaC_4H_4O_6.4H_2O)$ in 1921 by Joseph Valasek. Rochelle salt was the first crystal to be designated as a ferroelectric material, having two Curie temperatures, 297 K and 255 K. This salt shows a ferroelectric property with monoclinic symmetry at temperatures below 255K and above 297K but orthorhombic symmetry between 255 K and 297 K. According to Czech scientist Roger Valasek, electric displacement in one direction in the ferroelectric material depends on the previous values of the electric field, E, but if the electric field increases steadily up to a maximum value and then starts to decrease to reach the maximum value in the opposite direction, electric displacement traces a hysteresis loop. The ferroelectric material's hysteresis loop traced between applied electric fields resembles the ferromagnetic hysteresis loop traced between an applied magnetic field and magnetic induction, although it has no connection with iron (ferrum). It is this analogy with ferromagnetism that gives ferroelectricity its name. "Siegnett-electricity," an earlier name derived from its first discovery in Rochelle salt (Siegnett salt), and was introduced by Kurchatov (1933).

6.3   CRYSTAL SYMMETRY OF FERROELECTRIC MATERIALS

Ferroelectricity in crystal can be illuminated on the basis of symmetry operations. Crystals are classified into 32 classes (point groups) according to symmetry operations [6]. Of these 32 point groups, 11 are centrosymmetric and do not possess polar properties [7], as shown in Figure 1.1. The remaining 21 groups do not possess a symmetry center and are known as non-centrosymmetric; they can have one or more polar axes. Twenty point groups are piezoelectric and 10 have a unique polar axis (and the ability to develop an electrical charge when mechanical stress is applied). Crystals that belong to these 10 point groups are pyroelectric (able to change polarization with temperature). Ferroelectric crystals are pyroelectric and have the additional property that with the application of a suitable electric field, the spontaneous polarization direction can be reversed. Ferroelectric materials are a highly important research area due to a properties that make them useful in a number of applications. These properties are:

- High electromechanical coupling
- Relatively low dielectric loss
- Low coercive field
- Large remanent polarization
- Low leakage current
- Moisture insensitivity
- High electro-optic coefficient

These properties make ferroelectric materials useful and distinct in electronic devices such as small, high-capacitance capacitors, which have a variety of applications such as ferroelectric memories, piezoelectric/electrostatic transducers, pyroelectric sensors, high positive temperature coefficient thermistors, electro-optic devices, etc.

6.4 PIEZOELECTRICITY

The word piezoelectricity originates from the Greek word 'piezo,' to press. The properties of piezoelectric materials in ceramics are crucial in several applications. In ferroelectric ceramics there are two types of piezoelectric effects:

1. Direct effect of piezoelectricity
2. Converse effect of piezoelectricity.

The direct effect of piezoelectric was discovered in 1880 by Pierre Curie. Materials develop a potential across their boundaries when subjected to mechanical stress (or pressure). This effect is relevant in microphones, sensors, gas lighters, supersonic detectors, etc. [8–10].

The converse effect of piezoelectric is when an electric field is supplied to the material and causes a mechanical deformation. This property can be used to make actuators, crystal oscillators, crystal speakers, etc. Thus, piezoelectric materials can be used as sensors and actuators and hence they are considered to be smart materials

6.4.1 TECHNIQUES OF PIEZOELECTRICITY

In piezoelectric crystal, positive and negative electrical charges are separated and distributed symmetrically, so that the crystal is electrically neutral. When pressure is applied, the symmetry of the crystal is disturbed and charge asymmetry produces a potential difference. Converse piezoelectricity is also revealed in ferroelectrics when an electric field creates mechanical stress in the crystal.

Mathematically, this can be expressed as

$$D = Dt + \varepsilon'_r TE \tag{6.1}$$

$$S = s^E t + dE \tag{6.2}$$

where D stands for dielectric displacement, T for stress, E for the electric field strength, S for strain, d for the piezoelectric coefficient of the material, and ε'_r for the relative permittivity of the dielectric constant. The above equations are in matrix form and they describe piezoelectric materials with unequal orientation.

6.4.2 PIEZOELECTRIC PARAMETERS AND THEIR RELATIONS

a. Piezoelectric Charge Coefficient
   When a piezoelectric material is subjected to stress, the surface is generated due to electric charge. The charge generated per unit force is called the piezoelectric charge coefficient, denoted by d, and measured in pC/N. Piezoelectric charge coefficient has a directional property: subscripts are used to specify the conditions it is determined for (e.g., d_{33} or d_{31}). The first subscript corresponds to the direction of the applied stress and the second to the faces of the materials

on which charge is developed. Higher values are suitable for sonar, ultrasonic cleaner, transducers, etc.

b. Piezoelectric Voltage Coefficient

This coefficient is related to the electric field produced by a stress in piezoelectric materials. It is expressed as meter-volt/Newton. Piezoelectric voltage coefficient (g) and its charge coefficient (d) are interrelated:

$$g = d / (\varepsilon \varepsilon_0) \qquad (6.3)$$

c. Electromechanical Coupled Factor

The electromechanical coupled factor is another crucial parameter; it reflects the efficiency of piezoelectric materials. This factor measures the quantity of electrical energy converted into work, or vice versa, and is determined by the resonance method [19]. Such conversions are never complete so the value of the coupled factor is always less than unity.

$$k^2 = \frac{\text{Work converted into electrical energy}}{\text{Input work}} \qquad (6.4)$$

On the basiys of energy conversion, there are various electromechanical coupled factors; for example, k_p, k_t, k_{33}. Here, k_p is the planer coupling coefficient, which is related to energy conversion along the plane to generate mechanical vibrations when the applied electric field is perpendicular. k_t is the thickness coupling factor, which is related to energy conversion across the thickness of the material when the applied electric field is in the direction of generated mechanical vibrations.

d. Hydrostatic Charge Coefficient (d_h)

This coefficient measures the effect of development of charge when pressure is applied to the material.

It is measured in C/N and related to d_{33} and d_{31} piezoelectric charge coefficients by the relation

$$d_h = d_{33} + 2d_{31} \qquad (6.5)$$

6.5 PYROELECTRICITY

Pyroelectricity was discovered in 314 BC in tourmaline crystal. It depends on the spontaneous polarization of temperature and reduces the charge carrier for reparation of a reduction in dipole moment with the increase in temperature, and its reverse. The crystals exhibit ferroelectric behavior when the direction and magnitude of P_s are reversed The coefficient performance can be measured by a figure of merit and is proportional to temperature as well as displacement current.

$$p = \frac{P_t}{A \left(\dfrac{dT}{dt} \right)} \qquad (6.6)$$

where p is coefficient, P_t is current, a is electrode area, and dT/dt gives the rate of change of temperature.

6.6 EXPERIMENTAL TECHNIQUES FOR CHARACTERIZATION OF MATERIALS

The following section discusses instruments and techniques used for structural and electrical characterization of the samples. A brief theoretical introduction to the characterization parameters is also given.

6.6.1 STRUCTURAL CHARACTERIZATION

6.6.1.1 X-ray Diffraction (XRD)

X-ray crystallography is a technique in which the pattern produced by the diffraction of X-rays through the closely spaced atoms in a crystal is recorded and then analyzed to reveal the nature of that lattice (Figure 6.1). Geometrically, one may imagine that a crystal is made up of lattice planes and that the scattering from a given set of planes will only be strong if the X-ray reflected by each plane arrives at the detector in phase. This leads to a relationship between the order of diffraction pattern n, X-ray wavelength λ, the spacing between lattice planes d, and the angle of incidence θ known as Bragg's law [11, 12]:

$$2d \sin \theta = n \lambda \tag{6.7}$$

X-ray powder diffraction finds frequent use in materials science as it is non-destructive and easy to use.

6.6.1.2 Utility of the XRD Pattern

> ***Phase Identification:*** The most widespread use of this technique is the identification of crystalline solids. Each crystalline solid produces its own diffraction pattern, which can be compared with those in a database

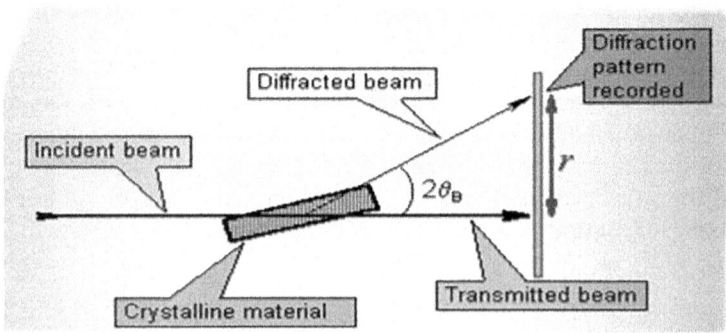

FIGURE 6.1   Working principle of X-ray diffraction

of known materials such as the Joint Committee of Powder Diffraction Standards database in order to identify the sample. The intensity of the lines are characteristic of that particular phase and its pattern thus provides a fingerprint of the material [13].

Crystallinity: In contrast to a crystalline pattern that consists of series of sharp peaks, amorphous materials (liquids, glasses, etc.) produce a broad background signal. XRD can be used to determine the degree of crystallinity by comparing the integrated intensity of the background pattern to that of the sharp peaks [14].

Crystallite Size: From the broadening of the XRD peak, it is possible to determine an average crystallite size by using the Debye–Scherrer formula [15]:

$$P_{hkl} = K \lambda / \beta_{1/2} \cos\theta \tag{6.8}$$

where $K = 0.89$, λ = radiation wavelength, $\beta_{1/2}$ is the half peak width, and θ the position of the peak.

Tetragonal Strain: The strain in a lattice along the c-axis is known as tetragonal strain, principal strain, or tetragonality and is expressed as (c/a) [16]. It is related to the phase transition temperature in ceramics and increase in tetragonal strain has been reported to increase the Curie temperature [17, 18].

XRD Analysis: X-ray diffractograms of ferroelectric materials were recorded for structural analysis and to determine the various phases present in the samples. An X-ray machine with CuKα radiation of wavelength (λ) 1.5405 Å was used. The samples were analyzed in the 2θ range from 10° to 120° at a scanning rate of 1°/min. The diffractometer was operated using software that evaluated the inter-planar spacing (d_{hkl}) values corresponding to all the observed peaks using Bragg's law, as shown in Figure 6.2.

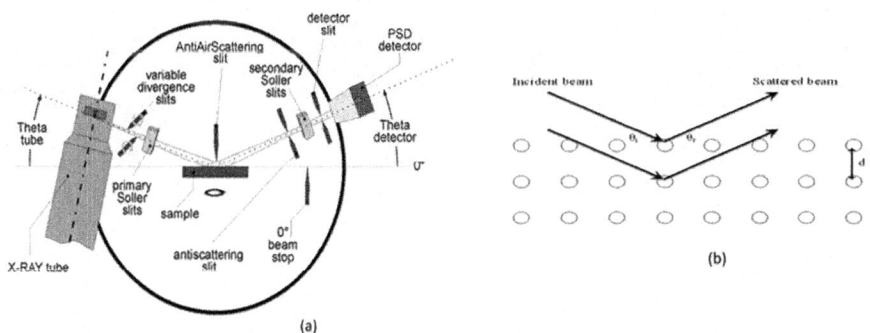

FIGURE 6.2   (a) Components diagram of XRD (b) Bragg's Law

Taking the intensity of the highest peak to be 100%, the relative intensity of other peaks is measured. The lattice parameters a, b, c were calculated from the observed d-values using the formula:

$$d_{hkl} = 1 / \left[\left(h/a \right)^2 + \left(k/b \right)^2 + \left(1/c \right)^2 \right]^{1/2} \tag{6.9}$$

where h, k, l are the Miller indices corresponding to different planes. The peaks were indexed and the lattice parameters deduced from the X-ray diffractogram data were refined using a least-squares refinement method using the computer program Powder X [19]. Structural distortion parameters like tetragonal strain were calculated using the lattice parameters and were investigated to observe the influence of substitution in the structure.

6.6.2 SCANNING ELECTRON MICROSCOPY

6.6.2.1 Electron Microscope

An electron microscope uses electrons rather than light to form an image. The scanning electron microscope (SEM) images the sample surface by scanning it with a high-energy beam of electrons in a raster scan pattern. The electrons interact with the atoms that make up the sample, producing signals that contain information about the sample's surface topography, composition, and other properties such as electrical conductivity. It has a large depth of field, which allows a large amount of the sample to be in focus at one time. The images produced have high resolution, which means that closely spaced features can be examined at a high magnification.

6.6.2.2 Working Principle of SEM

In a typical system (Figure 6.3), an electron beam is thermionically emitted from a tungsten or lanthanum hexaboride cathode (electron gun) or via field emission gun and accelerated toward an anode. The electron beam, which typically has an energy ranging from 0.5 keV to 40 keV, is focused by one or two condenser lenses to a spot 0.4 to 5 nm in diameter. The beam passes through pairs of scanning coils or deflector plates in the electron column, typically in the final lens, which deflect the beam in the *x* and *y* axes so that it scans in a raster fashion over a rectangular area of the sample surface.

When the primary electron beam interacts with the sample, the electrons lose energy by repeated random scattering and absorption within a teardrop-shaped volume of the specimen known as the interaction volume, which extends from less than 100 nm to around 5 μm into the surface. The interaction volume depends on the electron's landing energy, the atomic number of the specimen, and the specimen's density. Energy exchange between the electron beam and the sample results in the reflection of high-energy electrons by elastic scattering, emission of secondary electrons by inelastic scattering, and the emission of electromagnetic radiation, each of which can be detected by specialized detectors. The beam current absorbed by the specimen can also be detected and used to create images of the distribution of the specimen current. Electronic amplifiers of various types are used to amplify the

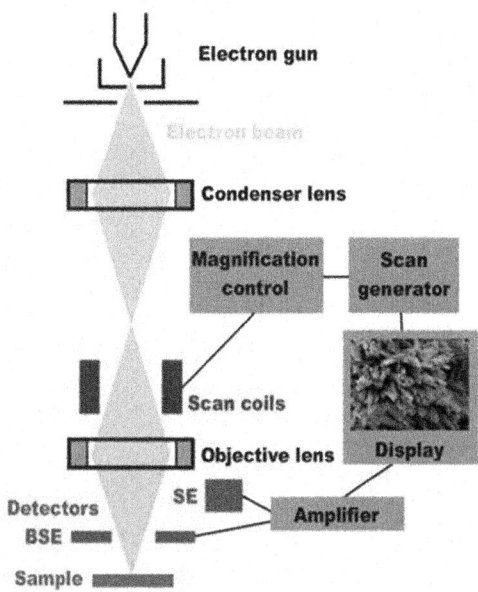

FIGURE 6.3   Working principle of SEM

signals, which are displayed as variations in brightness on a cathode ray tube. The raster scanning of the display is synchronized with that of the beam on the specimen, and the resulting images therefore make up a distribution map of the intensity of the signal being emitted from the scanned area of the specimen. The image may be captured by photography from a high-resolution cathode ray tube, but in modern machines is digitally captured and displayed on a computer monitor.

SEM is used to determine the average grain size and surface morphology of a sample. An accelerating voltage of 20 kV and a probe current of around 0.4 nA are used. Average grain size is calculated using the linear intercept method.

6.6.3   TRANSMISSION ELECTRON MICROSCOPY

The transmission electron microscope (TEM; Figure 6.4) is another type of electron microscope whereby a beam of electrons is transmitted through an ultra-thin specimen. An image is formed from the interaction of the electrons with the specimen; it is magnified and focused onto an imaging device such as a fluorescent screen, onto a layer of photographic film, or to a sensor such as a CCD camera. It is capable of imaging at a significantly higher resolution than SEM, enabling the instrument's user to examine fine details, even as small as a single column of atoms.

6.6.4   DENSITY MEASUREMENT

Density measurements are performed employing Archimedes' principle. The method is based on hydrostatic weighing and it is the best-known density measurement

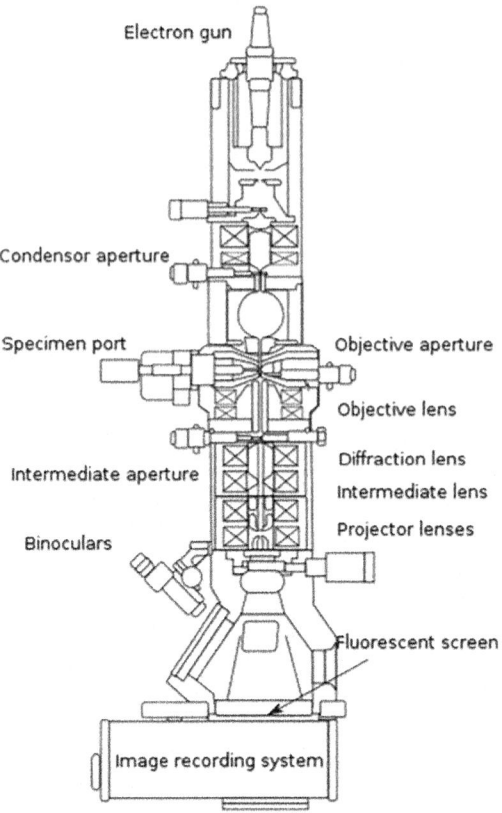

FIGURE 6.4 Working principle of TEM

technique [20]. The weight of a specimen is measured in two different media, e.g., air and liquid. To calculate the density of the specimen ρ_{exp}, the following equation is used:

$$\rho_{exp} = \frac{W_{air}\ \rho_{liq}}{W_{air} - W_{liq}} \tag{6.10}$$

where ρ_{liq} is the density of the weighing liquid. Weighing in air was done using a precision digital balance (Precisa, Switzerland) and the weight in liquid was determined by suspending the sample in ethylene glycol, which has a density of 1.112 g/cm^3.

6.7 ELECTRICAL CHARACTERIZATION

6.7.1 DIELECTRIC STUDIES

In these classes of materials, the dielectric constant and dielectric loss are two important parameters that provide a great deal of information for understanding the behavior and the mechanism of electric polarization [21–24]. Dielectric measurements are most commonly used for the identification of phase transitions. As the ability of these materials to polarize quickly is important for their use in memory applications, it is important to have information on the frequency dependence of the dielectric behavior [25, 26].

Capacitance (C), which is the ability of a body to hold electrical charge, is defined, in a parallel-plate capacitor where two parallel metal electrodes, each of area A square meters, separated by a distance d meters, filled with a dielectric material of permittivity ε, as

$$C = \varepsilon A / d \tag{6.11}$$

The permittivity ε, which is an intrinsic property of a dielectric material, is related to the permittivity of free space ε_0 as

$$\varepsilon = \varepsilon_0 \varepsilon_r \tag{6.12}$$

where ε_r is the relativity permittivity, which is simply a number.

6.7.2 COMPLEX PERMITTIVITY

The response of normal dielectrics to external fields generally depends on the frequency of the field. This frequency dependence reflects the fact that a material's polarization does not respond instantaneously to an applied field. The response is always causal (arising after the applied field), which can be represented by a phase difference. For this reason permittivity is often treated as a complex function of the angular frequency (ω) of the applied field and is represented as

$$\varepsilon_r^*(\omega) = \varepsilon_r'(\omega) - j\varepsilon_r''(\omega); \quad j = \sqrt{-1} \tag{6.13}$$

where $\varepsilon_r'(\omega)$ is the real part of the permittivity representing the lossless permittivity, which is related to the electrical energy stored within the medium and is known as dielectric constant, while $\varepsilon_r''(\omega)$ is the imaginary part of the permittivity, which is related to the loss of energy within the medium.

6.7.2.1 Phasor diagram

For sinusoidal applied voltage (V) , charge stored in the dielectric can be expressed as

$$Q = CV_0 e^{j\omega t} \tag{6.14}$$

and current flowing through the capacitor is given by

$$I = j\omega \varepsilon_r C_0 V \tag{6.15}$$

If the power were not dissipated at all in the dielectric of the capacitor ("ideal dielectric"), the phasor of current I through the capacitor would be ahead of the phasor of

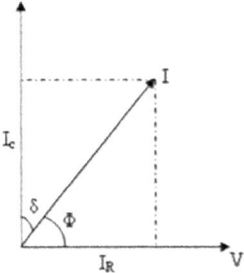

FIGURE 6.5   Phasor diagram of I and V in a dielectric

voltage V by precisely 90°. In real dielectrics, however, the phase angle φ is slightly less than 90°. The total current I through the capacitor can be resolved into two components as shown in Figure 6.5.

The current I_c is proportional to the charge stored in the capacitor and leads the voltage by 90°. The current I_R is an AC conduction current in phase with voltage V, which represents the energy loss or power dissipation in the dielectric.

Because the phase angle is very close to 90° in a capacitor with a high-quality dielectric, the angle δ is a more descriptive parameter that, when added to the angle φ, becomes equal to 90° [27].

$$\delta = 90° - \varphi \tag{6.16}$$

where the angle δ is called the dielectric loss angle, which is an important parameter for the material of a dielectric. All other conditions being equal, dielectric losses grow with this angle [27]. This parameter is usually described by the loss tangent or dissipation factor, tan δ.

On choosing the complex form of dielectric constant ε (Eq. 6.17) and using it in Eq. 6.19 we get

$$I = j\omega\varepsilon_r'\varepsilon_0 C_0 V + \omega\varepsilon_r'' C_0 V \tag{6.17}$$

$$I = I_C + I_R \tag{6.18}$$

From the magnitude of these currents, therefore, the dissipation factor tan δ can be defined as:

$$D\,(\text{dissipation factor})\tan\delta = \left|\frac{I_R}{I_c}\right| = \frac{\varepsilon_r''}{\varepsilon_r'} \tag{6.19}$$

The quality factor Q is defined as the ratio of the energy stored in a component to the energy dissipated by the component per cycle [27]:

$$Q = \frac{\text{Energy Stored}}{\text{Energy Dissipated}} = \frac{1}{\tan\delta} = \tan\varphi \tag{6.20}$$

The value of tan δ for a given material is not strictly constant and depends on various external factors.

6.7.2.2 Frequency Dependence of Permittivity

The dependence of ε on frequency is known as dielectric dispersion and is shown in Figure 6.6. The total polarization P, the total polarizability α, and the relative permittivity ε_r of a dielectric in an alternating field all depend on the ease with which the dipoles can reverse alignment with each reversal of the field [28, 29]. There are four frequency ranges in which different polarization mechanisms dominate. As the relaxation frequencies of all the four polarization processes differ, it is possible to separate contributions experimentally.

Each dielectric mechanism effect has a characteristic relaxation frequency. As the frequency becomes larger, the slower mechanisms drop off. This leaves only the faster mechanism to contribute to the dielectric constant, $\varepsilon_r'(\omega)$. The dielectric loss factor, $\varepsilon_r''(\omega)$ will correspondingly peak at each critical frequency.

6.7.2.3 Temperature Dependence of Permittivity

The effect of temperature on the dielectric constants of ionic and electronic materials is, in general, small at low temperatures but increases with increasing temperature. At elevated temperatures, ion mobility is appreciable [28].

The total permittivity of normal dielectrics increases with increasing temperature. However, in ferroelectric materials, the permittivity increases with increasing temperature, peaks at the Curie temperature (T_c), and then reduces as temperature is increased further. Above the Curie temperature, the dielectric constant follows the Curie–Weiss law [30].

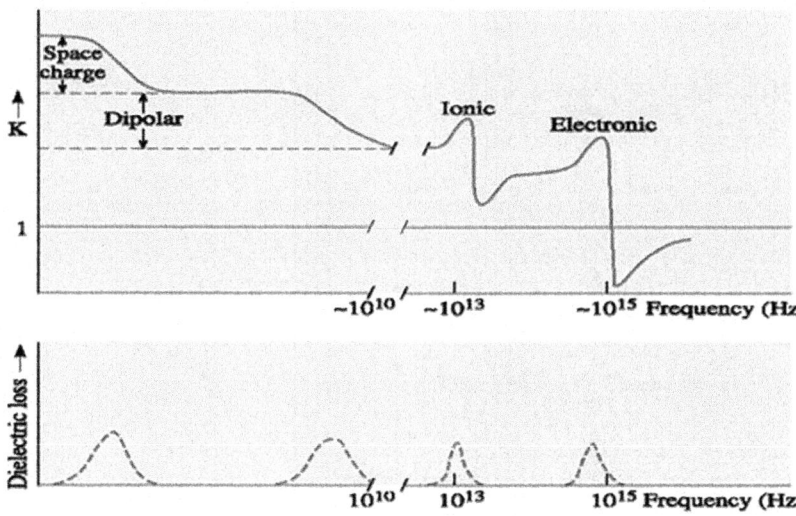

FIGURE 6.6 The frequency dependence of the real and imaginary parts of dielectric constant

Generally, dielectric loss appreciably increases when temperature rises. In polar dielectrics, apart from dipolar losses, there is also loss due to electrical conduction, which grows with an increase in temperature [31].

6.7.2.4   Measurement of Dielectric Parameters

A precision LCR (inductance [L], capacitance [C], and resistance [R]) meter was used for measuring the dielectric constant and dielectric loss in the present work. The frequency dependence of dielectric permittivity and loss was studied in the frequency range 20 Hz–1 MHz. The temperature dependence of dielectric properties was measured from room temperature to 400°C, except for a few samples for which the temperature was lowered using liquid nitrogen to get the Curie temperature, as a function of signal frequency (1 kHz, 10 kHz, and 100 kHz) at an oscillation amplitude of 1 V with no DC bias. The variation in dielectric constant and loss values with temperature were recorded at constant heating rate of 2°C/minute. For the measurements, the samples were mounted on a laboratory-made two-probe sample holder. A schematic diagram of the sample holder is shown in Figure 6.7.

6.7.3   ELECTRICAL CONDUCTION

Electrical conduction relates the current flow in a material to the electrical field applied across it at steady state [27]. Although dielectrics in principle do not allow current to flow under normal circumstances, stray charges and defects cause a small conduction. In other words, the conductivity of insulators is extremely small but not zero [32]. In 1926, Jaffe [31] put forward the idea that the conduction in ionic

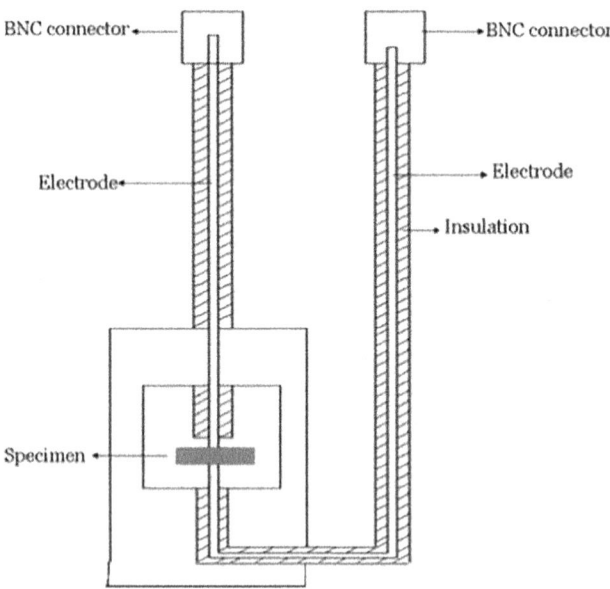

FIGURE 6.7   Schematic diagram of the laboratory made sample holder

crystals observed even at low temperature might be caused by ions tearing them-selves away from the lattice points as a result of thermal oscillations. This point of view was developed and quantitatively substantiated by Frenkel [33].

The current density in a parallel-plate capacitor can be defined in terms of material properties as [26]

$$J = \omega\varepsilon_0\varepsilon_r''E + j\omega\varepsilon_0\varepsilon_r'E \tag{6.21}$$

where E is the electric field strength. Thus we have

$$\frac{J}{E} = \sigma + j\omega\varepsilon_0\varepsilon_r' \tag{6.22}$$

which implies that the real part of the ratio J/E is defined as the electrical conductivity σ in the material, i.e.,

$$\sigma(\omega) = \omega\varepsilon_0\varepsilon_r''(\omega) \tag{6.23}$$

The measured loss of material ε_r'' can be expressed as a function of both dipolar loss (ε_d'') and conductivity (σ) [26],

$$\varepsilon_r'' = \varepsilon_d'' + \frac{\sigma}{\omega\varepsilon_0} \tag{6.24}$$

6.7.3.1 Mechanism of Electrical Conduction in Dielectrics

There are three mechanisms of electrical conduction in dielectrics depending on the type of carriers present in the material [32]:

i.   *Ionic conduction* occurs either through the migration of positive and negative ions in an external field or through the motion of ions in vacancies, which reflects the migration of vacancies.
ii. *Molionic conduction* occurs through the motion of molecular ions (molions) or groups of ionized molecules. This conduction is encountered, especially, in amorphous or liquid dielectrics.
iii. *Electronic conduction* is due to the motion of free charges (electrons and/or holes) in the solid.

Thus, electrical conductivity in ceramics can involve both electronic and ionic charge carriers. In a defect-free insulator, the electrons are either tightly bound or located in a filled electron band. Similarly, in a perfect crystal, where there is no mechanism for ions to diffuse, no current can result from ionic motion.

6.7.4 Ionic Conductivity

The migration of ionic defects (anion or cation vacancies) in the lattice gives rise to ionic conductivity, because the vacancies move through the lattice under the influence of an applied electric field.

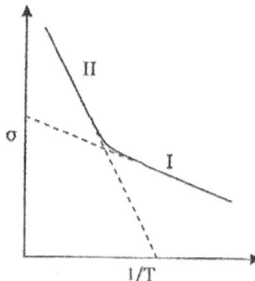

FIGURE 6.8 Ionic conductivity σ as a function of 1/T

It may be shown with the help of thermodynamics that the number of ionic defects increases rapidly at elevated temperatures. If conductivity is plotted against 1/T, one can usually distinguish between a low-temperature region (I) and a high-temperature region (II), as shown in Figure 6.8.

Low-temperature conduction results from the presence of impurity ions in the lattice and is described as the extrinsic region of conductivity [34]. The high-temperature region involves a considerably larger activation energy. It is called the intrinsic region of ionic conductivity and is caused by the movement of component ions in the lattice and/or thermal generation of new defects. This intrinsic conductivity is dominant at high temperatures.

The conductivity can be derived from the Arrhenius equation [32]

$$\sigma = \sigma_0 \exp\left(\frac{-E}{k_B T}\right) \tag{6.25}$$

$$\text{and } E = \frac{w}{2} + \mu \tag{6.26}$$

where w and µ are the activation energies for defect generation and migration, respectively. For extrinsic conduction w = 0 and E = µ; i.e., ionic mobility becomes the controlling factor in the conduction process of dielectrics.

6.7.5 ELECTRICAL CONDUCTIVITY

Electronic conduction is also known to occur in dielectric materials due to the ability of an ion to exist in more than one oxidation state, which leads to the generation of electrons. This process is a function of temperature and/or compositional fluctuation [32]. If present, then the electronic conduction is expected to dominate at low temperatures, as low thermal energies are sufficient to energize low-mass electrons.

6.7.5.1 Conductivity Measurement

Silver electrode pellets were used for the conductivity measurements using a two-probe method. DC conductivity was measured using a Keithley's 6517A electrometer. The measurement was performed at 1 V in the temperature range room temperature to 250°C; below room temperature, liquid nitrogen was used as coolant. Conductivity was calculated using the relation

$$\sigma = \frac{d}{AR} \tag{2.27}$$

where A is the cross-sectional area, d the thickness of the pellet, and R the measured resistance.

6.7.6 IMPEDANCE STUDIES

Impedance (Z) describes the measure of opposition a component offers to the flow of an alternating current at a given frequency. It is represented as a complex vector quantity (Figure 6.9). The impedance vector consists of a real part, resistance (R), and an imaginary part, reactance (X) [35]. The electrical properties of a material are often represented in terms of some complex parameters like impedance (Z*), complex admittance (Y*), complex permittivity (ε*), and dielectric loss (tan δ). These frequency-dependent parameters are related to each other by the following relations:

$$Z^*(\omega) = Z' - jZ'' = R_s - \frac{j}{\omega C_s} = \frac{1}{j\omega C_0 \varepsilon^*} \tag{6.28}$$

$$Y^*(\omega) = Y' + jY'' = \frac{1}{R_p} + j\omega C_p = j\omega C_0 \varepsilon^* = \frac{1}{Z^*} \tag{6.29}$$

$$\varepsilon^*(\omega) = \varepsilon' - j\varepsilon'' \tag{6.30}$$

$$\tan \delta = \frac{Z'}{Z''} = \frac{Y'}{Y''} = \frac{\varepsilon''}{\varepsilon'} \tag{6.31}$$

where R_s, C_s are the series resistance and capacitance; R_p, C_p are the parallel resistance and capacitance; C_0 is the geometrical capacitance; and Z', Y', ε' and Z'', Y'', ε'' are the real and imaginary components of impedance, admittance, and permittivity, respectively.

Complex impedance spectroscopy (CIS) is a powerful non-destructive technique for the characterization of electrical behavior of electro-ceramic materials. The

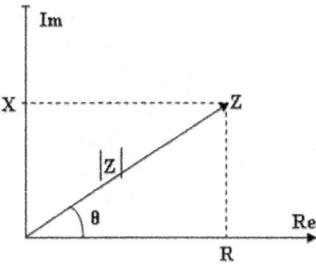

FIGURE 6.9 Graphical representation of the complex impedance plane

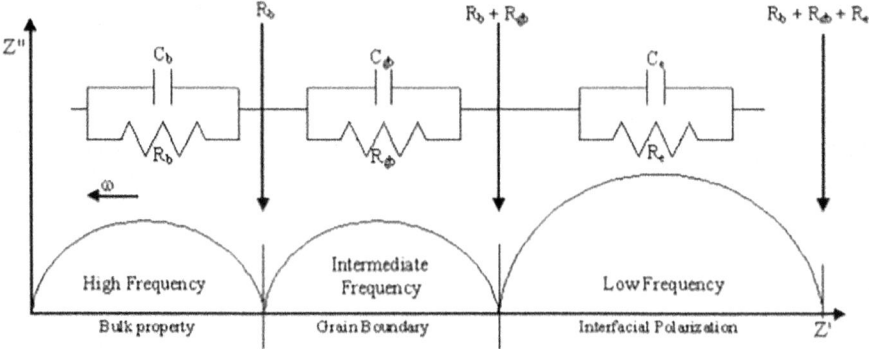

FIGURE 6.10   Various components in a complex impedance plane

technique is based on the principle of analyzing the response of a specimen to a sinu-soidal electrical signal and subsequent calculation of the resulting transfer function (impedance) with respect to the frequency of the applied signal. The output response, when plotted in a complex plane, appears in the form of a succession of semicircles (Nyquist or Cole–Cole plots) in the frequency domain. These arise as a result of the contribution of various components such as the bulk material, grain boundary effect, and interfacial polarization phenomenon (at the material electrode interface) to the electrical properties, as can be seen in Figure 6.10.

In view of this, the CIS technique enables us to separate the effects due to each component (bulk, grain boundary, and electrode interface) in a polycrystalline sample because of the different time constants of these phenomena. The first semi-circle in the high-frequency region manifests the effect of grain interior; the second semicircle represents the grain boundary effects. In some cases, a third semicircle that appears at very low frequencies can be represented by an equivalent circuit consisting of parallel resistance and capacitance (RC circuit) connected in series, each circuit being responsible for a semicircle in the experimental electrical response.

6.7.6.1   Impedance Measurement

Impedance analysis was performed on the samples, using an LCR meter in the fre-quency range 20 Hz to 1 MHz at an oscillation amplitude of 1 V. The studies were performed at temperatures much higher than the Curie temperature of the samples, as the semicircles starts manifesting only at these temperatures. The various parameters computed from the impedance plots are grain resistance (R_b), grain boundary resis-tance (R_{gb}), grain capacitance (C_b), grain boundary capacitance (C_{gb}), and relaxation time. The second intercept of each semicircle on the real (Z') axis gives the value of resistance (R_b and R_{gb}). The semicircles in the impedance spectrum have a character-istic peak occurring at a unique relaxation frequency ($\omega_r = 1/2\pi f_r$). It can be expressed as $\omega_r RC = \omega_r \tau = 1$. Therefore,

$$f_r = \frac{1}{2\pi\tau} = \frac{1}{2\pi RC} \tag{6.32}$$

where τ is relaxation time. The respective capacitances (C_b and C_{gb}) due to the grain and grain boundary effect can be calculated using this relation.

Bulk conductivity (σ_{bulk}) was also calculated from CIS data using the following equation [34]:

$$\sigma_{bulk} = \frac{d}{R_b A} \tag{6.33}$$

where d is the thickness of the pellet and A is the area of the conducting surface of the pellet.

6.8   FERROELECTRIC STUDIES

The hysteresis loop, commonly known as the polarization-electric (P-E) loop, analyzed using the Sawyer–Tower circuit (Figure 6.11), is widely used for the ferroelectric characterization of materials.

6.8.1   SAWYER–TOWER CIRCUIT

The ferroelectric device under test (DUT) is placed in series with a sense capacitance C_s and a resistor R_s. A computer-controlled waveform generator produces an AC signal, which is measured.

The capacitance of the sensing capacitor must be significantly larger than the capacitance of the DUT to minimize its effect on the measurement. The external charge on the sense capacitor, which is proportional to the external charge on the DUT, is calculated from the value of C_S and the voltage measured across C_S:

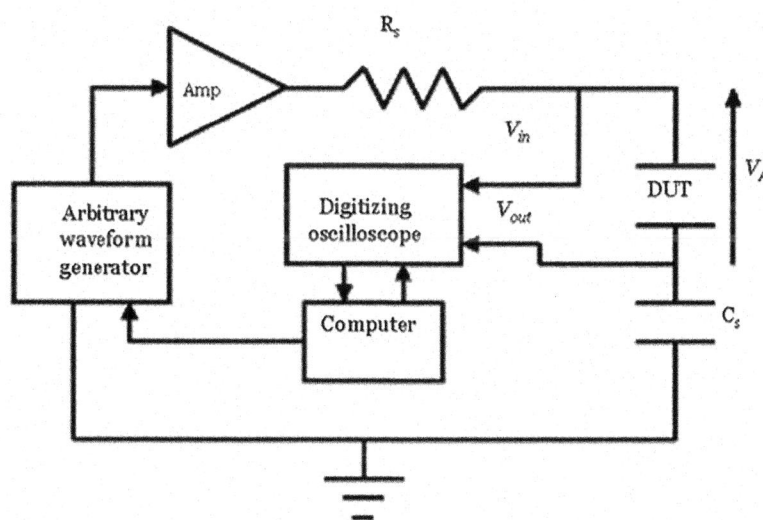

FIGURE 6.11   Block diagram of Sawyer –Tower circuit

$$q = CsV_{out} \qquad (6.34)$$

The voltage applied to the DUT is the difference between the voltage measured at the input and that at the output of the DUT,

$$V_A = V_{in} - V_{out} \qquad (6.35)$$

The polarization (or charge per unit area) is calculated by dividing the total charge by the area of the pellet. The electric field is calculated by dividing the applied voltage by the thickness of the pellet.

6.8.1.1 P-E Hysteresis Measurement

Hysteresis measurements were done at room temperature (expect for a few samples for which the temperature was lowered using liquid nitrogen as their Curie temperature is below room temperature) using an automatic P-E-loop tracer based on the Sawyer–Tower circuit at a switching frequency of 100 Hz. The pellet was placed in a silicon oil bath during the measurement. The measurement gives P-E hysteresis loops from which remanent polarization $(2P_r)$ and coercive field $(2E_c)$ were determined.

6.8.1.2 Poling

The piezoelectric properties manifest in a ferroelectric ceramic material after "poling". This means the domains are reoriented and aligned by application of a strong DC electric field, usually at an elevated temperature just below the Curie temperature. This is illustrated in Figure 6.12.

Through this polarizing (poling) treatment, domains that are almost aligned in the direction of the electric field expand at the expanse of domains that are not aligned with the field. When the electric field is removed, most of the dipoles are locked into a configuration of "near alignment" (Figure 6.12). The specimen now has a permanent polarization, called the remanent polarization.

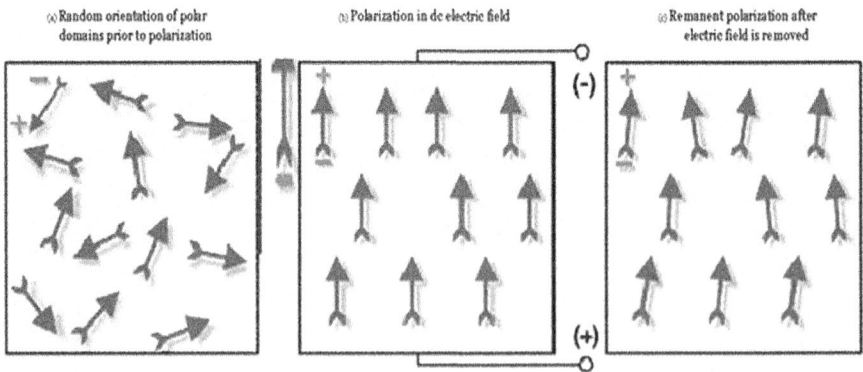

FIGURE 6.12   Poling of a ferroelectric ceramic

6.8.1.3   d$_{33}$ Measurement

The pellets were electrically poled by applying a high DC electric field (10–20 kV/cm) in a silicon oil bath at around 75–85°C for 6 h. The d$_{33}$ measurements of the samples used a piezometer system (PiezoTest PM300, UK) at room temperature.

6.8.1.4   Types of Ferroelectric

1. Normal ferroelectric
2. Relaxor ferroelectric

Diffuse Phase Transition

Various transition phases that are not clear are characterized by transition temperature in dielectric materials. In these ferroelectric ceramics, at $T_c = T_0$ the phase transition occurs and the polarization reduces to zero: the transition is tarnished by a firm temperature period. In this temperature region, physical properties transform steadily. The same kind of phenomenon is seen in different materials, but significant examples of diffuse phase transition (DPT) are established in ferroelectric materials. Important characteristics of DPT are:

a. Transition temperatures obtained by different techniques do not coincide
b. A gradual decrease of spontaneous and remanent polarization with rising temperature
c. Relaxation nature of dielectric properties in the transition region
d. Broadened maxima in the permittivity–temperature curve
e. Above the transition temperature there is no Curie–Weiss performance at firm temperature intervals.

Generally, DPT in ferroelectric materials is identified due to the occurrence of two types of fluctuations at large temperature intervals above the transition temperature:

a. Polarization (structural) fluctuations
b. Compositional fluctuations.

According to the thermodynamic point of view, polarization (structural) fluctuations are a result of little energy variation between low and high phases of temperature in the region of transition, and compositional fluctuations are present in ferroelectric solid solution. According to Xu [33], a substance of low stability is assumed to have the most-diffuse transition. For relaxor and other ferroelectric diffuse transition phase, the size of the transition area is significant for realistic applications. Xiang Ming Chen reported that in the contracted temperature region, in the region of the transition, materials with a perovskite-type structure split into paraelectric and ferroelectric micro-regions. Also, in the complex type of perovskite, ferroelectricity with an indistinct cation arrangement illustrates a DPT that is characterized by wide-ranging dielectric constant (ε') and dielectric diffusion temperature dependence in the transition region. For DPT, modified temperature dependence is followed by ε',

$$\frac{1}{\varepsilon'} - \frac{1}{\varepsilon m} = \left(T - T_m\right)^\gamma / C'$$ (6.33)

where T_m stands for the temperature at which ε' is highest, ε_m is ε' at T_m, C' stands for the modified Curie–Weiss constant, and γ is the exponential factor, which defines the diffusiveness of ceramics; $1 < \gamma < 2$ for the diffuse-phase transition. For normal ferroelectrics $\gamma = 1$ and for relaxor $\gamma = 2$. The dielectric constant versus temperature curve has been characterized in the micro-region with local compositions varying from 100 to 1000 Å over the average composition length scale. Ferroelectric systems with mechanical and dielectric properties below their T_c is the outcome of the state of strain and polarization. These DPT ferroelectric materials have key applications due to their electro-optic, hysteresis, and dielectric properties.

6.8.1.5 Relaxor Ferroelectrics

Relaxor ferroelectrics have stimulating and unusual dielectric properties. For instance:

a. The dependence of temperature on the real part of ε' is wide-ranging
b. A strong dispersion of relaxation of permittivity provides local polarization at temperatures above T_m when temperature is around T_m
c. The well-known Curie–Weiss law does not hold for temperatures around T_m

Relaxor ferroelectrics contain several properties related to spin and dipolar glass. Generally ferroelectric materials with relaxor behavior have compositionally induced disorder [34]. This type of behavior is observed in disordered perovskite-type (ABO_3) ferroelectrics and in various crystals with hydrogen-bonded ferroelectrics and antiferroelectrics, known as protonic glasses.

To examine relaxor ferroelectrics, it is essential to distinguish a few properties from those of normal ferroelectric materials.

1. The dielectric constant of normal ferroelectric materials has a sharp peak at the transition temperature. The peak in a single crystal is pointed and full width at half maxima (FWHM) is 10–20 K but in mixed oxide ferroelectrics, the peak shape varies and FWHM is 20–40 K.
2. The frequency response is independent of temperature in the audio range. In comparison, relaxor ferroelectric materials possess a wide-ranging peak of dielectric constant when a strong frequency dispersion of dielectric constant and temperature coexist.
3. Temperature dependence of the dielectric constant in normal ferroelectric materials follows the Curie–Weiss law, $\varepsilon' = \dfrac{\text{curie constant }(C)}{\text{Absolute temperature }(T) - \text{curie temperature }(To)}$ and varies linearly as $\dfrac{1}{\varepsilon'}$ vs T. Relaxor ferroelectric materials do not follow the Curie–Weiss law.
4. The ferroelectric transition shows strong optical anisotropy across the maximum transition temperature and shows a microscopic symmetry change at Curie temperature.

5. The hysteresis loop as mentioned for P-E of ferroelectricity is the mark of the low-temperature phase. Polarization P_r, having a large remanent material, has a ferroelectric nature; on the right-hand side, the relaxor type exhibits slim loops. For an extremely high electric field the nano domains of the relaxor may slope with the field to maintain larger polarization; with the removal of the field, a number of domains are oriented in random directions, resulting in small remanent polarization.

6. At the transition temperature, the saturation and remanent polarization of ferroelectric decreases with increasing temperature. The vanishing of spontaneous polarization at the transition temperature is nonstop for second order transitions and irregular for first order transition.

6.8.1.6 Multiferroic ferroelectric

Multiferroic materials are multifunctional and exhibit more than one ferroic property such as ferromagnetism, ferroelectricity, ferrotoroidicity, and ferroelasticity. Various approaches are considered to syndicate the above ferroic properties in single-phase system [35]. However, to explain the ferroelectricity in multiferroic materials the three basic mechanisms are:

a. It is induced by spin orders (antiferromagnetic in E phase and spiral), which affect the spatial inversion symmetry of the system
b. It originates from the states of charge ordered
c. Systems are ferrotoroidic.

Multiferroic ceramics are a probable explanation for the magnetoelectric effect. The magnetoelectric effect is analogous to the manifestation of dielectric polarization in the presence of a strong magnetic field. This can also be termed a straight magnetoelectric effect and is considered to be a magnetoelectric effect ($P = \alpha H$). The converse magnetoelectric effect can be explained as magnetization upon which an electric field is applied and considered a magnetoelectric effect ($M = \alpha E$). The above converse and direct magnetoelectric effects show an effectual conversion between energies of electric fields and magnetic fields.

6.9 SUMMARY

This chapter aimed to describe the technological importance of different types of ferroelectric ceramics. Structural properties are used to define the phase identification and internal structure of the materials. Dielectric constant and dielectric loss are two important parameters that provide a great deal of information for understanding the behavior and mechanisms of electric polarization. The effects of different metal ion substitutions are fascinating; in this way, dielectric and electrical properties can be enhanced and made suitable for device applications.

REFERENCES

1. Bussmann-Holder, A., H. Bilz, and G. Benedek, *Physical Review B* 39.13(1989) 9214.
2. Landauer, Rolf, *Journal of Applied Physics* 28.2(1957) 227–234.

3. Känzig, Werner, *Solid State Physics* 4 (1957) 1–197.
4. Safari, A., R. K. Panda, and V. F. Janas, *Engineering Materials* 122 (1996). 35–70
5. Känzig, Werner, *Solid State Physics* 4 (1957) 1–197.
6. Kimura, T., et al. *Nature* 426.6962 (2003) 55–58.
7. Safari, A., M. Allahverdi, and E. K. Androgen, *Frontiers of Ferro electricity*. Springer US (2006)177–198.
8. Jindal, S., Vasishth, A., Devi, S. and Singh, B., *Ferroelectrics*, *519*(1),2017 9–14.
9. Pratt, I. H., and S. Firestone 8.1(1971) 256–260.
10. S. O. Kasap, *Principles of Electronic materials and Devices*, 2nd Edition (Tata McGraw-Hill Inc., New York, 2002).
11. R. K. Pure and V. K. Babbar, *Solid State Physics* (S. Chand & CO. Ltd., New Delhi, 2001).
12. H. Nagata, N. Chikushi and T. Takenaka, *Jpn. J. Appl. Phy.* 38, 5498 (1999).
13. S. Luo, M. Miyayama and T. Kudo, *Mater. Res. Bull.* 36, 531 (2001).
14. V. Shrivastava, A. K. Jha and R. G. Mendiratta, *Solid State Comm.* 133, 125, (2005).
15. B. Su and W. Button, *J. Appl. Phys.* 95, 1382–(2004).
16. K. Uchino, *Ferroelectric Devices* (Marcel Dekker Inc., New York, 2000).
17. C. Dong, *J. Appl. Cryst.* 32, 838, (1999).
18. H. A. Bowman and R. M. Schoonover, *J. Res. Nat. Bur. Stand.*, 69 C, 217, (1965).
19. H. D. Megaw, *Ferroelectricity in Crystals* (Methen & Co. LTD., London, 1957).
20. W. Kanzig, *Ferroelectric and Antifreeoelectrics* (Academic Press, New York, 1957).
21. F. Jona and G. Shirane, *Ferroelectric Crystals* (Pergamon Press, Oxford, London, 1962).
22. J. C. Burfoot, *Ferroelectrics* (Van Nostrand, New York, 1967).
23. J. B. Grindlay, *An Introduction to Phenomenological Theory of Ferroelectricity* (Pergamon Press, Oxford, London, 1970).
24. J. C. Anderson and K. D. Leaver, *Materials Science* (Thomas nelson & Sons Ltd., London, 1969).
25. Y. H. Xu, *Ferroelectricity Materials* (Elsevier Science Publisher, Amsterdam, 1991).
26. R. C. Buchanan, *Ceramic Materials for Electronics: Processing, Properties & Applications* (Marcel Dekker inc., New York, 1991).
27. S. B. Lang, *Source Book of Pyroelectricity* (Gordon & breach Science Publishers, New York, 1965).
28. R. M. Rose, L. A. Shepard and J. Wulff, *Electronic properties* (Wiley eastern Pvt. Ltd., New York, 1965).
29. C. A. Kittle, *Introduction to solid State Physics*, 7th edition (John Wiley & Sons Inc., New Jersey, 1996).
30. B. Tareev, *Physics of Dielectic Materials* (Mir Publishers, Moscow, 1979).
31. I. Bunget and M. Popenscu, *Physics of Solid Dielectrics* (Elsevier, New York, 1984).
32. Y. I. Frenkel, *Elecktrichestro (Physics of Electrets)*, 8 (1974) 5.
33. I. S. Zheludev, *Physics of Crystalline dielectrics: Crystallography and Spontaneous polarization*, Vol. I (Plenum press, New York, 1971).
34. R. N. P. Choudhary and A. K. Thakur, *Complex impedance Analysis: A tool for ferro-electric Materials (Proceedings of NSFD – XIII)* Nov. 23–25, 2004.
35. E. Barsoukov and J. R. Macdonald, *Impedance Spectroscopy: theory, experiment and applications* (John Wiley & Sons Inc., New Jersey, 2005).

7 Thermal Characterization of Composites

Rajashri Priyadarshini

CONTENTS

7.1 WHAT IS THERMAL ANALYSIS AND WHY IS IT ESSENTIAL?

Thermal analysis or thermal characterization refers to the set of multicomponent techniques carried out to check material consistency and properties by measuring the changes in physical and/or reactive phenomena as a function of time and temperature when a sample is exposed to a controlled temperature program. Before using a manufactured composite or material, it is essential to check its durability and adaptability. [Paul Gabbort] Thermal analysis enables the user and manufacturer to look for possible issues, helps understand the material for its further development, guides the engineer in their selection of an optimized process for a tailor-made composite, and also suggests a suitable end use based on performance in the analysis. Thermal characterization is a primary analysis technique for composites that bridges the gap between manufacturing and the end use. [Brown, 1998]

Techniques available for thermal analysis might be dynamic or isothermal:

    Differential scanning calorimetry (DSC)

    Thermogravimetric analysis (TGA)

    Di-electric analysis (DEA)

    Thermo-mechanical analysis (TMA)

    Dynamic mechanical analysis (DMA)

149

7.2 DIFFERENTIAL SCANNING CALORIMETRY

For the direct determination of enthalpy, the only reliable method is calorimetry. DSC measures the properties of a material by analyzing the heat flowing into or out of the sample. Two major types of DSCs exist, both similar in principle: heat flux DSC and power compensation DSC.

7.2.1 HEAT FLUX DSC

This DSC set up consists of two pans, one containing the sample and the other empty, as a reference. The entire apparatus is enclosed in a furnace, which is heated at a linear rate. As the furnace is heated, a thermoelectric disk gets heated up, in turn heating the pans, which are stationed over it. As the pans contain different materials, they have different heat capacities (C_p) so heat up differently. The difference in the temperature between the sample and reference is measured by the area of thermocouples. The heat flow is given by $q = \dfrac{\Delta T}{R}$ where ΔT is the temperature difference and R the resistance of the thermoelectric disk.

7.2.2 POWER COMPENSATION DSC

This type of DSC uses a similar arrangement to the heat flux DSC except that the reference and sample units are heated by separate heaters in separate furnaces. The sample and references units should always be maintained at the same temperature. The difference in thermal power used to maintain the units at a temperature is plotted against temperature for the duration of the experiment. The primary equations in use are as follows:

$$\text{Heat flow} = \frac{q}{t} \tag{7.1}$$

$$\text{Heating rate} = \frac{\Delta T}{t} \tag{7.2}$$

$$Cp = \frac{q}{\Delta T} \tag{7.3}$$

where q is the heat flow in the sample, t is time and ΔT is the temperature difference.

Various data can be inferred from the graphs obtained from the DSC experiments: useful information such as curing rate, percentage crystallinity, melting point, specific heat capacity, and purity of sample.

A constant temperature will result in a uniform constant plot in the heat flow versus temperature plot but as the temperature rises in a polymeric composite, the graph shows a shift in the curve with an increase in heat flow, due to a thermal transition. It is proven that this is due to the glass transition temperature, taken as the temperature at the mid-point of the slope as shown in Figure 7.1 (DSC curve). When the temperature increases further, molecules form a crystalline order. This involves a release of heat (exothermic process), which can be observed in the graph as a dip in heat flow at a certain temperature, known as the crystallization temperature, T_c. The area of the dip is the latent energy of crystallization. The presence of a dip in the graph suggests that the given material can crystallize whereas a complete amorphous substance shows no dip.

Beyond the crystallization temperature occurs another transition called melting. Melting requires a huge amount of heat so this shows as a peak in the graph. The peak is called the melting temperature, T_m, and the area of the peak is the latent heat of melting. [Schave,J.E,K,1995]

7.3   THERMOGRAVIMETRIC ANALYSIS

TGA is another important thermal analysis technique. In this method, the change in the mass of the sample with a given change in temperature is observed. This provides many insights into the physical and chemical properties of the sample. Temperature, time, and mass are the base parameters for measurement. The technique requires a thermogravimetric analyzer or thermobalance, which measures the mass change of the sample with time and sends data for feedback to the programmable temperature control. The program is mostly run either by increasing the temperature or by controlling the temperature for a constant mass loss in a certain thermal reaction. The volatile samples can undergo any of the three types of TGA analysis:

Dynamic TGA: Involves a continuous linear increase in temperature with time.

Isothermal or static TGA: Maintains a constant temperature for a period of time.

Quasistatic TGA: Heats the sample in a series of increasing temperatures, to a constant weight.

A crucible contains the sample of initial known weight, enclosed in a furnace. The furnace is slowly heated. The null balance principle is used: a null sensor is balanced by a restoring force and the change in weight. Unequal light falls on the two photodiodes every time the balance is out of null position. A feedback system applies current to the meter for the reading to return it to its null position. The amount of current supplied is proportional to the loss or gain in weight. A thermogravimetric curve or thermogram of mass or percentage of initial mass versus temperature or time is plotted. An example graph follows that shows different inferences taken from the thermogram of increase in mass versus temperature. Weight losses can occur for various reasons such as decomposition, evaporation, or reduction reactions. Similarly, a gain in weight could be attributed to oxidation reactions in an oxidizing atmosphere, absorption, or adsorption reactions as shown in Figure 7.2 (TGA curve).

There are many factors influencing the thermogram. Considerations such as using a small amount of sample to avoid temperature gradients and having a uniform sample size are crucial in the measurements. Instrumental factors such as the heating rate and atmosphere of the furnace are also important.

The thermogravimetric curve is a gateway to a myriad of information. A material can be characterized by observing its decomposition patterns. If there is no loss in mass, the sample is thermally stable. This helps in identification of materials, quality control, or in understanding decomposition mechanisms. Oxidative stability or corrosion studies are carried out by oxidizing the materials in oxygen or another atmosphere. Reaction mechanisms can be formulated by checking the composition of multicomponent systems. Kinetic parameters for chemical reactions can also be calculated. The first derivative of the TGA curve, called the differential thermogravimetric curve (DTG curve), is plotted for inflection points, for in-depth interpretations and thermal analysis. The TGA curve also gives information about the purity of the sample, sublimation behavior, catalysts, etc. [Vassilikou-Dova,A,Kalogeras,I.M, 2008]

7.4   DI-ELECTRIC ANALYSIS

DEA or di-electric thermal analysis (DETA) is a thermal analysis technique that measures the changes with temperature or frequency in different physical properties in a polar material. When an AC field is applied to the sample, capacitance and conductance are measured with respect to time, temperature, and frequency, controlled by relaxation properties of the material.

Two electrodes acting as di-electric sensors are kept in contact with the sample under consideration. A sinusoidal voltage is applied to regulate excitation of charge carriers in one electrode and the response is checked using other electrode where the sinusoidal signal is attenuated or shifted in phase due to the mobility of ions and dipole alignment. Dipoles in the sample material try to align themselves with respect to the electric field, while ions move toward the electrode of opposite polarity. Di-electric properties such as permittivity or loss factor can be calculated from the amplitude and phase change recorded at the response electrode.

DEA is efficient in monitoring changes in viscosity and cure rate of resins, polymers, and composites via variations in the di-electric properties. Permittivity (ε'), loss factor (ε''), dissipation factor, conductivity, etc. can all be measured by DEA. DETA is helpful during the thermal transitions of polymeric substances for detecting their di-electric constants and polarizability. Gel point, diffusion properties, aging of materials, reactivity, and degree of cure are other properties that are understood and measured by di-electric analysis. [Peter T.Haines]

7.5   THERMO-MECHANICAL ANALYSIS

TMA is a thermal analysis technique that uses the stress applied to a material in a controlled temperature environment to analyze various mechanical properties. In a TMA analysis, the material's response to the changes in temperature, keeping in view the dimensional changes, is examined, rather than the response of the sample. TMA is predominantly known for its precise temperature control with small sizes of sample. The very first measurement of the glass transition temperature by TMA was done on inorganic glass; even today, the sensitivity of this method is higher by orders of magnitude than that of DSC.

Samples for measuring the coefficient of thermal expansion are the most difficult to prepare because they must be rectangular with parallel, flat ends. For anisotropic materials, a cubic shape is mostly used. Glass materials are made into samples 0.050–0.127 meters in length.[Alan T.Riga] The sample is placed in a furnace where it has contact with a probe, which is in turn touched by a linear variable differential transformer (LVDT) for detection of length, and a force generator. A thermocouple is located nearby for temperature measurement. The temperature of the sample is changed with the help of the furnace by applying force on the sample via the probe. Based on the temperature used and the purpose of the experiment, the probe can be made of different materials such as quartz, alumina, or any metal. There are three types of probe for various measurement of properties, as follows:

Expansion or compressive probe: This type of probe is used when measuring the deformation due to thermal expansions or any other transitions due to temperature.

Penetration probe: This probe is used while measuring softening temperature.

Tension probe: The tension probe is used for thermal expansion or shrinkage of fiber or film materials.

TMA calculates thermal expansivity from the data generated for T_g and is widely used for creep analysis, in accurate measurements of the coefficient of thermal analysis, and in monitoring the stress–strain response of a material. It is also used for shrinkage force testing, tensile testing, and stress relaxation analysis. [Simon Gaisford 2016]

7.6 DYNAMIC MECHANICAL ANALYSIS

DMA, also known as dynamic thermo-mechanical analysis or dynamic mechanical spectroscopy, is more sensitive than DSC or differential thermal analysis for measuring dynamic moduli and damping coefficients. Stress or strain is applied to a sample at certain frequencies and the response is analyzed to study bulk properties and thermal transitions. It is considered to be one of the most powerful tools for measuring polymer thermal transitions. A typical DMA instrument consists of an LVDT that acts as a displacement sensor, a temperature controlled furnace system, a drive motor system, a guidance system, a sample system, and an instrument probe that moves through a magnetic core resulting in a change in voltage. The probe applies a sinusoidal force to the sample to deform it. The deformation is detected to establish a relationship between its extent and the type of deformation that results from the force applied. There are different modes of operation based on the shape of the sample, its modulus, and the purpose of the experiment: tension, compression, dual cantilever bending, three-point bending, and shear modes. The stress or strain applied can be via torsional or axial analyzers. Torsional analyzers are mostly used for liquids or melts whereas axial analyzers are used for solids or semi-solids. Axial analyzers function with higher modulus materials than torsional ones. There are two types of analyzer based on resonance:

Free resonance analyzer: This measures the free oscillations of the damping of a suspended and swung sample. Free resonance analyzers are limited to rod-shaped or rectangular samples.

Forced resonance analyzer: This forces the sample to oscillate at a certain frequency in a temperature sweep experiment.

Based on stress or strain, two types of analyzer exist:

Strain control: The probe is displaced and the stress is measured by applying a force balance transducer using different shafts. This type has a shorter response time.

Stress control: A set force is applied to the sample and conditions such as temperature or frequency are varied. This has a single shaft, which it is hard to control, but is much less expensive than the strain control analyzer.

The most common testing modes are as follows:

Temperature sweep: The complex modulus is measured at a low constant frequency while the temperature of the sample varies. The most evident result from this is the $\tan(\Delta)$ peak that appears at T_g.

Frequency sweep: The temperature of the sample is fixed and the frequency is varied. Peaks in $\tan(\Delta)$ and E'' w.r.t frequency of the corresponding T_g are found.

Dynamic stress–strain: The amplitude of oscillations, storage, and loss moduli are gradually increased with stress. This characterizes the material and helps in finding the upper bound of a linear stress–strain regime.

The data obtained from a DMA help in calculating the damping, complex modulus, viscosity, and elasticity properties of viscoelastic materials. [Menard 2006]

REFERENCES

Brown, M.E., *Introduction to Thermal Analysis: Techniques and Applications*, Chapman and Hall, New York (1988)

Gabbott, P. (2008), *Principles and Applications of Thermal Analysis*, Wiley-Blackwell.

Simon Gaisford, V. K. (2016), *Principles of Thermal Analysis and Calorimetry*, The Royal Society of Chemistry, London.

Haines, P. J. (2012), *Thermal Methods of Analysis: Principles, Applications and Problems*, Springer Science & Business Media, Berlin.

Menard, K. P. (2006), '*Dynamic mechanical analysis, Encyclopedia of Analytical Chemistry: Applications*', *Theory and Instrumentation*, 1–25, Wiley Online Library.

Riga, A. T. (1991), *Materials Characterization by Thermomechanical Analysis*, ASTM Philadelphia.

Schawe, J. E. K. (1995) 'Principles for the interpretation of modulated temperature DSC measurements. Part 1. Glass transition', *Thermochimica Acta*, 261(C), pp. 183–194. doi: 10.1016/0040-6031(95)02315-S.

Vassilikou-Dova, A. and Kalogeras, I. M. (2008) *Dielectric Analysis (DEA), Thermal Analysis of Polymers: Fundamentals and Applications*. doi: 10.1002/9780470423837.ch6.

8 Mechanical Characterization Techniques for Composite Materials

Partha Pratim Das and Vijay Chaudhary

CONTENTS

8.1 INTRODUCTION

Some of our new technologies need materials with specific combinations of properties that cannot be matched by traditional metal alloys, ceramics, and polymeric materials. This is especially true for applications in aerospace, under water, and in transportation. For example, aircraft engineers are increasingly looking for low-density, solid, robust, etc. materials. Combinations of material properties have been and still are being expanded through the production of composite materials. In general, any multiphase material that displays a large proportion of the properties of both constituent phases is considered to be a composite [1]. The judicious combination of two or more distinct materials fashions properties according to this theory of joint action. The design of most composites requires trade-offs on properties.

Types of composite, including metal alloys, ceramics, and multiphase polymers, have already been discussed. For example, pearlitic steels have a microstructure composed of alternating layers of ferrite and cement. The ferrite phase is soft and ductile whereas the cement is hard and very fragile [2]. The combined mechanical properties of the pearlite—reasonably high ductility and resistance—are superior to those of each of the constituent phases alone.

There are also composites that arise naturally. For example, wood is made up of strong and flexible cellulose fibers that are surrounded and kept together by a more rigid material called lignin. Similarly, bone is a mixture of strong but soft protein collagen and thin, hard mineral apatite. A composite is a multi-phase material that is artificially produced, as opposed to one that exists or is shaped naturally. In addition, the constituent phases must be chemically dissimilar and separated by another material [3].

In the manufacture of composite materials, scientists and engineers have ingeniously combined complex metals, ceramics, and polymers to create a new generation of extraordinary materials. Some composites were engineered to improve combinations of mechanical properties such as rigidity, resilience, and ambient and high temperature resistance [4]. Many composite materials consist of only two phases: one is called the continuous matrix and covers the other, referred to as the scattered or reinforcing layer. The composite properties are a function of the properties of the constituent phases, their relative quantities, and the structure of the dispersed phase.

8.2   MECHANICAL CHARACTERIZATION TECHNIQUES

Characterization of composites using mechanical means is an important step before using the composites for various applications. Numerous experiments are conducted on a composite material to clarify its basic features (Figure 8.1) [5]. The standard tests to study the mechanical behavior of the composites are friction tests, flexural tests, impact tests, and compression tests under different loading conditions.

8.2.1   TENSILE TESTING

The mechanical properties of materials are calculated by carefully planned laboratory experiments that reproduce the conditions of operation as near as possible. In real life, there are many factors affecting the way loads are applied to a material [7]. Several specific examples of how to apply these loads are tensile, compressive, and shear, to name only a few. Those properties are essential to mechanical design when choosing materials. The test is a destructive method using a specimen of traditional shape and dimensions (prepared in compliance with D638/D3039 ASTM standards) [8].

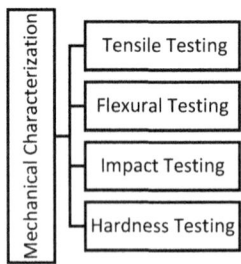

FIGURE 8.1   Various mechanical characterization techniques [6]

ASTM D638 is usually used during the manufacture of composite materials, but when the strength of polymer is high then ASTM D3039 is required.

Stress is usually expressed in N/m^2 or Pascal ($1\ N/m^2 = 1$ Pa). The stress value from the experiment is determined by dividing the amount of force (F) exerted by the device by its cross-sectional area (A) in the axial direction, which is measured before the experiment is carried out. Mathematically, it is expressed in Equation 8.1. Strain values that do not have units can be calculated using Equation 8.2. In the equation, L is the instantaneous length of the specimen and L_0 its original length.

$$\sigma = \frac{F}{A} \tag{8.1}$$

$$\varepsilon = \frac{L - L_0}{L_0} \tag{8.2}$$

The stress–strain curve is characteristic of ductile metallic constituents. Another interesting aspect is that we usually speak about the "engineering stress–strain" curve [9]. When a material approaches the stress-strain curve's maximum, it will significantly reduce its cross-sectional area, a phenomenon known as necking. The computer program assumes when plotting the stress–strain curve that the cross-sectional area will remain constant during the experiment, even throughout the necking process, allowing the curve to slope downwards. The "real" stress–strain curve could be plotted by directly installing a gauge to calculate the change in the specimen's cross-sectional area during the experiment.

When a force is released when the material is in its elastic zone, the material returns to its original form, the slope of the curve being a constant and an intrinsic property of a material, known as the elastic modulus, E (expressed in GPa). It gives a constant value and is given by:

$$E = \frac{\sigma}{\varepsilon} \tag{8.3}$$

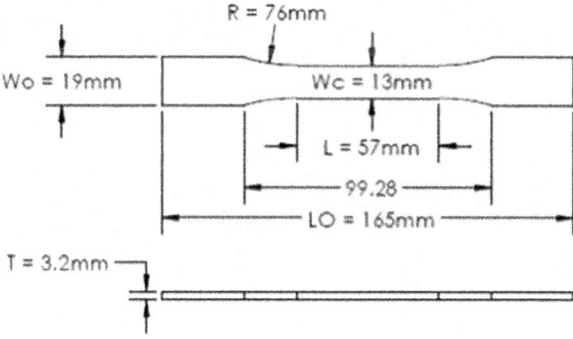

FIGURE 8.2   Standard specimen drawn using AutoCAD software

A drawing of the standard specimen is shown in Figure 8.2. The higher the value of the youth module, the higher the stiffness value, as the structure would deform less at a higher stress value. This property contributes to a concept called material stiffness, which is an indicator of resistance to deformation under the elastic limit. The elasticity modulus is a very important property that is used in formulas that deal with beams and columns where stiffness is an important criterion.

In theory, even without calculating the specimen's cross-sectional area during the tensile experiment, the real stress–strain curve could still be built by assuming the material volume remains the same. Using this definition, Equations 8.4 and 8.5 are used to measure both the true stress (σ_T) and the true pressure (ε_T). Within these equations, L_0 refers to the specimen's initial length, L the instant length, and γ the instant tension [10].

$$\sigma_T = \sigma \frac{L}{L_0} \tag{8.4}$$

$$\varepsilon_T = \ln \frac{L}{L_0} \tag{8.5}$$

A universal testing machine (UTM) is used for tensile testing.

8.2.2 FLEXURAL TESTING

Flexural work is used to assess the resistance or bending properties of a material. In a test referred to as a transverse beam test but also referred to as a three-point bend or four-point bending test, the sample is positioned between two points or supports. The composite's interlaminar strength gives very high resistance to the bending load applied during a bending operation. The ASTM D790 model is used for flexural testing of polymer-based composites [11].

Flexural strength is the property of the composites that reflects the interlaminar shear strength. This research can be conducted on a UTM computer like that used for the tensile testing. The flexural strength and modulus can be calculated using standard relations. The flexural stress, σ_f is calculated by [12]:

$$\sigma_f = \frac{3FL}{2bd^2} \tag{8.6}$$

$$\sigma_f = \frac{FL}{\pi R^3} \tag{8.7}$$

Equations 8.6 and 8.7 are used for rectangular and circular cross sections, respectively.

The flexural strain, ε_f is calculated by:

$$\varepsilon_f = \frac{6Dd}{L_2} \tag{8.9}$$

The flexural modulus of elasticity (MPa), E_f is calculated by:

$$E_f = \frac{mL^3}{4bd^3}$$ (8.10)

where,

F is the load at a given point on the load deflection curve (N)

L is the support span (mm)

b is the width of the test beam (mm)

d is the depth or thickness of the test beam (mm)

D is the maximum deflection of the center of the beam (mm)

m is the gradient (i.e., the slope) of the initial straight-line portion of the load deflection curve (N/mm)

R is the radius of the beam (mm)

8.2.3 IMPACT TESTING

Impact strength is the capability of the material to withstand a suddenly applied load and is expressed in terms of energy. It represents the material's behavior during impact at high speed. Impact behavior is the most important mechanical property of a material. Impact strength is tested through ASTM D256 using a testing sample of dimensions 127 mm × 13 mm × 3 mm. The unit of impact strength is J/cm. There are two types of impact test, Izod and Charpy methods [13].

Owing to its low cost and robustness Charpy has been used for several years to test composite material. In the Charpy method, the specimen is put in a vise so that at the ends it serves as a beam. Likewise, the specimen is put in a vise in the Izod test, so that its one end is free and behaves like a cantilever plate[14].

Composite materials may fail if the energy absorption capacity of the material is not sufficient to withstand the energy being transferred through the application of the impact charge. This test shows the ability of a material to absorb energy during impact on load on the specimen. Although the toughness of a material can be evaluated by area under the stress–strain diagram, an indication of relative toughness comes from the impact test. Usually notch-type specimens are used for impact studies. Two types of notches are used to test bending effect: V-note and keyhole note.

8.2.4 HARDNESS TEST

Hardness is another essential mechanical property to note. It measures part resistance to localized plastic deformation; for example, a slight dent or scrape. Hardness testing is widely used to inspect and verify composites. The idea of any hardness test method is to force an indenter into the sample surface, then measure the indentation (depth or actual indentation area). Hardness is not a fundamental property and the combination of yield force, tensile strength, and elasticity modulus determines its value. Depending on the magnitude of the loading force and the dimensions of the indentation, toughness is defined as macro-, micro-, or nano-hardness [15].

The macro-hardness tests (Rockwell, Brinell, Vickers) are the most widely used methods for fast, routine hardness measurements. The indenting forces for macro-hardness measurements are in the range from 50 to 30,000 N [16].

Plastic hardness is most measured via the Shore® (Durometer) or Rockwell hardness test. Both methods measure the indentation resistance of a plastic and have an objective hardness value that does not inherently fit well with certain properties or fundamental characteristics. Shore A or Shore D is used, which is also commonly used for softer plastics such as polyolefins, fluoropolymers, and vinyl. The Shore A scale is used for the softer rubbers whereas Shore D is used for firm ones. There are many other Shore hardness levels, such as Shore O and Shore H, but most plastics engineers rarely encounter them. The Shore hardness is measured using a Durometer proven tool, the indenter foot of which penetrates the sample, and is thus also known as Durometer hardness [17]. The indentation reading may change over time, so the indentation time and the hardness number are often registered. During hardness testing of composite materials based on polymers, the ASTM D785 standards are typically followed. Compared with other mechanical examinations, some advantages of the hardness test are as follows:

They are simple and cheap to perform – usually, no special specimen needs to be prepared [18].

They are non-destructive – the specimen is neither fractured nor excessively deformed.

The various hardness tests may be divided into three categories:

Elastic hardness.

Resistance to cutting or abrasion.

Resistance to indentation.

8.2.5 INDUSTRIAL APPLICATION OF MECHANICAL CHARACTERIZATION TECHNIQUES

Mechanical characterization testing is considered by academicians and researchers in the polymer and polymer composite-based industries as one of the most prominent, and popularly preferred methods for evaluating strength, rigidity, etc. Product research includes a wide range of high-performance, specialized industrial applications including textiles, packaging, building, aerospace, automotive, and many other products [19].

8.3 CONCLUSIONS

The strength, rigidity, and resilience of polymer composites is important to evaluate. All mechanical characterization services meet standardizations and common requirements. Mechanical testing includes tensile, flexural, and impact properties, giving a simple insight into the ability of materials to withstand sudden failure under the load or stress applied [20]. Nevertheless, the mechanical properties of the polymer composite materials depend on the quality of the fiber and polymer and the interfacial bond between the fiber and the matrix. This chapter covers various technical examinations available.

REFERENCES

1. Termonia Y. Tensile Strength of Discontinuous Fibre-Reinforced Composites. *J Mater Sci* 1990;259(11): 4644–4653.
2. Mohanty AK, Misra M, Drzal LT. Sustainable Bio-Composites from Renewable Resources: Opportunities and Challenges in the Green Materials World. *J Polymer Env* 2002; 10:19–26.
3. Goda K, Cao Y. (2007) Research and Development of Fully Green Composites Reinforced with Natural Fibers. *J Sol Mech Mater Eng* 2007; 1:1073–1084.
4. Ashik KP, Sharma RS. A Review on Mechanical Properties of Natural Fiber Reinforced Hybrid Polymer Composites. *J Miner Mater Charact Eng* 2015; 3: 420–426.
5. Kokta BV, Maldas D, Daneault C, Beland, P. Composites of Polyvinyl Chloride-Wood Fibers. III. Effect of Silane as Coupling Agent. *J Vinyl Technol* 1990; 12(3):146–153.
6. Salkind MJ. *Fatigue of Composites, Composite Materials: Testing and Design 2nd Conference, ASTM STP* 1972; 1:497:143.
7. Jena H, Pandit Ku. M, Pradhan, Ku. A. (2012) Study the Impact Property of Laminated Bamboo-Fiber Composite Filled with Cenosphere. *Int J Env Sci Dev* 2012; 3: 456–459.
8. Yousif BF, Ku H. Suitability of Using Coir fiber/Polymeric Composite for the Design of Liquid Storage Tanks. *Mater Design* 2012; 36: 847–853.
9. Nadir A, Songklod J, Vallayuth F, Piyawade B, White RH. Coir Fiber Reinforced Polypropylene Composite Panel for Automotive Interior Applications. *Fibers Polym* 2011; 12: 919–926.
10. Verma D, Gope PC, Shandilya A, Gupta A, Maheshwari MK. Coir Fiber Reinforcement and Application in Polymer Composites: A Review. *J Mater Environ Sci* 2013; 4: 263–276.
11. Abilash N, Shivaprakash M. Environmental Benefits of Eco-friendly Natural Fiber Reinforced Polymeric Composite Materials. *Int J Appl or Innov Eng Manag* 2013; 2 (1): 53–59. ISSN 2319–4847.
12. Siddiquee M, et al. Investigation of an Optimum Method of Biodegradation Process for Jute Polymer Composites. *Am J Eng Res* 2014; 31: 200–206.
13. Kim SW, Lee SH, Kang JS, Kang KH. Thermal Conductivity of Thermoplastics Reinforced with Natural Fibers. *Int J Thermophys* 2006; 27(6): 1873–1881.
14. Wambua P, Ivens J, Verpoest I. Natural Fibers: Can they Replace Glass in Fiber Reinforced Plastics?, *Compos Sci Technol* 2003; 63: 1259–1264.
15. Oksman K, Wallstrom L, Berglund LA, Filho RDT. Morphology and Mechanical Properties of Unidirectional Sisal–Epoxy Composites. *J Appl Polym Sci* 2002; 84:2358–2365.
16. Xuea, Y, Dub Y, Elderc S, Wang K, Zhang J. Temperature and Loading Rate Effects on Tensile Properties of Kenaf Bast Fiber Bundles and Composites. *Compos Part B Eng* 2009; 40(3):189–196.
17. Verma D, Gope PC, Shandilya A, Gupta A, Maheshwari MK. Coir Fiber Reinforcement and Application in Polymer Composites: A Review. *J Mat Environ Sci* 2013; 4:263–276.
18. Ku H, Wang H, Pattarachaiyakoop N, Trada M. A Review on the Tensile Properties of Natural Fiber Reinforced Polymer Composites. *Compos Part B Eng* 2011;42:856–873.
19. Wambua P, Ivens J, Verpoest I. Natural Fibres: Can They Replace Glass in Fibre Reinforced Plastics? *Compos Sci Technol* 2003;63:1259–1264.
20. Hu R, Lim J-K. Fabrication and Mechanical Properties of Completely Biodegradable Hemp Fiber Reinforced Polylactic Acid Composites. *J Compos Mater* 2007;41:1655–1669.

9 Humidity Sensor Based on Alum–Fly Ash Composite

Amit Sachdeva, Shri Prakash Pandey and Pramod K. Singh

CONTENTS

9.1   INTRODUCTION

Solid electrolytes characterized by exceptionally high ionic conductivity and relatively small electronic conductivity have attracted a great deal of attention because of their unique transport properties and potential applications in batteries, sensors, etc. [1, 2]. Although intensive research in the past decade has already resulted in a number of good solid electrolyte materials, efforts continue to discover new materials and methods for increasing the level of ionic conduction.

A number of recent investigations have reported significant enhancement in ionic conductivity by the dispersion of fine, insulating, insoluble fly ash (FA) particles; importantly, this does not appreciably alter the electronic conductivity. These so-called dispersed solid electrolyte systems (DSES) have become increasingly important to both experimentalists and theorists [3–5]. A comparison of the pure system and

163

the two-phase mixture (containing FA) reveals that the two-phase mixtures and/or DSES in general exhibit a higher conductivity than the starting materials. In this paper we report electrical and structural data on potash alum dispersed with FA. Additionally we have tried to develop a humidity sensor based on this composite electrolyte.

FA is a waste product produced by coal-fired thermal power stations during the combustion of coal. The many coal-fired power plants all over the world produce a large quantity of FA, causing serious environmental problems, as less than half of the ash is used as a raw material for concrete manufacturing and construction; the rest is directly dumped as landfill or simply piled up. Owing to environmental concerns, new ways of using FA have to be explored. Hence, there is considerable interest in FA as a raw material. Recently, materials scientists and engineers have suggested and devised various methods to use FA for the synthesis of some useful composites [6–8].

There have been many experimental analyses of FA's basic compositional, physical, and chemical properties for technical studies and applications. Raw FA consists of quartz and mullite as crystalline phases and some quantity of glassy phase. It is a gray, alkaline powder with pH 9–9.9.

The most common alum is the double sulfate of potassium and aluminum, $K_2Al_2(SO_4)_4.24H_2O$, a white crystalline powder that is readily soluble in water. It is used in curing animal skins. Other alums are used in papermaking and to fix dyes in the textile industry. The raw material of manufacture of common alums is alum rock, composed chiefly of alunite or alum stone. Alum is also made from alum shale, which is either allowed to decompose by exposure, or roasted. During the process, free sulfuric acid is formed, which acts upon the clay, producing aluminum sulfate, which is then dissolved out. Potassium sulfate or ammonium sulfate is added to the solution to produce potash alum or ammonia alum.

Through this work we have been successful in preparing potash alum–FA composites. We have developed a humidity sensor that gives a new, better way to use FA.

9.2   EXPERIMENTAL

Potash alum was purchased from market and used without further purification; FA of unknown purity and composition was collected from a local supplier. Appropriate amounts of potash alum and FA were weighed and thoroughly mixed in an agate mortar and pestle (~2 hours) followed by pulverization and pelletization in a nickel-plated steel die at pressures of 2.5 tons using a hydraulic pelletizer machine. The circular pellets thus obtained were 0.15 cm^2 in area and 4–6 mm in thickness. Silver paste was coated onto both surfaces of the pellets and dried under room environment to produce pellets that were ready for electrical measurement.

9.2.1   COMPLEX IMPEDANCE SPECTROSCOPY

The electrical conductivity of the composite pellets was measured at 1 KHz frequency using a Hioki 3522-50LCR Hi Tester (Hioki, Japan). The electrical conductivity (σ) was evaluated using the formula

$$\sigma = R_b \left(1/A \right)$$

where σ is ionic conductivity, R_b is the bulk resistance, l is the thickness of the pellet, and A is the area of the sample.

9.3 RESULTS AND DISCUSSION

9.3.1 ELECTRICAL

9.3.1.1 Complex Impedance Spectroscopy

The calculated values of ionic conductivity are listed in Table 9.1 and plotted in Figure 9.1. It is clear that the electrical conductivity of the potash alum–FA system increases initially with increasing content of FA to a conductivity maximum obtained at 65% FA, then decreases. To interpret the results of conductivity measurements in dispersed solid electrolytes, several theoretical models have been proposed. For ionic transport in dispersed ionic conductors, a percolation model was proposed, assuming that highly conducting paths are created along the interface between the host electrolyte and the dispersoid. Electrical conductivity enhancement in two-phase composites was shown to be strongly dependent on sample preparation conditions, and higher conductivities are expected if a better contact between the solid electrolyte and dispersoid can be obtained [9].

The high ionic conductivity can be quantitatively explained by a space-charge effect induced by internal cation adsorption at the FA surfaces [10–14]. The observed conductivity maximum could be explained by the percolation model as water of crystallization present in alum is adsorbed on the surface of the composite, where movement of H^+ and OH^- ions are responsible for the increase in conductivity. The decrease in conductivity after 65% FA composition was due to achievement of the percolation threshold [15, 16].

9.3.1.2 Temperature Dependence of Conductivity

In DSES the temperature dependence can either be Arrhenius (linear plots) or Vogel–Tammann–Fulcher type. Figure 9.2 shows the temperature dependence of the electrical conductivity of a typical composite system (maximum conductivity composition).

TABLE 9.1
Room Temperature Conductivities in the Potash Alum–fly Ash Composite System

Sr No	Fly Ash (wt%)	Potash Alum (wt%)	Conductivity (S/cm)
1	95	5	3.2×10^{-6}
2	85	15	5.3×10^{-6}
3	75	25	9.2×10^{-6}
4	65	35	1.5×10^{-5}
5	60	40	1.2×10^{-5}
6	40	60	5.4×10^{-6}
7	10	90	4.6×10^{-6}

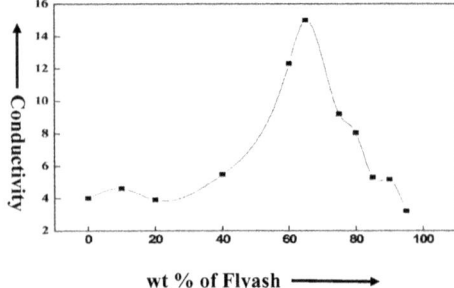

FIGURE 9.1   Variation in conductivity with concentration of fly ash in the potash alum–fly ash composite system

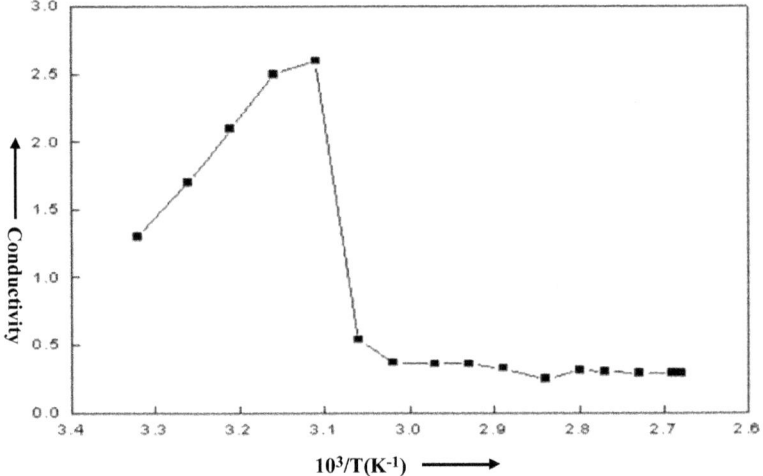

FIGURE 9.2   Variation of conductivity with temperature (maximum conductivity)

It follows Arrhenius behavior in which conductivity can be explained using the formula $\sigma = \sigma_0 \exp(-E_a/kT)$, where σ is the conductivity, σ_0 the pre-exponential factor that depends upon the concentration, the attempt frequency, and the jump distance of the atomic defects, E the overall activation energy, k the Boltzmann constant, and T the absolute temperature (in K).

From the figure we can clearly observe that conductivity rises with increasing temperature up to 48°C and then falls before attaining a constant value. This is because physisorbed water present in potash alum, which plays an important role in conductivity enhancement, is lost between 45 and 50°C.

9.3.2   STRUCTURAL

9.3.2.1   Scanning Electron Microscopy

Scanning electron microscopy (SEM) has been used to examine the surface morphology of composite electrolytes. We have recorded SEM micrographs using a Hitachi S-570 SEM instrument, which are shown in Figure 9.3.

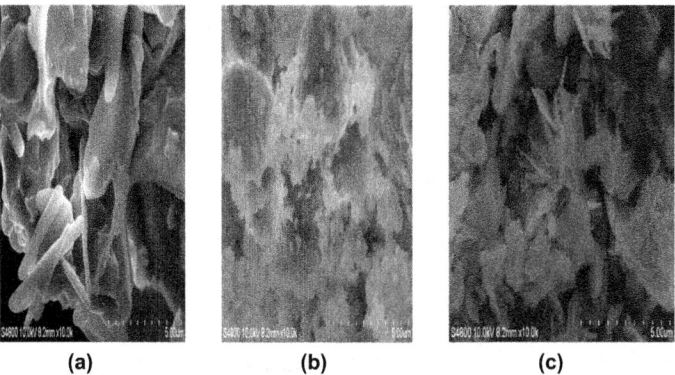

(a)   (b)   (c)

FIGURE 9.3 SEM micrographs of (a) Pure Potash Alum (b) Potash Alum Flyash composite and (c) pure Flyash

It is clear from the figure that the uncrushed potash alum shows a needle-like morphology (Figure 9.3a) whereas the FA of unknown purity shows an irregular structure (Figure 9.3b). In the composite, morphology is mixed: white FA is uniformly mixed with darker potash alum grains (Figure 9.3c).

9.3.2.2 Infrared Spectroscopy

Infrared spectroscopy was carried out to study the nature and functional groups present in the composite. Figure 9.4 shows the recorded infrared spectra (Perkin Elmer 883) of pure potash alum, pure FA, and potash alum doped with FA (maximum conductivity composition).

It is obvious that the IR spectrum of composite materials (Figure 9.4c) contains the peaks from pure potash alum (Figure 9.4a) or FA (Figure 9.4b). The absence of any new peaks clearly confirms its composite nature.

The characteristic broad band at 3422 cm^{-1} in the composite spectrum is due to N-H stretching of aniline. The peak observed at 1094 cm^{-1} in the spectrum of FA (Figure 9.4b) may be attributed to the presence of silica [17]. This peak confirms the highest % of alum in the FA, which is confirmed by chemical analysis [8]. Several peaks are observed between 1400 and 1300 cm^{-1}, which correspond to various metal oxides present in the FA. Absorptions at 1297.49 ± 10, 1139.71 ± 10, 1089.48 ± 10, and 794.81 ± 20 cm^{-1} correspond to FA constituents in the potash alum–FA composite.

Figure 9.4a shows the spectrum of potash alum. Infrared spectra of alums based on one monovalent and one trivalent cation have been published [18]. Ross interpreted the infrared spectrum of potassium alum as 981 cm^{-1}, 465 cm^{-1}, 1200 cm^{-1}, 1105 cm^{-1}, 618 cm^{-1}, and 600 cm^{-1} for $(SO4)^{2-}$. Water stretching modes were reported at 3400 and 3000 cm^{-1}, bending modes at 1645 cm^{-1}, and librational modes at 930 and 700 cm^{-1} [19].

9.3.2.3 X-Ray Diffraction

To further investigate the composite, we recorded its X-ray diffraction (XRD) patterns along with those of the host materials (Figure 9.5). Peaks in the XRD pattern of

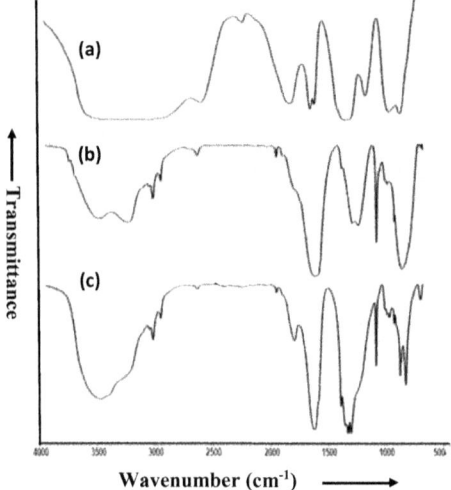

FIGURE 9.4   Infrared spectra of (a) pure potash alum, (b) pure fly ash (c) potash alum–fly ash composite

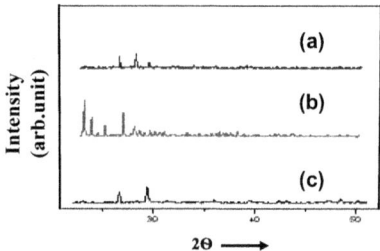

FIGURE 9.5   X-ray diffraction patterns of (a) pure fly ash, (b) pure potash alum, and (c) potash alum–fly ash composite

the composite (Figure 9.5c) also appear in the XRD patterns of FA (Figure 9.5a) or potash alum (Figure 9.5b), which affirms its composite nature, supporting the IR data.

9.3.3   HUMIDITY SENSOR

Based on the maximum electrical conductivity sample, we tried to fabricate a humidity sensor in our laboratory (Figure 9.6). We developed a finger-type electrode on the surface of a composite pellet using vacuum-coated silver paint as electrode material, deposited using a vacuum-coating unit (Hind High Vac, India) at 10^{-3} Torr pressure. Conducting copper wire was used for contact and sensing behavior. To measure the

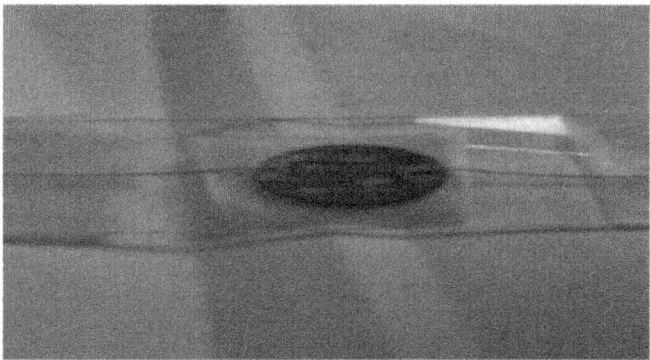

FIGURE 9.6 Photograph of humidity sensor based on potash alum–fly ash composite

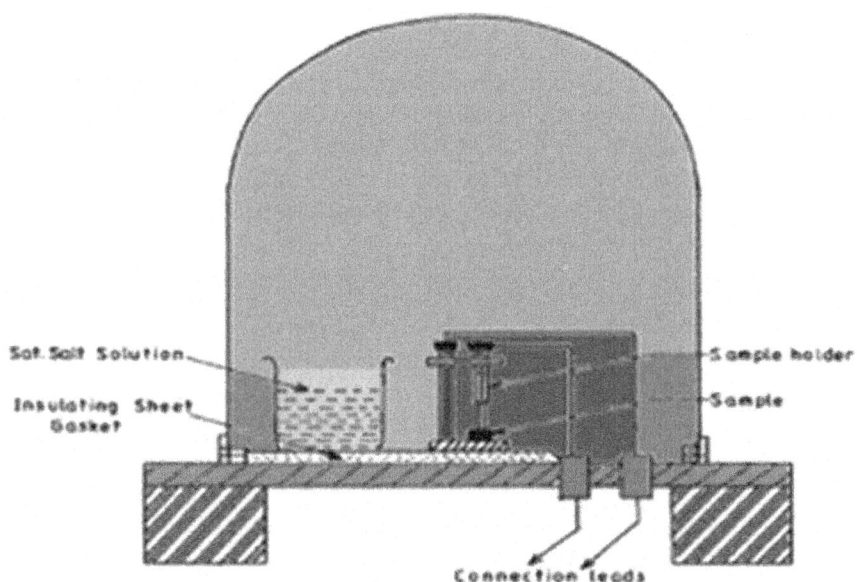

FIGURE 9.7 Experimental setup to maintain constant humidity

response of the humidity sensor we have also developed a constant humidity chamber in our laboratory (Figure 9.7), as designed by Chandra et al. [20].

A very simple approach was adopted for creating different constant humidities for in situ measurement. It is known that the water vapor pressure of over-saturated solutions of different salts gives different relative humidities.

The response of the sensor (voltage vs. humidity) is shown in Figure 9.8. It is clear that the sensor works well and has a quick response, rapidly showing an exponential decrease in voltage with increase in humidity level.

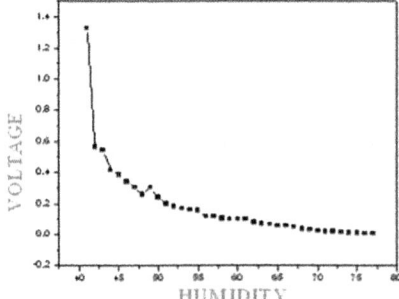

FIGURE 9.8 Response of humidity sensor based on potash alum–fly ash composite (maximum conductivity sample)

9.4 CONCLUSION

A solid-state composite based on potash alum and FA has been developed and well characterized using various techniques. Electrical conductivity measurement shows that the addition of FA enhances electrical conductivity to a maximum 1.5×10^{-5} S/cm at 65% FA composition, then decreases. Infrared spectroscopy as well as XRD confirmed the composite nature of the material while SEM affirmed the homogeneous mixing. A humidity sensor has been fabricated (with maximum σ sample) that shows stable and good performance.

ACKNOWLEDGMENT

This work was supported by the DST project, Government of India (SR/S2/CMP-0065/2010).

REFERENCES

1. Jow, T. and Wagner, J. B. (1979). The effect of dispersed alumina particles on the electrical conductivity of cuprous chloride, *Journal of the Electrochemical Society.*, 126: 1963–1972.
2. Gupta, R.K. and Agarwal, R. C. (1999). Superionic solids:composite electrolyte phase –an overview, *Journal of Material Science* ,34:1131–1162.
3. Vaidehi, N., Akila, .R. , Shukla, A.K. and Jacob, K.T. (1986). Enhanced ionic conduction in dispersed solid electrolyte systems $CaF_2:Al_2O_3$ and $CaF_2:CeO_2$, *Materials Research Bulletin*, 21: 909–916
4. Reddy, S .N. , Chary, A.S. , Saibabu, K. and Chiranjivi, T. (1989). Enhancement of dc ionic conductivity in dispersed solid electrolyte system - $Sr(NO_3)_2:\gamma\text{-}Al_2O_3$, *Solid State Ionics*, 34:73–77.
5. Dudney, N. J. (1989). Composite electrolytes, *Annual Review of Materials Science*, 19:103–120
6. Owens, B.B., Skarstad, P.M. (1992). Ambient temperature solid state batteries, *Solid State ionics*, 53–56:665–672.
7. Narayan, H. , Alemu, H.M., Somerset, V.S., Iwuoha, E.I., Hernández, M.L., Hernández, J.A., Montaño, A.M., and Henao, J.A. (2009) . Chapter 5 In: *Fly Ash: Reuse,*

Environmental Problems and Related Issues, Ed: Peter H. Talone, 111–136, Nova Science Publishers, Inc.,New York.

9. Bose, D.N. and Majumdar, D. (1984). Lithium solid electrolytes and their applications, *Bulletin of Materials Science*, 6: 223–230

10. Shahi, K. and Wagnor Jr., J.B. (1982).Enhanced ionic conduction in dispersed solid electrolyte system (DSES) and/or multiphase systems:AgI-Al$_2$O$_3$,AgI-SiO$_2$,AgI-Flyash,AgI-AgBr, *Journal of Solid State Chemistry* 42:107–119.

11. Ahmad, K., Pan, W. and Shi, S.L. (2006). Electrical conductivity and dielectric properties of multiwalled carbon nanotube and alumina composites, *Applied Physics Letters*, 89: 133122–133124.

12. Stuffer, D. (1985). *Introduction to Percolation Theory*, Taylor and Francis, London.

13. Maier, J.(1987), Defect chemistry and conductivity effects in heterogeneous solid electrolytes, *The Journal of the Electrochemical Society*, 134:1524–1535.

14. Maier, J.(1994), Defect chemistry at interfaces. *Solid State Ionics*, 70/71: 43–51.

15. Maier, J.(1985), Space charge regions in solid two-phase systems and their conduction contribution- I. Conductance enhacement in the system ionic conductor-'inert' phase and application on AgCl:Al$_2$O$_3$ and AgCl:SiO$_2$, *Journal of Physics and Chemistry of Solids*, 46: 309–320.

16. Stoneham, A. M. , Wade, E. and Kilner, J. A.(1979), A model for the fast ionic diffusion in alumina-doped LiI, *Materials Research Bulletin*, 14: 661–666.

17. Wagner, J. B.(1980), Transport in compounds containing a dispersed second phase, *Materials Research Bulletin*, 15: 1691–1701.

18. Raghavendra, S.C.,Khasim Syed, Revanasiddapa, M., Prasad Ambika, M.V.N., Kulkarni, A.B.,(2003). Synthesis, characterization and low frequency a.c. conduction of polyaniline/fly ash composites, *Indian Academy of Sciences*,26, 739.

20. Ross, S.D. (1974) Chapter 18 in: Sulphatee and other oxy-anions of Group VI, *The Infrared Spectra of Minerals*, V.C. Farmer editor, 423–444,The Mineralogical Society, London.

21. Ross, S.D. (1972). *Inorganic Infrared and Raman Spectra*, McGraw-Hill Book Company Ltd, London.

22. Srivastava, N. and Chandra, S. (2000). Studies on a new proton conducting polymer system: poly(ethylene succinate) + NH4ClO4. *European Polymer Journal* 36: 421–433.

10 Applications of Graphene-Based Composite Materials

Gorkem Memisoglu, Burhan Gulbahar and Canan Varlikli

CONTENTS

10.1   INTRODUCTION

Graphene has been known for over 150 years (Schafhaeutl 1840), (Brodie 1859), (Boehm et al., 1962), (Boehm et al., 1986), (Bommel and Van Crombeen, 1975), (Boehm et al., 1994), (Lu et al. 1999), (Novoselov et al. 2004). In 1840, for the first time, graphite was chemically intercalated and exfoliated using nitric and sulfuric acids (Schafhaeutl 1840). In addition, the first oxide layers of graphene were formed in 1859 (Brodie 1859). Since the discovery of the transport properties in micromechanically exfoliated layers with the isolation of single-layer free-standing graphene (Novoselov et al. 2004), graphene has been the target of great attention and has been investigated for its optical, mechanical, thermal, and electrical properties. It has extraordinary properties such as high optical transparency, wavelength-independent broad-spectrum absorption, mechanical strength, chemical stability, high carrier mobility, high electrical conductivity, high thermal conductivity, high optical damage threshold, ultra-low weight, large Young's modulus (1 TPa for single-layer graphene), cost-effectivity, low residual stress, and large-scale, single-, or multilayer production potential (Bolotin et al. 2008), (Balandin et al. 2008), (Morozov et al. 2008), (Cai et al. 2009), (Zhu et al. 2010), (Park and Ruoff 2009), (Wu et al. 2017), (Park et al. 2014), (Du et al. 2014), (Mohan et al. 2018), (Zhang et al. 2016), (Mak et al. 2008), (Zhou et al. 2006), (Hasan et al. 2011), (Weiss et al. 2012), (Li et al. 2014), (Sadasivuni et al. 2015), (Ferrari et al. 2015) (Luo et al. 2015), (Nemade et al., 2018), (Ghayoor et al. 2019).

Figure 10.1 shows the history of graphene development showing selected events like the first reports of chemical intercalation of graphite, preparation of graphene oxide layers, and mechanically exfoliated graphene preparation. The figure also illustrates some applications in sensors, solar cells, lighting or display applications, optic modulators, bioelectronics, or batteries. Growth and preparation methods and the integration technique are crucial in order to enhance the performance of graphene-based devices or develop new graphene-related applications by changing the optical, thermal, or electrical properties of the graphene-based layers (graphene, reduced graphene oxide, or graphene oxide) (Colombo et al. 2013), (Novoselov et al. 2004) (Bleu et al. 2018), (Gulbahar and Memisoglu 2017). The performance of graphene-based devices can be improved by improving the preparation method. As an example, if the graphene-based layers are thinner, the performance of the device can be improved by means of more electrical conductivity and more optic transmittance (Galindo et al. 2014).

There are two main techniques for graphene growth: top down and bottom up. The bottom-up technique involves chemical vapor deposition (CVD), epitaxial growth, or pyrolysis, whereas the top-down techniques are exfoliation processes such as mechanical or liquid exfoliation, or chemical synthesis (Bhuyan et al. 2016), (Coros et al. 2020).

The preparation method affects the graphene product's quality in terms of its electrical or thermal conductivity, optic transparency, and charge carrier mobility. CVD, liquid exfoliation, mechanical exfoliation, epitaxial growth, nanotube slicing growth, and intercalation are some widely used methods for graphene production (Geim and MacDonald 2007), (Batzill 2012), (Colombo et al. 2013), (Whitener Jr and Sheehan 2014), (Bianco et al. 2015), (Ionita et al. 2017), (Bleu et al. 2018). In addition, wet exfoliation, sonication, or Hummer's method are used to prepare the oxide or reduced oxide form of graphene with a controllable large surface area (Hummers Jr and Offeman 1958), (Chen et al. 2013), (Toh et al. 2014). Of these techniques, CVD provides high-quality, large, flat surfaces, and single or multi-layer graphene growth on metal host such as nickel or copper. However, there can be integrating the CVD-grown graphene to a system by transferring the graphene from its metal host to the target substrate, because of the strong interaction between the growth graphene layer(s) and host (Whitener Jr and Sheehan, 2014).

Other relatively facile, up-scalable, and economic graphene-based layer growth methods include the mechanical or liquid exfoliations which are useful for systems requiring graphene-based layer(s) that are a few microns wide. The drawbacks of such methods are their size limit, and the manipulation process to integrate them to a photonic or optoelectronic device application. Transfer and integration of the growth graphene layer(s) to a device application are generally achieved by dry-stamp transfer with poly-dimethyl siloxane, wet-stamp transfer, or wet etching with chemicals (Castellanos-Gomez et al. 2014), (Frisenda et al. 2018). Figure 10.2 shows examples of the facile and low-cost preparation of a graphene-based layer from the exfoliation of graphite flakes (a–c), and solution-processed graphene hydrosol drop-casting and annealing processes (d–g). Graphite flakes with 1 mm width are bought from HQ-Graphene Inc. (Groningen, the Netherlands), and monolayer dispersed graphene oxide hydrosol is from Graphenea Inc., (San Sebastian, Spain). In Figure 10.2 (a), graphite flakes are shown.

Scotch-tape-based mechanical exfoliation cycle-dependent graphene many layer (GML) preparation is shown in Figure 10.2(b): the GML is obtained after many mechanical exfoliation steps. GML with diameter less than 50 μm and thickness 80 nm on glass substrate is shown in Figure 10.2(c). Figure 10.2(d) and (e) show

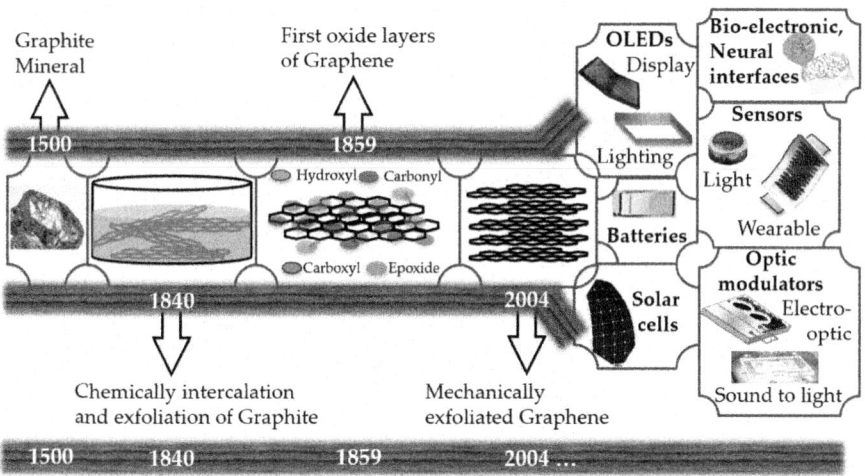

FIGURE 10.1   Timeline for graphene preparation and some selected applications

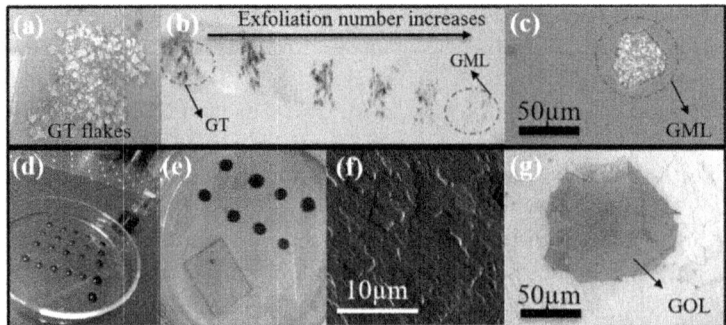

FIGURE 10.2   Graphene-based layer preparation examples from the mechanical exfoliation of graphite (GT) flakes (a–c), and the solution-processed graphene hydrosol drop-cast and annealing processes (d–g). Experiments were performed at the University of the Basque Country laboratories, Bilbao, Spain

graphene oxide layer (GOL) preparation by drop-casting from commercial graphene hydrosol before and after the annealing process, respectively. Atomic force microscopy pictures of graphene single-layer flakes with width less than 10 μm are presented in Figure 10.2(f). Furthermore, an optical microscopy image used to estimate the diameter (~100 μm) of the prepared GOL sample on a hole in glass is shown in Figure 10.2(g).

For visible-range photonic applications including photosensors, photovoltaics, and energy conversion or storage devices, graphene plays an important role as the organic, transparent, flexible, thermally or electrically conductive counterpart with an additive material such as polymer, quantum dots, or metal oxide (Memisoglu et al. 2015), (Raccichini et al. 2015), (Gulbahar and Memisoglu 2017a), (Gulbahar and Memisoglu 2017b). Graphene is also highly useful in lighting and display devices (Li et al. 2019), (Diker et al. 2019), optic modulators, Förster resonance energy transfer (FRET)-based photosensors or point-of-care devices (Li et al. 2014), (Luo et al. 2015), (Memisoglu et al. 2018), (Memisoglu et al. 2020), (Memisoglu, 2020), or biosensors (Shao et al. 2010), (Pan et al. 2015), (Sha et al. 2017).

Pure graphene layers have high thermal and electrical conductivities, whereas the oxide form of graphene is known as an insulator and the reduced graphene oxide form can have electrical conductivity with low thermal conductivity (Zeng et al. 2018). However, additive materials like polymers, quantum dots, or metal oxides can enhance the optical, thermal, electrical, and mechanical properties of a graphene-based composite device system (Shao et al. 2010), (Li et al. 2014), (Pan, Gu, Lan, Sun, and Gao 2015), (Luo et al. 2015), (Memisoglu et al., 2015), (Raccichini et al. 2015), (Gulbahar and Memisoglu 2017a), (Gulbahar and Memisoglu 2017b), (Memisoglu et al. 2018), (Li et al. 2019), (Sha et al., 2017), (Diker et al., 2020), (Memisoglu et al., 2020), (Memisoglu, 2020). It is critical to choose appropriate materials and preparation methods to optimize both the quality and functional properties like light sensing, mechanical strength, durability, electrical or thermal conductivity, or bioactivity of the graphene-based composite. Graphene-based composites are used in many varied areas such as photonics, biotechnology and medical, aerospace, automobile, flexible wearable or consumer electronics, or the defense industries (Basavaraja, Kim, and Do Kim 2011), (Du and Cheng 2012), (Zhang et al. 2016), (Stankovich et al. 2006). Quality and functional properties of the graphene-based composites can be affected by many factors, such as the physical or chemical properties of the additives, the interfacial relationship between additive and graphene materials, and the network between materials in the composite (Colombo et al., 2013).

This chapter is organized as follows. Section 10.2 reviews the photonic applications of graphene-based composites, which are essential for understanding graphene-polymer composites, graphene–quantum dot composites, and graphene metal oxide composites in terms of photosensor, solar cell, lighting, and biological applications. Then Section 10.3 summarizes applications related to graphene-composite-based visible-range photonics, and their future prospects.

10.2 PHOTONIC APPLICATIONS OF GRAPHENE-BASED COMPOSITES

Nanotechnology and nanomaterial developments play crucial roles in revolutionizing energy conversion, lighting, biological or photo-sensing device applications. Photonic-related nanotechnological applications of graphene-based nanomaterials have been under investigation for more than a decade. In addition, when additive materials like polymers, quantum dot nanocrystals, or metal oxides are combined with graphene, the functional properties of the resulted composite nanomaterial can improve their optical, electric, or thermal performance, making them functional composite materials to be used in photosensors, photovoltaics, lighting, and biological applications (Lu et al. 2016), (Stankovich et al. 2006), (Bao et al. 2010), (Rafiq et al. 2010), (Basavaraja, Kim, and Do Kim 2011), (Hasan et al. 2011), (Kim, Lee, and Lee 2011), (Bai, Zhai, and Zhang 2011), (Pang et al. 2011), (Song et al. 2011), (Zhang et al. 2012), (Konstantatos et al. 2012), (Du and Cheng 2012), (Huang et al. 2014), (Memisoglu et al., 2015), (Ferrari et al. 2015), (Tang et al., 2017), (Lee and Shiang Wang 2019), (Han et al. 2019), (Hyeong et al. 2019), (Zhao et al. 2020), (Park et al. 2020).

The optical, thermal, mechanical, and electrical properties of composites combining graphene with polymer, quantum dots, or metal oxides can be developed. As an example, indium tin oxide (ITO) is a transparent conductive electrode material widely used in photonics or optoelectronics applications. ITO is more expensive than graphene due to the rare earth element indium in its structure. A graphene-based composite layer is a candidate to take the place of ITO due to its similar work function (4.5 eV), high transparency to light, and high flexibility (Zhang et al. 2006), (Wu et al. 2009), (Kumar and Zhou 2010), (Keersmaecker et al. 2018). It is very important and valuable to develop low-cost, easily prepared, high-quality graphene-based composites to use instead of ITO in organic lighting applications like organic light-emitting diodes (OLEDs) or solar cells and other related applications (Wu et al. 2009), (Chang et al. 2010), (De et al. 2010), (Kumar and Zhou 2010), (Keersmaecker et al. 2018).

In this section, various visible-range photonic applications of graphene-based composite materials with polymers, quantum dots, and metal oxides are detailed, presenting applications in photosensors, solar cells, lighting, and biology.

10.2.1 GRAPHENE-POLYMER COMPOSITES

The future of the electronics industry is expected to lie mostly in flexible electronics (Han et al. 2019), (Costa et al. 2019). Polymers and their derivatives have important roles in the development of photonic-related technologies thanks to their multifunctional properties like flexibility, mechanical strength, thermal or electrical conductivity, light weight, and optical absorption and emission. The presence of polymers in the graphene-based composite can improve functional properties like photo-sensing, light emission, or bioactivity. However, graphene-polymer composites have wide applications such as solar cells, photosensors, lighting, displays, biological applications, and in the automobile and defense industries, thanks to specific properties of the polymer medium such as high absorption coefficient, specific strength, modulus, structure preservation, and easy processing and fabrication (Bao et al. 2010), (Rafiq et al. 2010), (Basavaraja et al., 2011), (Hasan et al. 2011), (Kim, Kyoo Lee, and Sang Lee, 2011), (Bai et al., 2011), (Pang et al. 2011), (Sengupta et al. 2011), (Song et al. 2011), (Zhang et al. 2012), (Du and Cheng 2012), (Huang et al. 2014), (Ferrari et al. 2015), (Lee and Shiang Wang 2019) (Park et al. 2020), (Meng et al. 2020). In addition, the performance of devices based on graphene-polymer composites can be improved by optimizing the preparation and characterization processes of the composite. Issues and challenges such as graphene and polymer types, graphene dispersion in in polymer, photooxidation, or interaction between the graphene and the polymer interface can affect the device's performance (Bao et al. 2010), (Zhao et al. 2010), (Ates 2013), (Stankovich et al. 2006), (Memisoglu et al., 2015), (Tang et al. 2017), (Mistretta et al. 2019), (Biswas and Oh 2019).

Polymers' optical and electrical properties can be tuned by doping with another component (Nalwa 2000). Graphene and polymer are good alternatives to be used in the composite form with their own functional properties.

Various polymers like poly(3,4-ethylenedioxythiophene)-poly(styrene sulfonate) (PEDOT:PSS), polythiophene (P3HT), polyurethane (PU), polyethylene glycol (PEG), polysulfone (PlS), polytetrafluoroethylene (PTFE), polypropylene (PPl), polyfluorene (PF), polyvinyl carbazole (PVK), polyphenylene vinylene (PPV), polyphenylene sulfide, polyaniline, poly-pyrrole, polyacetylene, polythiophene, polymethylmethacrylate (PMMA), polystyrene (PS), or polycarbonate (PC) and their derivatives are commonly used with graphene-based materials for photosensor, solar cell, lighting, or biological applications (Lange et al. 2008), (Shao et al. 2010), (Bao et al. 2010), (Memisoglu et al. 2015), (Tang et al. 2017).

Applications for graphene-polymer composites are detailed in the following sections.

10.2.1.1 Graphene-Polymer Composites for Photosensor Applications

Graphene-polymer composite materials are used in photosensor applications to increase the optical sensing, flexibility, and durability of a membrane, or the flexibility, durability, or thermal strength of an electrode (Bao et al. 2010), (Li et al. 2013), (Yan et al. 2015), (Memisoglu et al. 2018), (Zeranska-Chudek et al. 2018), (Aydın et al. 2018). As an example, the optical sensitivity of a PVA polymer material was increased tenfold by adding graphene material into the polymer (Bao et al. 2010). In addition, the graphene in the polymer can improve the photo-resistance of the composite; this

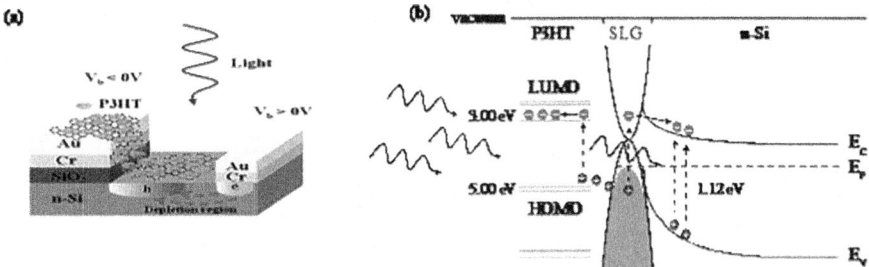

FIGURE 10.3 (a) Structure of a photosensor based on a graphene-P3HT composite and (b) the corresponding energy-band diagram. (Aydın et al., 2018)

effect increases as the graphene amount in the composite increases (Mistretta et al. 2019). The presence of graphene in the polymer can improve the photocatalytic reaction of the system (Wu et al. 2020). Another application example for graphene-polymer composites is strain sensors, which are useful in vibrating Förster resonance energy transfer (VFRET)-based acousto-optic sensor applications (Eswaraiah et al. 2011), (Kim, Yun, et al., 2011), (Al-Solamy et al., 2012), (Kuang et al. 2013), (Goncalves et al. 2014), (Memisoglu et al. 2018). Furthermore, a specific example of a graphene-polymer composite used in photosensor applications is the graphene-PEDOT:PSS composites that work as hole-dominant electrodes (Wu et al. 2014).

Another specific example of a graphene-polymer composite is the graphene-P3HT composite that is used in the electrode of a Schottky junction photosensor, as presented in Figure 10.3. The spectral photoresponsivity of the Schottky junction photosensor increased more than three orders of magnitude when the P3HT polymer was added to the graphene-based electrode (Aydin et al. 2018).

10.2.1.2 Graphene-Polymer Composites for Solar Cell Applications

Graphene-polymer composites have received immense interest for improving the performance of photovoltaic applications. They are lightweight and flexible and suitable for the solar cells that serve as a clean and sustainable energy resource for today and for the future. Various types of graphene-polymer composite materials have been studied and reported as potential material for solar cells (Tung et al. 2011), (Dai et al. 2012), (Hsu et al. 2012), (Yu et al. 2014), (Memisoglu et al. 2015). The excellent mechanical properties and exceptional electrochemical properties, electrical conductivity, thermal conductivity, and surface functionality of polymers add value to graphene-based solar cells with regard to generating electricity efficiently. The most common polymers used in solar cells are P3HT, PEDOT:PSS, PF, PPV, and PVK.

In a photovoltaic application, device performance can be improved by using graphene and a polymer, e.g., fluorene-conjugated polymer can be covalently grafted on graphene sheets, and charge transfer (photo-induced) from polymer to graphene was found (He et al. 2017). A graphene-P3HT composite increased the power conversion efficiency of a solar cell device (Gao et al. 2011).

The hole-transport property of PEDOT:PSS can be improved by graphene, and graphene-PEDOT:PSS composites provide better energy conversion efficiency than

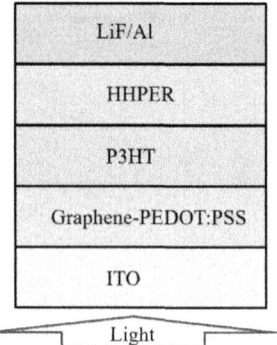

FIGURE 10.4 Structure of a solar cell based on the graphene-PEDOT:PSS composite (adapted from Memisoglu et al. 2015). ITO, transparent electrode; P3HT, donor; HHPER, acceptor; LiF/Al, cathode electrode

PEDOT:PSS alone, for solar cell applications (Tung et al. 2011), (Yu et al. 2014), (Memisoglu et al. 2015). The structure of a solar cell device with a graphene-PEDOT:PSS hole-transport layer is given in Figure 10.4. Moreover, it was reported that the graphene reduces charge transfer resistance and enhances device performance when used as a hole-transport layer of a solar cell by means of increasing the short circuit current and open circuit voltage of the device (Memisoglu et al. 2015).

10.2.1.3 Graphene-Polymer Composites for Lighting Applications

Polymers are used in lighting applications thanks to their easy processing, chemical resistance, transparency to light, and low cost (Mittal et al. 2014). For lighting applications, PF, PVK, PPV, PMMA, PS, epoxy resins, or PC and their derivatives are some commonly used polymers (Ma, Jen and Dalton 2002), (Hasan et al. 2011).

In optoelectronic applications like OLEDs, graphene composites are placed in the active layer or the transparent conductive electrode material. In addition, graphene-polymer composites are used to provide or improve the mechanical strength, flexibility, lightness of weight, low sheet resistance, high transparency to light, and durability of a layer, or the flexibility, durability, transparency, low sheet resistance, mechanical strength, or thermal strength of an electrode (Wu et al. 2009), (Chang et al. 2010), (Kumar and Zhou 2010), (Hasan et al. 2011), (Wu et al. 2014), (Dehsari et al. 2014), (Liu et al. 2015), (Xu, Li, and Tang 2016) (Keersmaecker et al. 2018), (Memisoglu et al. 2018), (Diker et al. 2019), (Diker et al. 2020).

As an application example of the graphene-polymer composites, PEDOT:PSS polymer material has investigated widely in the graphene-polymer composite structure that works as a transparent electrode instead of ITO in the OLED (Wu et al. 2009), (Chang et al. 2010), (Kumar and Zhou 2010), (Wu et al. 2014), (Dehsari et al. 2014), (Liu et al. 2015), (Xu et al., 2016) (Keersmaecker et al. 2018), (Diker et al., 2019). Another example is given in Figure 10.5 with a similar graphene-polymer composite structure that is used as the hole-transport layer in an OLED where ITO is the transparent electrode, Al is the cathode electrode, ADS231BE is the emitter, and Cs_2CO_3 is the electron transport layer (Diker et al. 2020).

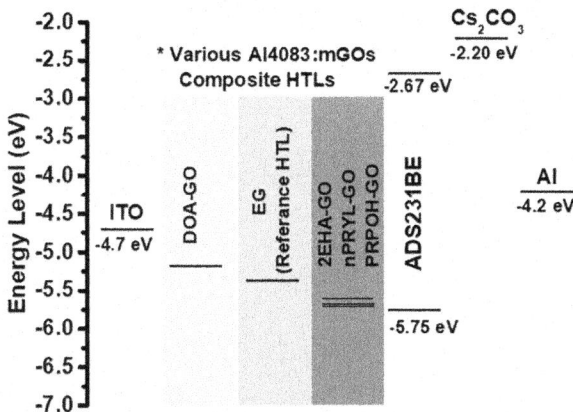

FIGURE 10.5 Energy diagram scheme of the OLED based on modified graphene oxide-polymer composite (Diker, Bozkurt, and Varlikli 2020). HTL, hole transport layer

10.2.1.4 Graphene-Polymer Composites for Biological Applications

Biocompatible polymers such as PEDOT:PSS, PU, PEG, PlS, PTFE, PPl, epoxy, or PVA are used in biological applications such as biosensors, bio-actuators, neural probes, drug-delivery devices, electrodes for electroactive polymer actuators, or membranes for point-of-care devices (Stankovich et al. 2006), (Bao et al. 2010), (Feng and Liu 2011), (Eswaraiah et al. 2011), (Guo, Han, et al. 2012), (Ates 2013), (Pandele et al. 2014), (Park et al. 2020), (Memisoglu et al. 2020). In addition, graphene material is known to be biocompatible, and graphene-polymer composites are perfectly fit to be used *in vivo* (Feng and Liu 2011), (Pandele et al. 2014). Furthermore, graphene-polymer composites find applications as polymer strip sensors for detecting biological materials *in vitro* (Al-Graiti et al. 2019).

Polymers such as epoxy derivatives are grafted onto the graphene hydrosol while the graphene is synthesized to increase the stretching and elasticity effects (Que et al. 2012).

An example of a biological application based on a graphene-polymer composite suspended membrane is given in Figure 10.6. The graphene-polymer membranes are

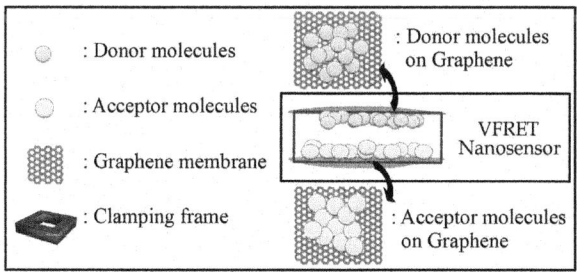

FIGURE 10.6 A biological application where donor and acceptor materials are attached to graphene-polymer composite membranes (Memisoglu et al. 2020)

suspended on frames for use as the nanoscale deflection medium by acoustic strain (Memisoglu et al. 2020).

10.2.2 GRAPHENE–QUANTUM DOT COMPOSITES

Quantum dots are semiconductor nanomaterials of diameter 1.5–8 nm whose optical and electronic properties change depending on their size or type because of quantum confinement. Quantum dot materials have a high extinction coefficient and a bright and stable fluorescence emission character (Leatherdale et al. 2002), (Zhang et al. 2015), (2017).

Core type quantum dot materials like cadmium selenide (CdSe), cadmium sulfide (CdS), and cadmium telluride (CdTe) are used for visible photosensors, whereas indium arsenide (InAs), lead sulfide (PbS), lead selenide (PbSe), indium phosphide (InP), or core/shell-type quantum dots such as cadmium selenide/zinc sulfide (CdSe/ZnS), cadmium sulfide/cadmium selenide (CdS/CdSe), or cadmium sulfide/cadmium telluride (CdS/CdTe) are some of the quantum dot materials used for photosensor, solar cell, lighting, or biological applications (Yan et al. 2012), (Gao et al. 2012), (Sevim et al. 2014), (Song et al. 2016), (Memisoglu et al. 2020), (Memisoglu 2020).

The first graphene–quantum dot composite was synthesized in 2010 (Geng et al. 2010), (Guo et al. 2010); since then, various sizes and types of core or core-shell quantum dots with graphene material have been studied in the literature (Murray et al. 1993), (Leatherdale et al. 2002), (Li et al. 2011), (Keuleyan et al. 2011), (Sun et al. 2012), (Konstantatos et al. 2012), (Gao et al. 2012), (Yan et al. 2012), (Ick Son et al. 2013), (Klekachev et al. 2013), (Rana et al., 2014), (Sevim et al. 2014), (Zheng et al. 2014), (Zhang et al. 2015), (Li et al. 2015), (Yogamalar et al. 2015), (Song et al. 2016), (Gulbahar and Memisoglu 2017a), (Zhang et al. 2017), (Tong et al. 2017), (Ma et al. 2018), (Sun et al. 2019), (Memisoglu et al., 2020). Graphene–quantum dot composites can provide flexibility, light weight, high conductivity, high optic absorption, strong light emission, low cost, an easy fabrication process, and improvement in electric and thermal properties for potential applications in optoelectronic and light-harvesting devices (Murray et al. 1993), (Li et al. 2011), (Gao et al. 2012), (Huang et al. 2013), (Gromova et al. 2015), (Li et al. 2015), (Tong et al. 2017).

Applications of graphene–quantum dot composites are detailed in the following sections.

10.2.2.1 Graphene–Quantum Dot Composites for Photosensor Applications

Graphene's low light absorbance (~2% for single-layer graphene) can be improved by adding quantum dots into the graphene composite (Sun et al. 2012). In addition, photonic applications based on graphene–quantum dot composites, like ultrafast photosensors, optical modulators, or tunable surface plasmon polariton devices, receive much attention and have experienced rapid development thanks to their high photo-responsivities (Geim and Novoselov 2007), (Wang et al. 2008), (Mak et al. 2008), (Xia et al. 2009), (Park et al., 2009), (Keuleyan et al. 2011), (Konstantatos et al. 2012), (Sun et al. 2012), (Ick Son et al. 2013), (Zhang et al. 2015), (Lu et al. 2016), (Song et al. 2016), (Zhao et al. 2020). As an example, graphene–quantum dot composites such as graphene-colloidal zinc oxide (ZnO) quantum dots have been prepared on a flexible poly(ethylene terephthalate) (PET) substrate to investigate the photoresponse characteristics of flexible photosensors (Ick Son et al. 2013).

In photosensor applications where the quantum dots are attached on the graphene membrane layer, or dispersed in graphene, the light absorbance and charge separation can be improved by increasing contact area between the two materials by the dispersion of the quantum dots on the graphene layer or in the graphene (Huang et al. 2013), (Gromova et al. 2015), (Tong et al. 2017), (Gulbahar and Memisoglu 2017), (Memisoglu et al. 2018), (Diker et al., 2020).

10.2.2.2 Graphene–Quantum Dot Composites for Solar Cell Applications

Quantum dots can be advantageous to be used in the solar cell applications because of their optic and physical properties like photo- and chemical stability, and their high light absorption coefficients. Photocatalytic activity, stability, and photovoltaic device applications of graphene–quantum dot composites have been investigated in the literature (Guo et al. 2010), (Gao et al. 2011), (Yan et al. 2012), (Li et al. 2015). In a photovoltaic application, photocatalytic activity, stability, light absorption, and color quality of quantum dots can be improved by using graphene in the composite (Gao et al. 2011), (Li et al. 2015). When CdS quantum dot particles form a composite material with graphene, the dispersed particles cover a smaller volume and larger surface than bulk CdS. Thus, the optical absorption of the CdS particles increases thanks to the increased surface area of the quantum dots (Li et al. 2011), (Yogamalar et al. 2015), (Li et al. 2015).

10.2.2.3 Graphene–Quantum Dot Composites for Lighting Applications

Quantum dot materials find applications in lighting thanks to their tunable emission wavelength, high intrinsic stability, sharp and pure bright color emission, and high photoluminescence quantum yield (Huang et al. 2013), (Ma et al. 2018), (Sun et al. 2019). Graphene–quantum dot composites are generally used in the active layer or as the transparent anode electrode of flexible optoelectronic devices (Klekachev et al. 2013), (Huang et al. 2013), (Rana et al., 2014), (Gromova et al. 2015), (Geng et al. 2015), (Tong et al. 2017). However, their utilization in lighting applications may involve down-conversion of energy as well and may present an extremely high color rendering index accompanied by tunable correlated color temperature values.

10.2.2.4 Graphene–Quantum Dot Composites for Biological Applications

In biological applications of graphene–quantum dot composite materials, the quantum dots generally chosen are biocompatible and water-soluble or dispersible, such as surface functionalized water-soluble biocompatible CdTe, CdSe/ZnS, CdSe, or InP (Rosenthal et al. 2011), (Zhou, Yang, and Zhang 2015), (Chinnathambi et al. 2017), (Tong et al. 2017), (Erogbogbo et al. 2010), (Erogbogbo et al. 2011), (Liu et al. 2013).

Graphene–quantum dot composites are used in the fluorescent imaging of biological molecules; blood plasma protein-binding studies; immunosensing; the detection of alpha fetoprotein, nucleic acid, carbohydrate antigen, chlorinated phenol, and cytochrome C; and FRET-based biological applications (Li L.L. et al. 2011), (Rosenthal et al. 2011), (Wang T. et al. 2012), (Guo Z. et al. 2012), (Yang et al. 2013), (Liu et al.

2013), (Zheng et al. 2014), (Zhang X. et al. 2014), (Liang et al. 2014), (Razali et al. 2016), (Tong et al. 2017), (Chinnathambi et al. 2017), (Gulbahar and Memisoglu 2017a), (Memisoglu et al. 2018), (Memisoglu et al. 2018), (Memisoglu et al., 2020).

10.2.3 GRAPHENE METAL OXIDE COMPOSITES

The versatile and tailor-made performances of the metal oxide materials have received attention thanks to their transparency to light, flexibility, and high quality. Titanium dioxide (TiO_2), zinc oxide (ZnO), tungsten trioxide (WO_3), magnesium oxide (MgO), stannic oxide (SnO_2), tri-manganese tetroxide (Mn_3O_4), indium oxide (In_2O_3), zinc ferrite ($ZnFe_2O_4$), calcium carbonate ($CaCO_3$), ferric oxide (Fe_2O_3), and gallium oxide (Ga_2O_3) are some metal oxide materials used with graphene in photosensors, solar cells, lighting, and biological applications (Stankovich et al., 2006), (Lu et al., 2006), (Chen et al. 2010), (Fu and Wang 2011) (Liu et al. 2012), (Khurana et al. 2013), (Zhang et al. 2014), (Fan et al. 2014), (Wang et al. 2014), (Eshaghi and Aghaei 2015), (Zubair et al. 2015), (Ma et al. 2018), (Wu et al. 2020). In addition, TiO_2, and ZnO are the metal oxide materials that are the most investigated, highly transparent, roll to roll process compatible, easily synthesizable, high quality, low cost, and have the best-defined crystal structure; they are used in graphene-based composite systems (Huang Q. et al. 2013), (Chen et al. 2010), (Liu et al. 2012), (Ma et al. 2018).

Applications of graphene metal oxide composites are described in the following sections.

10.2.3.1 Graphene–Metal Oxide Composites for Photosensor Applications

Graphene metal oxide material composites are useful for photosensors, with their mechanical strength, high electrical and thermal conductivities, and low-cost, large-scale production availability (Lu, Chang, and Fan 2006), (Fu et al. 2012), (Shu et al. 2013), (Nie et al. 2013), (Zheng et al. 2013), (Ick Son et al. 2013), (Liang et al. 2016). TiO_2 and ZnO are the most used metal oxide materials in graphene-based composite systems for photosensor applications ((Fu et al. 2012), Huang Q. et al. 2013), (Nie et al. 2013), (Zheng et al. 2013), (Ick Son et al. 2013), (Liang et al. 2016). As an example, in the graphene-TiO_2 composite, where the graphene works as the electron acceptor material, a synergetic effect occurs between the graphene and the TiO_2 nanoparticles, which improves the photocatalytic activity of the device (Huang Q. et al. 2013).

10.2.3.2 Graphene–Metal Oxide Composites for Solar Cell Applications

In solar cell applications, graphene metal oxide composites are used as photoanode, counter electrode, and photocatalytic reaction support (Stankowich et al. 2006), (Lu, Chang, and Fan 2006), (Khurana et al. 2013), (Zhang et al. 2014), (Wang et al. 2014), (Shu et al. 2013), (Beliatis et al. 2014), (Eshaghi and Aghaei 2015), (Wu et al. 2020).

Graphene-TiO_2 composites are mostly used as photoanode or counter electrode in solar cell applications (Stankovich et al., 2006), (Lu et al., 2006), (Khurana et al. 2013), (Zhang et al. 2014), (Wang et al. 2014), (Shu et al. 2013), (Beliatis et al. 2014), (Eshaghi and Aghaei 2015), (Wu et al. 2020). However, graphene can reduce interface resistance and improve the electron transport and charge-collection efficiency of a metal-oxide-based solar cell where TiO_2 is the metal oxide (Han et al. 2015). In addition, the photocatalytic reaction rate is enhanced with polymer-supported graphene–TiO_2 (Wu et al. 2020), and $ZnFe_2O_4$–graphene composite photocatalysts (Fu and Wang 2011). Furthermore, ZnO is used with graphene in the solar cell to improve photovoltaic performance (Khurana et al. 2013). Moreover, in the dye-sensitized solar cell, Mn_3O_4-graphene composites are used as the counter electrode (Wang et al. 2014).

10.2.3.3 Graphene–Metal Oxide Composites for Lighting Applications

Graphene metal oxide composite materials are used in lighting applications as an electron transport layer, or as an active layer to enhance either the charge-transport rate of the application or electrical conductivity (Lu, Chang, and Fan 2006), (Fan et al. 2014), (Zubair et al. 2015), (Liang et al. 2018).

As an example, a graphene metal oxide composite system was used to improve the emission quality of a light-emitting device, in terms of improving the current density or charge transport and resulting in enhancement of the brightness (Lu, Chang, and Fan 2006), (Zubair et al. 2015), (Liang et al. 2018). In addition, electrical conductivity improvement is reported where a metal oxide is used in a graphene-based composite medium; electrical conductivity is increased with an increasing proportion of graphene in a composite medium where metal oxide is used (Fan et al. 2014).

10.2.3.4 Graphene–Metal Oxide Composites for Biological Applications

Biocompatible metal oxides like ZnO, TiO_2, MgO, $CaCO_3$, or Fe_2O_3 are used in biological applications for drug delivery, bioimaging, antimicrobial adhesion, cellular compatibility investigations, or medical diagnostics (Lu, Chang, and Fan 2006), (Acosta-Torres et al. 2011), (Cheng et al. 2017). Graphene-based composites are of vital importance in bio-applications. For example, chemically modified graphene-based composites are used as drug carriers for cancer therapy (Ma et al. 2017). In another example, graphene composites doped with metal oxides are used for photocatalytic activity and disinfection investigations; performance is improved with metal nanoparticles like iron, nickel, and copper (Acosta-Torres et al. 2011).

10.3 FUTURE PHOTONICS-RELATED APPLICATIONS OF GRAPHENE-BASED COMPOSITES

The unique nature of graphene and its composite forms will allow improvement in photonics technology. By using additives like polymers, metal oxides, or quantum dots, there is a growing graphene composite-based photonics sector.

The future prospects of graphene-based composite technology lie mostly in the area of photonics-related biological applications, energy management systems, optical nanoscale tagging and signal generation, environmental monitoring, high-resolution imaging, point-of-care devices, smart flexible wearables, energy harvesting and storage, microfluidic particle tracking, and real-time, nanoscale-resolution air and liquid flow monitoring and sensing (Gulbahar and Memisoglu 2017a), (Memisoglu et al. 2018), (Memisoglu et al. 2018), (Dideikin and Vul 2019), (Memisoglu, Gulbahar, and Bello 2020), (Memisoglu, 2020). Graphene-based composite material development has shown great improvements and will go on extending its applications and improving modern industry with photon-counting units, space applications, photonic-integrated circuits, bolometers, and quantum experiments.

10.4   CONCLUSION

In this chapter, a brief summary on the visible-light photosensor, solar cell, lighting and biological applications of graphene composites with polymers, quantum dots, and metal oxides is presented. Graphene material filled or coated with matrix-like polymers, quantum dots, or metal oxides forms composites with improved mechanical strength, electrical or thermal conductivity, light absorption or emission, or large-scale production. The future of graphene-based composite materials in photonic-related technology is predicted to be flexible and transparent substrates, with unlimited sizes and long lifetimes with lower costs and facile preparation. Consequently, it is expected that the application of graphene-based composites will continue to be a hot topic for the near future.

ACKNOWLEDGMENTS

Gorkem Memisoglu is thankful to the European Union's Horizon 2020 research and innovation program supports under Marie Sklodowska-Curie grant agreement No 713694. Canan Varlikli thanks the project support funds of Scientific and Technological Research Council of Turkey (TUBITAK) (Project #:114M508).

REFERENCES

Acosta-Torres, L. S., Lopez-Marin, L. M., Nunez-Anita, R. E., Hernandez-Padron, G., & Castano, V. M. 2011. 'Biocompatible metal-oxide nanoparticles: nanotechnology improvement of conventional prosthetic acrylic resins'. *Journal of Nanomaterials* 2011. https://doi.org/10.1155/2011/941561.

Al-Graiti, W., Foroughi, J., Liu, Y., & Chen, J. 2019. 'Hybrid Graphene/Conducting Polymer Strip Sensors for Sensitive and Selective Electrochemical Detection of Serotonin'. *ACS Omega*. https://doi.org/10.1021/acsomega.9b03456.

Al-Solamy, F. R., Al-Ghamdi, A. A., & Mahmoud, W. E. 2012. 'Piezoresistive behavior of graphite nanoplatelets based rubber nanocomposites'. *Polymers for Advanced Technologies* 23: 478–482. https://doi.org/10.1002/pat.1902.

Ates, M. 2013. 'A review study of bio sensor systems based on conducting polymers'. *Materials Science and Engineering: C* 33 (4): 1853–1859. https://doi.org/10.1016/j.msec.2013.01.035.

Aydın, H., Kalkan, S. B., Varlikli, C., & Çelebi, C. 2018. 'P3HT–graphene bilayer electrode for Schottky junction photodetectors'. *Nanotechnology* 29 (14), 145502. https://doi.org/10.1088/1361-6528/aaaaf5.

Bai, X., Zhai, Y., & Zhang, Y. 2011. 'Green Approach to Prepare Graphene-Based Composites with High Microwave Absorption Capacity'. *The Journal of Physical Chemistry C* 115 (23): 11673–11677. https://doi.org/10.1021/jp202475m.

Balandin, A.A., Ghosh, S., Bao, W., Calizo, I., Teweldebrhan, D., Miao, F., & Lau, C.N. 2008. 'Superior Thermal Conductivity of Single Layer Graphene'. *Nano Letters* 8 (3): 902–907. https://doi.org/10.1021/nl0731872.

Bao, B. Q., Zhang, H., Yang, J. X., Wang, S., Tang, D. Y., Seeram Ramakrishna, R. J., Lim, C. T., & Loh, K. P. 2010. 'Graphene – Polymer Nanofiber Membrane for Ultrafast Photonics'. *Advanced Functional Materials* 782–791. https://doi.org/10.1002/adfm.200901658.

Basavaraja, C., Kim, W. J., & Do Kim, Y. 2011. 'Synthesis of Polyaniline-Gold/Graphene Oxide Composite and Microwave Absorption Characteristics of the Composite Films'. *Materials Letters* 65 (19–20): 3120–3123. https://doi.org/10.16194/j.cnki.31-1059/g4.2011.07.016.

Batzill, M. 2012. 'The surface science of graphene: Metal interfaces, CVD synthesis, nanoribbons, chemical modifications, and defects'. *Surface Science Reports* 67 (3-4): 83–115. https://doi.org/10.1016/j.surfrep.2011.12.001.

Beliatis, M. J., Gandhi, K. K., Rozanski, L. J., Rhodes, R., McCafferty, L., Alenezi, M. R., ... & Silva, S. R. P. 2014. 'Hybrid Graphene-metal oxide solution processed electron transport layers for large area high-performance organic photovoltaics'. *Advanced Materials* 26 (13): 2078–2083. https://doi.org/10.1002/adma.201304780.

Bianco, G. V., Losurdo, M., Giangregorio, M. M., Sacchetti, A., Prete, P., Lovergine, N., Capezzuto, P., & Bruno, G. 2015. 'Direct epitaxial CVD synthesis of tungsten disulfide on epitaxial and CVD graphene'. *RSC Advances* 5 (119): 98700–98708. https://doi:10.1039/C5RA19698A.

Biswas, M. R. U. D., & Oh, W. C. 2019. 'Comparative study on gas sensing by a Schottky diode electrode prepared with graphene–semiconductor–polymer nanocomposites'. *RSC Advances* 9 (20): 11484–11492. https://doi.org/10.1039/C9RA00007K.

Bhuyan, M. S. A., Uddin, M. N., Islam, M. M., Bipasha, F. A., & Hossain, S. S. 2016. 'Synthesis of graphene. *International Nano Letters* 6 (2): 65–83'. https://doi.org/10.1007/s40089-015-0176-1.

Bleu, Y., Bourquard, F., Tite, T., Loir, A. S., Maddi, C., Donnet, C., & Garrelie, F. 2018. 'Review of graphene growth from a solid carbon source by pulsed laser deposition (PLD)'. *Frontiers in Chemistry* 6: 572. https://doi.org/10.3389/fchem.2018.00572.

Boehm, H.P., Clauss, A., Fischer, G.O., & Hofmann, U. 1962. 'The Adsorption Behavior of Very Thin Carbon Films'. *Anorg. Allg. Chem* 316 (3–4): 119–127.

Boehm, H. P., Setton, R., & Stumpp, E. 1986. 'Nomenclature and Terminology of Graphite Intercalation Compounds.' *Carbon* 24 (2): 241–245. https://doi.org/10.1016/0008-6223(86)90126-0

Boehm, H. P., Setton, R., & Stumpp, E. 1994. 'Nomenclature and Terminology of Graphite Intercalation Compounds'. *Pure and Applied Chemistry* 66 (9): 1893–1901. https://doi.org/10.1351/pac199466091893.

Bolotin, K. I., Sikes, K. J., Jiang, Z., Klima, M., Fudenberg, G., Hone, J., Kim, P., & Stormer H. L. 2008. 'Ultrahigh Electron Mobility in Suspended Graphene'. *Solid State Communications* 146 (9–10): 351–355. https://doi.org/10.1016/j.ssc.2008.02.024.

Bommel, A. J., & Crombeen, J. E. Van. 1975. 'LEED and Auger Electron Observations of the SiC (0001) Structure'. *Surface Science* 48 (2): 463–472.

Brodie, B. C. 1859. 'On the Atomic Weight of Graphite'. *Royal Society of London* 149: 249–259. https://doi.org/10.1098/rstb.2010.0374.

Cai, W., Zhu, Y., Li, X., Piner, R. D., & Ruoff, R. S. 2009. 'Large Area Few-Layer Graphene/Graphite Films as Transparent Thin Conducting Electrodes'. *Applied Physics Letters* 95: 123115. https://doi.org/10.1063/1.3220807.

Castellanos-Gomez, A., Buscema, M., Molenaar, R., Singh, V., Janssen, L., Van Der Zant, H. S., & Steele, G. A. 2014. 'Deterministic transfer of two-dimensional materials by all-dry viscoelastic stamping'. *2D Materials* 1: 011002. https://doi.org/10.1088/2053-1583/1/1/011002.

Chakraborty, I., Bodurtha, K. J., Heeder, N. J., Godfrin, M. P., Tripathi, A., Hurt, R. H., Shukla, A., & Bose, A. 2014. 'Massive Electrical Conductivity Enhancement of Multilayer Graphene/Polystyrene Composites Using a Nonconductive Filler'. *ACS Applied Materials and Interfaces* 29: 8–11. https://doi.org/10.1021/am5044592.

Chang, H., Wang, G., Yang, A., Tao, X., Liu, X., Shen, Y., & Zheng, Z. 2010. A transparent, flexible, low-temperature, and solution-processible graphene composite electrode. *Advanced Functional Materials* 20 (17): 2893-2902. https://doi.org/10.1002/adfm.201000900.

Chen, C., Cai, W., Long, M., Zhou, B., Wu, Y., Wu, D., & Feng, Y. 2010. Synthesis of visible-light responsive graphene oxide/TiO$_2$ composites with p/n heterojunction. *ACS Nano* 4 (11): 6425–6432. https://doi.org/10.1021/nn102130m.

Chen, J., Yao, B., Li, C., & Shi, G. 2013. 'An improved Hummers method for eco-friendly synthesis of graphene oxide'. *Carbon* 64: 225–229. https://doi.org/10.1016/j.carbon.2013.07.055.

Cheng, C., Li, S., Thomas, A., Kotov, N. A., & Haag, R. 2017. Functional graphene nanomaterials-based architectures: biointeractions, fabrications, and emerging biological applications. *Chemical Reviews* 117 (3): 1826–1914. https://doi.org/10.1021/acs.chemrev.6b00520.

Chinnathambi, S., Abu, N., & Hanagata, N. 2017. Biocompatible CdSe/ZnS quantum dot micelles for long-term cell imaging without alteration to the native structure of the blood plasma protein human serum albumin. *RSC Advances* 7 (5): 2392–2402. https://doi.org/10.1039/C6RA26592H.

Colombo, L., Wallace, R. M., & Ruoff, R. S. 2013. 'Graphene growth and device integration'. *Proceedings of the IEEE* 101 (7): 1536–1556. https://doi.org/10.1109/JPROC.2013.2260114.

Coros, M., Pruneanu, S., & Stefan-van Staden, R. I. 2019. 'Recent Progress in the Graphene-Based Electrochemical Sensors and Biosensors'. *Journal of the Electrochemical Society* 167 (3): 037528. https://doi.org/10.1149/2.0282003JES.

Costa, P., Gonçalves, S., Mora, H., Carabineiro, S. A., Viana, J. C., & Lanceros-Mendez, S. 2019. 'Highly Sensitive Piezoresistive Graphene-Based Stretchable Composites for Sensing Applications'. *ACS Applied Materials & Interfaces* 11 (49): 46286–46295. https://doi.org/10.3390/ma13030528.

Dai, L., Chang, D. W., Baek, J. B., & Lu, W. 2012. 'Carbon nanomaterials for advanced energy conversion and storage'. *Small* 8 (8): 1130–1166. https://doi.org/10.1002/smll.201101594.

De, S., King, P. J., Lotya, M., O'Neill, A., Doherty, E. M., Hernandez, Y., ... & Coleman, J. N. 2010. 'Flexible, transparent, conducting films of randomly stacked graphene from surfactant-stabilized, oxide-free graphene dispersions'. *Small* 6 (3): 458–464. https://doi.org/10.1002/smll.200901162.

Dehsari, H. S., Shalamzari, E. K., Gavgani, J. N., Taromi, F. A., & Ghanbary, S. 2014. 'Efficient preparation of ultra large graphene oxide using a PEDOT: PSS/GO composite layer as hole transport layer in polymer-based optoelectronic devices'. *RSC Advances* 4 (98): 55067-55076. https://doi.org/10.1039/C4RA09474C.

Dideikin, A. T., & Vul, A. Y. 2019. 'Graphene oxide and derivatives: the place in graphene family'. *Frontiers in Physics* 6, 149. https://doi.org/10.3389/fphy.2018.00149.

Diker, H., Yesil, F., & Varlikli, C. 2019. 'Contribution of O_2 plasma treatment and amine modified GOs on film properties of conductive PEDOT: PSS: Application in indium tin oxide free solution processed blue OLED'. *Current Applied Physics* 19 (8): 910–916. https://doi.org/10.1016/j.cap.2019.04.018.

Diker, H., Bozkurt, H., & Varlikli C. 2020. 'Dispersion stability of amine modified graphene oxides and their utilization in solution processed blue OLED'. *Chemical Engineering Journal* 381 (1): 122716. https://doi.org/10.1016/j.cej.2019.122716.

Du, J. & Cheng, H. M. 2012. 'The Fabrication, Properties, and Uses of Graphene/Polymer Composites'. *Macromolecular Chemistry and Physics* 213: 1061–1077. https://doi.org/10.1002/macp.

Du, J., Pei, S., Ma, L., & Cheng, H. M. 2014. '25th anniversary article: carbon nanotube-and graphene-based transparent conductive films for optoelectronic devices'. *Advanced Materials* 26 (13): 1958–1991. https://doi.org/10.1002/adma.201304135.

Erogbogbo, F., Yong, K. T., Hu, R., Law, W. C., Ding, H., Chang, C. W., ... & Swihart, M. T. 2010. 'Biocompatible magneto fluorescent probes: luminescent silicon quantum dots coupled with superparamagnetic iron (III) oxide'. *ACS Nano* 4 (9): 5131–5138. https://doi.org/10.1021/nn101016f.

Erogbogbo, F., Yong, K. T., Roy, I., Hu, R., Law, W. C., Zhao, W., ... & Prasad, P. N. 2011. 'In vivo targeted cancer imaging, sentinel lymph node mapping and multi-channel imaging with biocompatible silicon nanocrystals'. *ACS Nano* 5 (1): 413–423. https://doi.org/10.1021/nn1018945.

Eshaghi, A., & Aghaei, A. A. 2015. 'Effect of TiO_2–graphene nanocomposite photoanode on dye-sensitized solar cell performance'. *Bulletin of Materials Science* 38 (5): 1177–1182. https://doi.org/10.1007/s12034-015-0998-5.

Eswaraiah, V., Balasubramaniam, K., & Ramaprabhu, S. 2011. 'Functionalized graphene reinforced thermoplastic nanocomposites as strain sensors in structural health monitoring'. *Journal of Materials Chemistry* 21: 12626–12628. https://doi.org/10.1039/C1JM12302E.

Fan, Y., Kang, L., Zhou, W., Jiang, W., Wang, L., & Kawasaki, A. 2015. 'Control of doping by matrix in few-layer graphene/metal oxide composites with highly enhanced electrical conductivity'. *Carbon* 81: 83–90. https://doi.org/10.1016/j.carbon.2014.09.027.

Feng, L., & Liu, Z. 2011. 'Graphene in biomedicine: opportunities and challenges'. *Nanomedicine* 6 (2): 317–324. https://doi.org/10.2217/nnm.10.158.

Ferrari, A. C., Bonaccorso, F., Falko, V., Novoselov, K. S., Roche, S., Boggild, P., Borini, S. et al. 2015. 'Science and Technology Roadmap for Graphene, Related Two-Dimensional Crystals, and Hybrid Systems'. *Nanoscale* 7 (11): 4598–4810. https://doi.org/10.1039/c4nr01600a.

Fu, Y., & Wang, X. 2011. Magnetically separable $ZnFe_2O_4$–graphene catalyst and its high photocatalytic performance under visible light irradiation. *Industrial & Engineering Chemistry Research* 50 (12): 7210–7218. https://doi.org/10.1021/ie200162a.

Fu, X. W., Liao, Z. M., Zhou, Y. B., Wu, H. C., Bie, Y. Q., Xu, J., & Yu, D. P. 2012. Graphene/ZnO nanowire/graphene vertical structure based fast-response ultraviolet photodetector'. *Applied Physics Letters* 100 (22): 223114. https://doi.org/10.1063/1.4724208.

Frisenda, R., Navarro-Moratalla, E., Gant, P., De Lara, D. P., Jarillo-Herrero, P., Gorbachev, R. V., & Castellanos-Gomez, A. 2018. 'Recent progress in the assembly of nanodevices and van der Waals heterostructures by deterministic placement of 2D materials'. *Chem. Soc. Rev.* 47: 53–68. https://doi.org/10.1039/C7CS00556C.

Galindo, B., Alcolea, S. G., Gomez, J., Navas, A., Murguialday, A. O., Fernandez, M. P., & Puelles, R. C. 2014. 'Effect of the number of layers of graphene on the electrical properties of TPU polymers'. *IOP Publishing, In IOP Conference Series: Materials Science and Engineering* 64 (1): 012008. https://doi.org/10.1088/1757-899X/64/1/012008.

Gao, Y., Yip, H. L., Chen, K. S., O'Malley, K. M., Acton, O., Sun, Y., ... & Jen, A. K. Y. 2011. 'Surface doping of conjugated polymers by graphene oxide and its application for organic electronic devices'. *Advanced Materials* 23 (16): 1903-1908. https://doi.org/10.1002/adma.201100065.

Gao, Z., Liu, N., Wu, D., Tao, W., Xu, F., & Jiang, K. 2012. 'Graphene–CdS composite, synthesis and enhanced photocatalytic activity'. *Applied Surface Science* 258 (7): 2473–2478. https://doi.org/10.1016/j.apsusc.2011.10.075.

Geim, A. K., & MacDonald, A. H. 2007. 'Graphene: Exploring carbon flatland'. *Physics Today* 60 (8): 35–41. https://doi:10.1063/1.2774096.

Geim, A. K. & Novoselov, K. S. 2007. 'The rise of graphene. *Nature Mater.* 6: 183–191.

Geng, X., Niu, L., Xing, Z., Song, R., Liu, G., Sun, M., ... & Sun, L. 2010. 'Aqueous-processable noncovalent chemically converted graphene–quantum dot composites for flexible and transparent optoelectronic films'. *Advanced Materials* 22 (5): 638–642. https://doi.org/10.1002/adma.200902871.

Ghayoor, R., Keshavarz, A., & Soltani Rad, M. N. 2019. 'Facile Preparation of TiO_2 Nanoparticles Decorated by the Graphene for Enhancement of Dye-Sensitized Solar Cell Performance'. *Journal of Materials Research* 34 (12): 2014–2023. https://doi.org/10.1557/jmr.2019.142.

Goncalves, V., Brandao, L., & Mendes, A. 2014. 'Development of porous polymer pressure sensors incorporating graphene plate qlets'. *Polymer Testing* 37: 129–137. https://doi.org/10.1016/j.polymertesting.2014.05.010.

Gromova, Y., Alaferdov, A., Rackauskas, S., Ermakov, V., Orlova, A., Maslov, V., ... & Fedorov, A. 2015. 'Photoinduced electrical response in quantum dots/graphene hybrid structure'. *Journal of Applied Physics* 118 (10): 104305. https://doi.org/10.1063/1.4929970.

Gulbahar, B., & Memisoglu, G. 2017a. 'CSSTag: Optical Nanoscale Radar and Particle Tracking for In-Body and Microfluidic Systems with Vibrating Graphene and Resonance Energy Transfer'. *IEEE Transactions on Nanobioscience* 16 (8). https://doi.org/10.1109/TNB.2017.2785226.

Gulbahar, B., & Memisoglu, G. 2017b. 'Nanoscale Optical Communications Modulator and Acousto-Optic Transduction with Vibrating Graphene and Resonance Energy Transfer'. *In IEEE International Conference on Communications.* https://doi.org/10.1109/ICC.2017.7997036.

Guo, C. X., Yang, H. B., Sheng, Z. M., Lu, Z. S., Song, Q. L., & Li, C. M. 2010. 'Layered graphene/quantum dots for photovoltaic devices'. *Angewandte Chemie International Edition* 49 (17): 3014–3017. https://doi.org/10.1002/anie.200906291.

Guo, Y. J., Han, Y. J., Shuang, S. M., & Dong, C. 2012. 'Rational synthesis of graphene-metal coordination polymer composite nanosheet as enhanced materials for electrochemical biosensing'. *Journal of Materials Chemistry* 22: 13166–13173. https://doi.org/10.1039/C2JM31997G.

Guo, Z., Hao, T., Duan, J., Wang, S., & Wei, D. 2012. 'Electrochemiluminescence immunosensor based on graphene–CdS quantum dots–agarose composite for the ultrasensitive detection of alpha fetoprotein'. *Talanta* 89: 27–32. https://doi.org/10.1016/j.talanta.2011.11.017.

Han, G. S., Song, Y. H., Jin, Y. U., Lee, J. W., Park, N. G., Kang, B. K., ... & Jung, H. S. 2015. 'Reduced graphene oxide/mesoporous TiO_2 nanocomposite based perovskite solar cells'. *ACS Applied Materials & Interfaces* 7 (42): 23521–23526. https://doi.org/10.1021/acsami.5b06171.

Han, S., Chand, A., Araby, S., Cai, R., Chen, S., Kang, H., ... & Meng, Q. 2019. 'Thermally and electrically conductive multifunctional sensor based on epoxy/graphene composite'. *Nanotechnology* 31 (7): 075702. https://doi.org/10.1088/1361-6528/ab5042.

Hasan, T., Scardaci, V., Tan, P. H., Bonaccorso, F., Rozhin, A. G., Sun, Z., & Ferrari, A. C. 2011. 'Nanotube and graphene polymer composites for photonics and optoelectronics'. *In Molecular-and Nano-Tubes* 279–354. Springer, Boston, MA. https://doi.org/10.1007/978-1-4419-9443-1.

He, J., Bao, F., Yan, S., Weng, F., Ma, R., Liu, Y., & Ding, H. 2017. 'Soluble fluorene–benzothiadiazole polymer-grafted graphene for photovoltaic devices'. *RSC Advances* 7 (57): 35950–35956. https://doi.org/10.1039/C7RA05937J.

Helgesen, M., Jorgensen, M., Nielsen, T. D., & Krebs, F. C. 2012. 'Printed polymer solar cells'. *In Printed Films*, Woodhead Publishing 550–574. https://doi.org/10.1533/9780857096210.2.550.

Hu, C., Lu, T., Chen, F., & Zhang, R. 2013. 'A brief review of graphene–metal oxide composites synthesis and applications in photocatalysis'. *Journal of the Chinese Advanced Materials Society* 1 (1): 21–39. https://doi.org/10.1080/22243682.2013.771917.

Hu, S., Huang, Q., Lin, Y., Wei, C., Zhang, H., Zhang, W., ... & Hao, A. 2014. 'Reduced graphene oxide-carbon dots composite as an enhanced material for electrochemical determination of dopamine'. *Electrochimica Acta* 130: 805–809. https://doi.org/10.1016/j.electacta.2014.02.150.

Huang, Y. Q., Zhu, R. J., Kang, N., Du, J., & Xu, H. Q. 2013. 'Photoelectrical response of hybrid graphene-PbS quantum dot devices'. *Applied Physics Letters* 103 (14): 143119. https://doi.org/10.1063/1.4824113.

Huang, Q., Tian, S., Zeng, D., Wang, X., Song, W., Li, Y., ... & Xie, C. 2013. 'Enhanced photocatalytic activity of chemically bonded TiO_2/graphene composites based on the effective interfacial charge transfer through the C–Ti bond'. *ACS Catalysis* 3 (7), 1477–1485. https://doi.org/10.1021/cs400080w.

Huang, J., Xie, G., Zhou, Y., Xie, T., & Yang, G. 2014. '*NO2 Gas Sensor Based on Polyvinylpyrrolidone/ Reduced Graphene Oxide Nanocomposite*', *In 2014 IEEE Workshop on Advanced Research and Technology in Industry Applications (WARTIA)* (2): 1059–1063. https://doi.org/10.1109/WARTIA.2014.6976459.

Hummers Jr, W. S., & Offeman, R. E. 1958. 'Preparation of graphitic oxide'. *Journal of the American Chemical Society* 80 (6): 1339–1339. https://doi.org/10.1021/ja01539a017.

Hyeong, S. K., Choi, K. H., Park, S. W., Bae, S., Park, M., Ryu, S., … & Lee, S. K. 2019. 'Review of the Direct Laser Synthesis of Functionalized Graphene and its Application in Sensor Technology'. *Applied Science and Convergence Technology* 28 (5): 148–154. https://doi.org/10.5757/ASCT.2019.28.5.148.

Hsu, C. L., Lin, C. T., Huang, J. H., Chu, C. W., Wei, K. H., & Li, L. J. 2012. 'Layer-by-Layer Graphene/TCNQ Stacked Films as Conducting Anodes for Organic Solar Cells'. *ACS Nano* 6: 5031-5039. https://doi.org/10.1021/nn301721q.

Ick Son, D., Yeon Yang, H., Whan Kim, T., & Il Park, W. 2013. 'Photoresponse mechanisms of ultraviolet photodetectors based on colloidal ZnO quantum dot-graphene nanocomposites'. *Applied Physics Letters* 102 (2): 021105. https://doi.org/10.1063/1.4776651.

Ionita, M., Vlasceanu, G. M., Watzlawek, A. A., Voicu, S. I., Burns, J. S., & Iovu, H. 2017. 'Graphene and functionalized graphene: Extraordinary prospects for nanobiocomposite materials'. *Composites Part B: Engineering* 121: 34–57. http://dx.doi.org/10.1016/j.compositesb.2017.03.031.

Keersmaecker, M. D., Lang, A. W., Osterholm, A. M., & Reynolds, J. R. 2018. 'All Polymer Solution Processed Electrochromic Devices: A Future without Indium Tin Oxide?'. *ACS Applied Materials & Interfaces* 10 (37): 31568–31579. https://doi.org/10.1021/acsami.8b10589.

Keuleyan, S., Lhuillier, E., Brajuskovic, V. & Guyot-Sionnest, P. 2011. 'Mid-infrared HgTe colloidal quantum dot photodetectors'. *Nature Photonics* 5 (8): 489–493. https://doi.org/10.1038/nphoton.2011.142.

Khurana, G., Sahoo, S., Barik, S. K., & Katiyar, R. S., 2013. 'Improved photovoltaic performance of dye sensitized solar cell using ZnO–graphene nano-composites'. *Journal of Alloys and Compounds* 578: 257–260. https://doi.org/10.1016/j.jallcom.2013.05.080.

Kim, H. M., Kyoo Lee, J., & Sang Lee, H. 2011. 'Transparent and High Gas Barrier Films Based on Poly(Vinyl Alcohol)/Graphene Oxide Composites'. *Thin Solid Films* 519 (22): 7766–7771. https://doi.org/10.1016/j.tsf.2011.06.016.

Kim, J. S., Yun, J. H., Kim, I., & Shim, S. E. 2011. 'Electrical properties of graphene/SBR nanocomposite prepared by latex heterocoagulation process at room temperature'. *Journal of Industrial and Engineering Chemistry* 17: 325-330. https://doi.org/10.1016/j.jiec.2011.02.034.

Klekachev, A. V., Kuznetsov, S. N., Asselberghs, I., Cantoro, M., Hun Mun, J., Jin Cho, B., … & De Gendt, S. 2013. 'Graphene as anode electrode for colloidal quantum dots based light emitting diodes'. *Applied Physics Letters* 103 (4): 043124. https://doi.org/10.1063/1.4816745.

Konstantatos, G., Badioli, M., Gaudreau, L., Osmond, J., Bernechea, M., De Arquer, F. P. G., … & Koppens, F. H. 2012. 'Hybrid graphene–quantum dot phototransistors with ultrahigh gain'. *Nature Nanotechnology* 7 (6): 363–368. https://doi.org/10.1038/nnano.2012.60

Kuang, J., Liu, L. Q., Gao, Y., Zhou, D., Chen, Z., Han, B.H., & Zhang, Z., 2013. 'A hierarchically structured graphene foam and its potential as a large-scale strain-gauge sensor'. *Nanoscale* 5: 12171–12177. https://doi.org/10.1039/C3NR03379A.

Kumar, A., & Zhou, C. 2010. 'The race to replace tin-doped indium oxide: which material will win?'. *ACS Nano* 4 (1): 11–14. https://doi.org/10.1021/nn901903b.

Lange, U. Roznyatovskaya, N. V., & Mirsky, V. M. 2008. 'Conducting polymers in chemical sensors and arrays'. *Analytica Chimica Acta* 614 (1): 1–26. https://doi.org/10.1016/j.aca.2008.02.068.

Leatherdale, C. A., Woo, W. K., Mikulec, F. V., & Bawendi, M. G. 2002. 'On the Absorption Cross Section of CdSe Nanocrystal Quantum Dots'. *The Journal of Physical Chemistry B* 106 (31): 7619–7622. https://doi.org/10.1021/jp025698c.

Lee, C. T., & Shiang Wang, Y. 2019. 'High-Performance Room Temperature NH_3 Gas Sensors Based on Polyaniline-Reduced Graphene Oxide Nanocomposite Sensitive Membrane'. *Journal of Alloys and Compounds* 789: 693–696. https://doi.org/10.1016/j.jallcom.2019.03.124.

Li, Q., Guo, B., Yu, J., Ran, J., Zhang, B., Yan, H., & Gong, J. R. 2011. 'Highly efficient visible-light-driven photocatalytic hydrogen production of CdS-cluster-decorated graphene nanosheets'. *Journal of the American Chemical Society* 133 (28): 10878–10884. https://doi.org/10.1021/ja2025454.

Li, L. L., Liu, K. P., Yang, G. H., Wang, C. M., Zhang, J. R., & Zhu, J. J. 2011. 'Fabrication of graphene–quantum dots composites for sensitive electrogenerated chemiluminescence immunosensing'. *Advanced Functional Materials* 21 (5): 869–878. https://doi.org/10.1002/adfm.201001550.

Li, M., Gao, C., Hu, H., & Zhao, Z. 2013. 'Electrical conductivity of thermally reduced graphene oxide/polymer composites with a segregated structure'. *Carbon* 65: 371–373. https://doi.org/10.1016/j.carbon.2013.08.016.

Li, W., Chen, B., Meng, C., Fang, W., Xiao, Y., Zhifang Hu, X. L. et al. 2014. 'Ultrafast All-Optical Graphene Modulator'. *Nano Letters* 14 (2): 955–959. https://doi.org/10.1021/nl404356t.

Li, Q., Li, X., Wageh, S., Al-Ghamdi, A. A., & Yu, J. 2015. 'CdS/graphene nanocomposite photocatalysts'. *Advanced Energy Materials* 5 (14): 1500010. https://doi.org/10.1002/aenm.201500010.

Li, H., Liu, Y., Su, A., Wang, J., & Duan, Y. 2019. 'Promising Hybrid Graphene-Silver nanowire composite electrode for flexible organic Light-emitting Diodes'. *Scientific Reports* 9 (1): 1–10. https://doi.org/10.1038/s41598-019-54424-3.

Liang, J., Yang, S., Luo, S., Liu, C., & Tang, Y. 2014. 'Ultrasensitive electrochemiluminescent detection of pentachlorophenol using a multiple amplification strategy based on a hybrid material made from quantum dots, graphene, and carbon nanotubes'. *Microchimica Acta* 181 (7–8): 759–765. https://doi.org/10.1007/s00604-013-1081-9.

Liang, F. X., Zhang, D. Y., Wang, J. Z., Kong, W. Y., Zhang, Z. X., Wang, Y., & Luo, L. B. 2016. 'Highly sensitive UVA and violet photodetector based on single-layer graphene-TiO_2 heterojunction'. *Optics Express* 24 (23): 25922–25932. https://doi.org/10.1364/OE.24.025922.

Liang, F. X., Gao, Y., Xie, C., Tong, X. W., Li, Z. J., & Luo, L. B. 2018. 'Recent advances in the fabrication of graphene–ZnO heterojunctions for optoelectronic device applications'. *Journal of Materials Chemistry C* 6 (15): 3815–3833. https://doi.org/10.1039/C8TC00172C.

Liu, X., Pan, L., Zhao, Q., Lv, T., Zhu, G., Chen, T., … & Sun, C. 2012. 'UV-assisted photocatalytic synthesis of ZnO–reduced graphene oxide composites with enhanced photocatalytic activity in reduction of Cr (VI)'. *Chemical Engineering Journal* 183: 238–243. https://doi.org/10.1016/j.cej.2011.12.068.

Liu, W., Zhang, Y., Ge, S., Song, X., Huang, J., Yan, M., & Yu, J. 2013. 'Core–shell Fe3O4–Au magnetic nanoparticles based nonenzymatic ultrasensitive electrochemiluminescence immunosensor using quantum dots functionalized graphene sheet as labels'. *Analytica Chimica Acta* 770: 132–139. https://doi.org/10.1016/j.aca.2013.01.039.

Liu, Y. F., Feng, J., Zhang, Y. F., Cui, H. F., Yin, D., Bi, Y. G., … & Sun, H. B. 2015. 'Improved efficiency of indium-tin-oxide-free organic light-emitting devices using PEDOT: PSS/graphene oxide composite anode'. *Organic Electronics* 26: 81–85. https://doi.org/10.1016/j.orgel.2015.06.031.

Lu, X., Yu, M., Huang, H., & Ruoff, R. S. 1999. 'Tailoring graphite with the goal of achieving single sheets'. *Nanotechnology* 10 (3): 269. https://doi.org/10.1088/0957-4484/10/3/308.

Lu, J. G., Chang, P., & Fan, Z. 2006. 'Quasi-one-dimensional metal oxide materials—Synthesis, properties and applications'. *Materials Science and Engineering: R: Reports* 52 (1–3): 49–91. https://doi.org/10.1016/j.mser.2006.04.002.

Lu, Y., Wu, Z., Xu, W., & Lin, S. 2016. 'ZnO quantum dot-doped graphene/h-BN/GaN-heterostructure ultraviolet photodetector with extremely high responsivity'. *Nanotechnology* 27 (48): 48LT03.

Luo, S., Wang, Y., Tong, X., & Wang, Z. 2015. 'Graphene-Based Optical Modulators'. *Nanoscale Research Letters* 10 (1): 1–11. https://doi.org/10.1186/s11671-015-0866-7.

Ma, N., Liu, J., He, W., Li, Z., Luan, Y., Song, Y., & Garg, S. 2017. 'Folic acid-grafted bovine serum albumin decorated graphene oxide: An efficient drug carrier for targeted cancer therapy'. *Journal of Colloid Interface Science* 490: 598–607. https://doi.org/10.1016/j.jcis.2016.11.097.

Ma, X., Xiang, Q., Liao, Y., Wen, T., & Zhang, H. 2018. 'Visible-light-driven CdSe quantum dots/graphene/TiO_2 nanosheets composite with excellent photocatalytic activity for E. coli disinfection and organic pollutant degradation'. *Applied Surface Science* 457: 846–855. https://doi.org/10.1016/j.apsusc.2018.07.003.

Mak, K. F., Sfeir, M. Y., Wu, Y., Lui, C. H., Misewich, J. A., & Heinz, T. F. 2008. 'Measurement of the Optical Conductivity of Graphene'. *Physical Review Letters* 101 (19): 2–5. https://doi.org/10.1103/PhysRevLett.101.196405.

Memisoglu, G. 2020. 'Vibrating FRET based Nanomechanical Sensor Preparation and Characterization for Environmental Monitoring Applications'. *IEEE Sensors Journal*. 8 pages. https://doi.org/10.1109/JSEN.2020.3028269.

Memisoglu, G., Diker, H., & Varlikli, C. 2015. 'Solution Processable Graphene Oxide Hole Transport Layers and Their Application in P3HT:HHPER Active Layer Based BHJSC'. *Turkish Journal of Physics* 39 (3). https://doi.org/10.3906/fiz-1504-5.

Memisoglu, G., Gulbahar, B. Zubia, J., & Villatoro, J. 2018. 'Theoretical Modeling of Viscosity Monitoring with Vibrating Resonance Energy Transfer for Point-of-Care and Environmental Monitoring Applications'. *Micromachines* 10 (1): 3. https://doi.org/10.3390/mi10010003.

Memisoglu, G., Gulbahar, B., & Fernandez Bello, R. 2020. 'Preparation and Characterization of Freely-Suspended Graphene Nanomechanical Membrane Devices with Quantum Dots for Point-of-Care Applications'. *Micromachines* 11 (1): 104 https://doi.org/10.3390/mi11010104.

Meng, Q., Kenelak, V., Chand, A., Kang, H., Han, S., & Liu, T. 2020. 'A highly flexible, electrically conductive, and mechanically robust graphene/epoxy composite film for its self-damage detection'. *Journal of Applied Polymer Science* 48991. https://doi.org/10.1002/app.48991.

Mittal, V., Luckachan, G. E., & Matsko, N. B. 2014. 'PE/Chlorinated-PE Blends and PE/Chlorinated- PE/Graphene Oxide Nanocomposites: Morphology, Phase Miscibility, and Interfacial Interactions'. *Macromol. Chem. Phys.* 215, 255–268. https://doi.org/10.1002/macp.201300613.

Mohan, V. B., Lau, K. T., Hui, D., & Bhattacharyya, D. 2018. 'Graphene-Based Materials and Their Composites: A Review on Production, Applications and Product Limitations'. *Composites Part B: Engineering*. https://doi.org/10.1016/j.compositesb.2018.01.013.

Morozov, S. V., Novoselov, K. S., Katsnelson, M. I., Schedin, F., Elias, D. C., Jaszczak, J. A., & Geim, A. K. 2008. 'Giant Intrinsic Carrier Mobilities in Graphene and its Bilayer'. *Physical Review Letters* 100 (1): 11–14. https://doi.org/10.1103/PhysRevLett.100.016602.

Murray, C., Norris, D. J., & Bawendi, M. G. 1993. 'Synthesis and characterization of nearly monodisperse CdE (E= sulfur, selenium, tellurium) semiconductor nanocrystallites'. *Journal of the American Chemical Society* 115 (19), 8706–8715. https://doi.org/10.1021/ja00072a025.

Nalwa, H. S. 2000. '*Handbook of Nanostructured Materials and Nanotechnology*'. New York, USA: Academic Press 5: 501–575. https://10.1016/B978-012513760-7/50070-8.

Nemade, K., Priyanka D., & Pradip T. 2018. 'Enhancement of Photovoltaic Performance of Polyaniline/Graphene Composite-Based Dye-Sensitized Solar Cells by Adding TiO_2 Nanoparticles'. *Solid State Sciences* 83: 99–106. https://doi.org/10.1016/j.solidstatesciences.2018.07.009.

Nie, B., Hu, J. G., Luo, L. B., Xie, C., Zeng, L. H., Lv, P., … & Yu, Y. Q. 2013. 'Monolayer Graphene Film on ZnO Nanorod Array for High-Performance Schottky Junction Ultraviolet Photodetectors'. *Small* 9 (17): 2872–2879. https://doi.org/10.1002/smll.201203188.

Novoselov, K. S., Geim, A. K., Morozov, S. V., Jiang, D., Zhang, Y., Dubonos, S. V., … & Firsov, A. A. 2004. 'Electric Field Effect in Atomically Thin Carbon Films'. *Science* 306: 666–670. https://doi.org/10.1126/science.1102896.

Pan, D., Gu, Y., Lan, H., Sun, Y., & Gao, H. 2015. 'Functional Graphene-Gold Nano-Composite Fabricated Electrochemical Biosensor for Direct and Rapid Detection of Bisphenol A'. *Analytica Chimica Acta* 853: 297–302. https://doi.org/10.1016/j.aca.2014.11.004.

Pandele, A. M., Ionita, M., Crica, L., Dinescu, S., Costache, M., & Iovu, H. 2014. 'Synthesis, characterization, and in vitro studies of graphene oxide/chitosan–polyvinyl alcohol films'. *Carbohydrate Polymers* 102: 813–820. https://doi.org/10.1016/j.carbpol.2013.10.085.

Pang, H., Chen, C., Zhang, Y. C., Ren, P. G., Yan, D. X., & Li, Z. M. 2011. 'The Effect of Electric Field, Annealing Temperature and Filler Loading on the Percolation Threshold of Polystyrene Containing Carbon Nanotubes and Graphene Nanosheets'. *Carbon* 49 (6): 1980–1988. https://doi.org/10.1016/j.carbon.2011.01.023.

Park, S., & Ruoff, R.S. 2009. 'Chemical Methods for the Production of Graphene.' *Nature Nanotechnology* 4 (4): 217. https://doi.org/10.1038/nnano.2009.58.

Park, J., Ahn, Y. H., & Ruiz-Vargas, C. 2009. 'Imaging of photocurrent generation and collection in single-layer graphene'. *Nano Letters* 9: 1742–1746. https://doi.org/10.1021/nl8029493.

Park, C. Y., Choi, J. G., Ghosh, T., Meng, Z. D., Zhu, L., & Oh, W. C. 2014. 'Preparation of ZnS-Graphene / TiO_2 Composites Designed for Their High Photonic Effect and Photocatalytic Activity Under Visible Light'. *In Nanotubes and Carbon Nanostructures* 22: 630–642. https://doi.org/10.1080/1536383X.2012.717556.

Park, M., Kim, S., Sohn, K. Y., Kim, S., & Jeon, M. 2020. 'Poly (3, 4-ethylene dioxythiophene): Poly (styrene sulfonate)-Functionalized Reduced Graphene Oxide Electrode for Ionic Electroactive Polymer Actuators'. *Science of Advanced Materials* 12 (3): 313–318. https://doi.org/10.1166/sam.2020.3642.

Que, R., Qi, S., Qinliang, L., Mingwang, S., Shiduan, C., Suidong, W., & Shuit-Tong, L. 2012. 'Flexible nanogenerators based on graphene oxide films for acoustic energy harvesting'. *Angew. Chem. Int. Ed.* 51: 5418–5422. https://doi.org/10.1002/anie.201200773.

Raccichini, R., Varzi, A., Passerini, S., & Scrosati, B. 2015. 'The Role of Graphene for Electrochemical Energy Storage'. *Nature Materials* 14 (3): 271–279. https://doi.org/10.1038/nmat4170.

Rafiq, R., Cai, D., Jin, J., & Song, M. 2010. 'Increasing the Toughness of Nylon 12 by the Incorporation of Functionalized Graphene'. *Carbon* 48 (15): 4309–4314. https://doi.org/10.1016/j.carbon.2010.07.043.

Rana, K., Singh, J., & Ahn, J. H. 2014. 'A graphene-based transparent electrode for use in flexible optoelectronic devices'. *Journal of Materials Chemistry C* 2 (15): 2646–2656. https://doi.org/10.1039/C3TC32264E.

Razali, W. A., Sreenivasan, V. K., Bradac, C., Connor, M., Goldys, E. M., & Zvyagin, A. V. 2016. 'Wide-field time-gated photoluminescence microscopy for fast ultrahigh-sensitivity imaging of photoluminescent probes'. *Journal of Biophotonics* 9 (8): 848–858. https://doi.org/10.1002/jbio.201600050.

Rosenthal, S. J., Chang, J. C., Kovtun, O., McBride, J. R., & Tomlinson, I. D. 2011. 'Biocompatible quantum dots for biological applications'. *Chemistry & Biology* 18 (1): 10–24. https://doi.org/10.1016/j.chembiol.2010.11.013.

Sadasivuni, K. K., Ponnamma, D., Kim, J., & Thomas, S. 2015. *'Graphene-Based Polymer Nanocomposites in Electronics'*. Switzerland: Springer. https://doi.org/10.1007/978-3-319-13875-6.

Schafhaeutl, C. 1840. 'Ueber Die Verbindungen Des Kohlenstoffes Mit Silicium, Eisen Und Andern Metallen, Welche Die Verschiedenen Arten von Gusseisen, Stahl Und Schmiedeeisen Bilden'. *Journal Für Praktische Chemie* 19 (1): 159–174. https://doi.org/10.1017/CBO9781107415324.004.

Sengupta, R., Bhattacharya, M., Bandyopadhyay, S. & Bhowmick, A. K. 2011. 'A review on the mechanical and electrical properties of graphite and modified graphite reinforced polymer composites'. *Prog. Polym. Sci.* 36: 638–670. https://doi.org/10.1016/j.progpolymsci.2010.11.003.

Sevim, S., Memisoglu, G., Varlikli, C., Dogan, L. E., Tascioglu, D., & Ozcelik, S. 2014. 'An ultraviolet photodetector with an active layer composed of solution processed polyfluorene: ZnO. 71CdO. 29S hybrid nanomaterials'. *Applied Surface Science* 305: 227–234. https://doi.org/10.1016/j.apsusc.2014.03.042.

Sha, R., Komori, K., & Badhulika, S. 2017. 'Graphene–Polyaniline Composite Based Ultra-Sensitive Electrochemical Sensor for Non-Enzymatic Detection of Urea'. *Electrochimica Acta* 233: 44–51. https://doi.org/10.1016/j.electacta.2017.03.043.

Shao, Y., Wang, J., Wu, H., Liu, J., Aksay, I. A., & Lin, Y. 2010. 'Graphene based electrochemical sensors and biosensors: a review'. *Electroanalysis: An International Journal Devoted to Fundamental and Practical Aspects of Electroanalysis* 22 (10): 1027–1036. https://doi.org/10.1002/elan.200900571.

Shu, W., Liu, Y., Peng, Z., Chen, K., Zhang, C., & Chen, W. 2013. 'Synthesis and photovoltaic performance of reduced graphene oxide–TiO_2 nanoparticles composites by solvothermal method'. *Journal of Alloys and Compounds* 563: 229–233. https://doi.org/10.1016/j.jallcom.2013.02.086.

Stankovich, S., Dikin, D.A., Dommett, G.H., Kohlhaas, K.M., Zimney, E.J., Stach, E. A., ... & Ruoff, R.S. 2006. 'Graphene-based composite materials'. *Nature* 442 (7100): 282–286. https://doi.org/10.1038/nature04969.

Song, P., Cao, Z., Cai, Y., Zhao, L., Fang, Z., & Fu, S. 2011. 'Fabrication of Exfoliated Graphene-Based Polypropylene Nanocomposites with Enhanced Mechanical and Thermal Properties'. *Polymer* 52 (18): 4001–4010. https://doi.org/10.1016/j.polymer.2011.06.045.

Song, X., Zhang, Y., Zhang, H., Yu, Y., Cao, M., Che, Y., ... & Yao, J. 2016. 'Improved photoelectronic performance of graphene, polymer and PbSe quantum dot infrared photodetectors'. *Materials Letters* 178: 52–55. https://doi.org/10.1016/j.matlet.2016.04.202.

Sun, Z., Liu, Z., Li, J., Tai, G. A., Lau, S. P., & Yan, F. 2012. 'Infrared photodetectors based on CVD-grown graphene and PbS quantum dots with ultrahigh responsivity'. *Advanced Materials* 24 (43): 5878–5883. https://doi.org/10.1002/adma.201202220.

Sun, Y., Jiang, Y., Sun, X. W., Zhang, S., & Chen, S. 2019. 'Beyond OLED: Efficient Quantum Dot Light-Emitting Diodes for Display and Lighting Application'. *The Chemical Record* 19 (8): 1729-1752. https://doi.org/10.1002/tcr.201800191.

Sung, H., Ahn, N., Jang, M. S., Lee, J. K., Yoon, H., Park, N. G., & Choi, M. 2016. 'Transparent conductive oxide-free graphene-based perovskite solar cells with over 17% efficiency'. *Advanced Energy Materials* 6 (3): 1501873. https://doi.org/10.1002/aenm.201501873.

Tang, L.C., Zhao, L., & Guan, L.Z. 2017. 'Graphene/Polymer Composite Materials: Processing, Properties and Applications'. Advanced Composite Materials: Properties and Applications', *Sciendo Migration* 7: 349–419.

Toh, S. Y., Loh, K. S., Kamarudin, S. K., & Daud, W. R. W. 2014. 'Graphene production via electrochemical reduction of graphene oxide: synthesis and characterisation'. *Chemical Engineering Journal* 251: 422–434. https://doi.org/10.1016/j.cej.2014.04.004.

Tong, L., Qiu, F., Zeng, T., Long, J., Yang, J., Wang, R., ... & Yang, Y. 2017. 'Recent progress in the preparation and application of quantum dots/graphene composite materials'. *RSC Advances* 7 (76): 47999–48018. https://doi.org/10.1039/C7RA08755A.

Tung, V. C., Kim, J., Cote, L. J., & Huang, J. 2011. 'Sticky interconnect for solution-processed tandem solar cells'. *Journal of the American Chemical Society* 133 (24): 9262–9265. https://doi.org/10.1021/ja203464n.

Wang, F., Zhang, Y., Tian, C., Girit, C., Zettl, A., Crommie, M., & Shen, Y. R. 2008. 'Gate-variable optical transitions in graphene'. *Science* 320 (5873): 206–209. https://doi.org/10.1126/science.1152793.

Wang, T., Zhang, S., Mao, C., Song, J., Niu, H., Jin, B., & Tian, Y. 2012. 'Enhanced electrochemiluminescence of CdSe quantum dots composited with graphene oxide and chitosan for sensitive sensor'. *Biosensors and Bioelectronics* 31 (1): 369–375. https://doi.org/10.1016/j.bios.2011.10.048.

Wang, X., Cheng, Z., Xu, K., Tsang, H. K., & Xu, J. B. 2013. 'High-responsivity graphene/silicon-heterostructure waveguide photodetectors'. *Nature Photonics* 7 (11): 888. https://doi.org/10.1038/nphoton.2013.241.

Wang, J. T. W., Ball, J. M., Barea, E. M., Abate, A., Alexander-Webber, J. A., Huang, J., ... & Nicholas, R. J. 2014. 'Low-temperature processed electron collection layers of graphene/TiO$_2$ nanocomposites in thin film perovskite solar cells'. *Nano Letters* 14 (2): 724–730. https://doi.org/10.1021/nl403997a.

Weiss, N. O., Zhou, H., Liao, L., Liu, Y., Jiang, S., Huang, Y., & Duan, X. 2012. 'Graphene: An Emerging Electronic Material'. *Advanced Materials* 24 (43): 5782–5825. https://doi.org/10.1002/adma.201201482.

Whitener Jr, K. E., & Sheehan, P. E. 2014. 'Graphene synthesis'. *Diamond and Related Materials* 46: 25–34. https://doi.org/10.1016/j.diamond.2014.04.006.

Wu, J., Agrawal, M., Becerril, H. C. A., Bao, Z., Liu, Z., Chen, Y. & Peumans, P. 2009. 'Organic light-emitting diodes on solution-processed graphene transparent electrodes'. *ACS Nano* 4: 43–48. https://doi.org/10.1021/nn900728d.

Wu, Z. S., Zhou, G., Yin, L. C., Ren, W., Li, F., & Cheng, H. M. 2012. 'Graphene/metal oxide composite electrode materials for energy storage'. *Nano Energy* 1 (1): 107–131. https://doi.org/10.1016/j.nanoen.2011.11.001.

Wu, X., Liu, J., Wu, D., Zhao, Y., Shi, X., Wang, J., ... & He, G. 2014. 'Highly conductive and uniform graphene oxide modified PEDOT: PSS electrodes for ITO-Free organic light emitting diodes'. *Journal of Materials Chemistry C* 2 (20): 4044–4050. https://doi.org/10.1039/C4TC00305E.

Wu, Y., Yu, C., Wu, F., Li, C., Zhou, J., Gong, Y., Rao, Y., & Chen, Y. 2017. 'A Highly Sensitive Fiber- Optic Microphone Based on Graphene Oxide Membrane'. *Journal of Lightwave Technology* 35 (19): 4344–4349.

Wu, Y., Mu, H., Cao, X., & He, X. 2020. 'Polymer-supported graphene–TiO$_2$ doped with non-metallic elements with enhanced photocatalytic reaction under visible light'. *Journal of Materials Science* 55 (4): 1577–1591. https://doi.org/10.1007/s10853-019-04100-8.

Xia, F., Mueller, T., Lin, Y.-M., Valdes-Garcia, A. & Avouris, P. 2009. 'Ultrafast graphene photodetector. *Nature Nanotechnology* 4: 839–843.

Xu, R. P., Li, Y. Q., & Tang, J. X. 2016. Recent advances in flexible organic light-emitting diodes'. *Journal of Materials Chemistry C* 4 (39): 9116–9142. https://doi.org/10.1039/C6TC03230C.

Yan, J., Ye, Q., Wang, X., Yu, B., & Zhou, F. 2012. 'CdS/CdSe quantum dot co-sensitized graphene nanocomposites via polymer brush templated synthesis for potential photo-voltaic applications'. *Nanoscale* 4 (6): 2109–2116. https://doi.org/10.1039/C2NR11893A.

Yan, D. X., Pang, H., Li, B., Vajtai, R., Xu, L., Ren, P. G., ... & Li, Z. M. 2015. 'Structured reduced graphene oxide/polymer composites for ultra-efficient electromagnetic interference shielding'. *Advanced Functional Materials* 25 (4): 559–566. https://doi.org/10.1002/adfm.201403809.

Yang, S., Liang, J., Luo, S., Liu, C., & Tang, Y. 2013. 'Supersensitive detection of chlorinated phenols by multiple amplification electrochemiluminescence sensing based on carbon quantum dots/graphene'. *Analytical Chemistry* 85 (16): 7720-7725. https://doi.org/10.1021/ac400874h.

Yogamalar, N. R., Sadhanandam, K., Bose, A. C., & Jayavel, R. 2015. 'Quantum confined CdS inclusion in graphene oxide for improved electrical conductivity and facile charge transfer in hetero-junction solar cell'. *RSC Advances* 5 (22): 16856–16869. https://doi.org/10.1039/C4RA13061H.

Yu, J. C., Jang, J. I., Lee, B. R., Lee, G. W., Han, J. T., & Song, M. H. 2014. 'Highly efficient polymer-based optoelectronic devices using PEDOT: PSS and a GO composite layer as a hole transport layer'. *ACS Applied Materials & Interfaces* 6 (3), 2067–2073. https://doi.org/10.1021/am4051487.

Zeng, W., Tao, X. M., Lin, S., Lee, C., Shi, D., Lam, K. H., ... & Zhao, Y. 2018. 'Defect-engineered reduced graphene oxide sheets with high electric conductivity and controlled thermal conductivity for soft and flexible wearable thermoelectric generators'. *Nano Energy* 54: 163–174. https://doi.org/10.1016/j.nanoen.2018.10.015.

Zeranska-Chudek, K., Lapinska, A., Wroblewska, A., Judek, J., Duzynska, A., Pawlowski, M., ... & Zdrojek, M. 2018. 'Study of the absorption coefficient of graphene-polymer composites'. *Scientific Reports* 8 (1): 1–8.

Zhang, D., Ryu, K., Liu, X., Polikarpov, E., Ly, J., Tompson, M. E. & Zhou, C. 2006. 'Transparent, conductive, and flexible carbon nanotube films and their application in organic light-emitting diodes'. *Nano Letters* 6: 1880–1886. https://doi.org/10.1021/nl0608543.

Zhang, H. B., Wen G. Z., Qing Y., Zhi G. J., & Zhong Z. Y. 2012. 'The Effect of Surface Chemistry of Graphene on Rheological and Electrical Properties of Polymethylmethacrylate Composites'. *Carbon* 50 (14): 5117–5125. https://doi.org/10.1016/j.carbon.2012.06.052.

Zhang, Q., Liu, Y., Duan, Y., Fu, N., Liu, Q., Fang, Y., Sun, Q., & Lin, Y., 2014. 'Mn₃O₄/graphene composite as counter electrode in dye-sensitized solar cells'. *RSC Advances* 4 (29): 15091–15097. https://doi.org/10.1039/C4RA00347K.

Zhang, X., Xu, F., Zhao, B., Ji, X., Yao, Y., Wu, D., ... & Jiang, K. 2014. 'Synthesis of CdS quantum dots decorated graphene nanosheets and non-enzymatic photoelectrochemical detection of glucose'. *Electrochimica Acta* 133: 615–622. https://doi.org/10.1016/j.electacta.2014.04.089.

Zhang, B., Wang, Y., Yang, C., Hu, S., Gao, Y., Zhang, Y., Wang, Y., Demir, H. V., Liu, L., & Yong, K. T. 2015. 'The composition effect on the optical properties of aqueous synthesized Cu-In-S and Zn-Cu-In-S quantum dot nanocrystals'. *Phys. Chem. Chem. Phys.* 17: 25133–25141. https://doi.org/10.1039/C5CP03312H.

Zhang, Y., Zhang, C., Wang, W., Du, X., Dong, W., Han, B., & Chen, Q. 2016. 'One-Step Synthesis of Polyvinylpyrrolidone-Reduced Graphene Oxide-Pd Nanoparticles for Electrochemical Sensing'. *Journal of Materials Science* 51 (13): 6497–6508. https://doi.org/10.1007/s10853-016-9949-9.

Zhang, B., Yang, C., Gao, Y., Wang, Y., Bu, C., Hu, S., Liu, L., Demir, H. V., Qu, J., & Yong, K.T. 2017. 'Engineering quantum dots with different emission wavelengths and specific fluorescence lifetimes for spectrally and temporally multiplexed imaging of cells'. *Nano-Theranostics* 1: 131–140. https://doi.org/10.7150/ntno.18989.

Zhang, X., Wang, N., Liu, R., Wang, X., Zhu, Y., & Zhang, J. 2018. 'SERS and the photo-catalytic performance of Ag/TiO$_2$/graphene composites'. *Optical Materials Express* 8 (4): 704–717. https://doi.org/10.1364/OME.8.000704.

Zhao, X., Zhang, Q., Chen, D., & Lu, P. 2010. 'Enhanced mechanical properties of graphene-based poly (vinyl alcohol) composites'. *Macromolecules* 43 (5): 2357–2363. https://doi.org/10.1021/ma902862u.

Zhao, X., Zhou, S., Jiang, L. P., Hou, W., Shen, Q., & Zhu, J. J. 2012. 'Graphene–CdS Nanocomposites: Facile One-Step Synthesis and Enhanced Photoelectrochemical Cytosensing'. *Chemistry–A European Journal* 18 (16): 4974–4981. https://doi.org/10.1002/chem.201102379.

Zhao, W. R., Kang, T. F., Xu, Y. H., Zhang, X., Liu, H., Ming, A. J., ... & Wei, F. 2020. 'Electrochemiluminescence solid-state imprinted sensor based on graphene/CdTe@ ZnS quantum dots as luminescent probes for low-cost ultrasensing of diethylstilbestrol'. *Sensors and Actuators B: Chemical* 306: 127563. https://doi.org/10.1016/j.snb.2019.127563.

Zheng, K., Meng, F., Jiang, L., Yan, Q., Hng, H. H., & Chen, X. 2013. 'Visible Photoresponse of Single-Layer Graphene Decorated with TiO$_2$ Nanoparticles'. *Small* 9 (12): 2076–2080. https://doi.org/10.1002/smll.201202885.

Zheng, X. T., Ananthanarayanan, A., Luo, K. Q., & Chen, P. 2015. 'Glowing graphene quantum dots and carbon dots: properties, syntheses, and biological applications'. *Small* 11 (14): 1620–1636. https://doi.org/10.1002/smll.201402648.

Zhou, S. Y., Gweon, G. H., Graf, J., Fedorov, A. V., Spataru, C. D., Diehl, R. D., Kopelevich, Y., Lee, D. H., Louie, S. G., & Lanzara, A. 2006. 'First Direct Observation of Dirac Fermions in Graphite'. *Nature Physics* 2 (9): 595–599. https://doi.org/10.1038/nphys393.

Zhou, J., Yang, Y., & Zhang, C. Y. 2015. 'Toward biocompatible semiconductor quantum dots: from biosynthesis and bioconjugation to biomedical application'. *Chemical Reviews* 115 (21): 11669–11717. https://doi.org/10.1021/acs.chemrev.5b00049.

Zhu, Y., Murali, S., Cai, W., Li, X., Suk, J. W., Potts, J. R., & Ruoff, R. S. 2010. 'Graphene and Graphene Oxide: Synthesis, Properties, and Applications'. *Advanced Materials* 3906–3924. https://doi.org/10.1002/adma.201001068.

Zubair, M., Mustafa, M., Ali, A., Doh, Y. H., & Choi, K. H. 2015. 'Improvement of solution based conjugate polymer organic light emitting diode by ZnO–graphene quantum dots'. *Journal of Materials Science: Materials in Electronics* 26 (5): 3344–3351. https://doi.org/10.1007/s10854-015-2837-2.

11 Low Power Ge-Si$_{0.7}$Ge$_{0.3}$ nJLTFET and pJLTFET Design and Characterization in Sub-20 nm Technology Node

Suman Lata Tripathi, Sobhit Saxena,
Yogesh Kumar Verma and Manoj Singh Adhikari

CONTENTS

11.1 INTRODUCTION

Multi-gate MOSFET structures have emerged as a potential candidates for low power SoC or integrated circuits [1–5] to overcome the thermal limits imposed due to the scaling of transistors. A vertical p-channel and n-channel hetero-junction tunnel field effect transistor (TFET) has been proposed with the steep subthreshold slope and very small OFF-state current that are important design considerations while device scaling in sub 45 nm technology (Figure 11.1) [6, 7]. Narrow bandgap materials such as SiGe, Ge, and InAs reduce tunneling distance and enhance the band to band tunneling (BTBT) effect in TFET with a better ON- and OFF-state current ratio [8]. The band-to-band tunneling in TFET follows the charge-plasma concept that depends upon the work function difference of metallic gate contact to source/drain region. Also, the silicon body thickness must be less than the Debye length [4].

Debye length can be expressed mathematically by

$$L_D = \sqrt{\frac{V_{Th} \in_{Si}}{qN}} \tag{11.1}$$

199

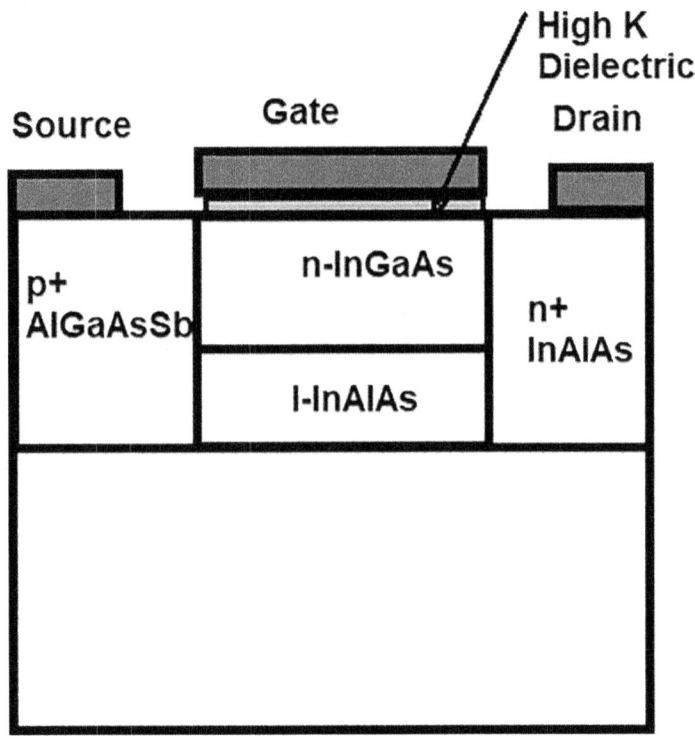

FIGURE 11.1   Hetero-junction TFET based on III-V compounds

where N denotes carrier concentration, ϵ_{Si} the dielectric constant of silicon, and V_{Th} is constant with temperature and termed "thermal voltage."

A thin SiGe pocket region is incorporated to reduce tunneling distance, which increases the BTBT of electrons, resulting in steep subthreshold characteristics [9, 10]. Longer channel lengths are preferred in TFET [11–17] for better ON-state current. Junction-less tunnel FET with a silicon-on-insulator (SOI) box region shows an increase in ON and OFF to improved I_{ON}/I_{OFF} ratio [18–21]. A small pocket $Ge\text{-}Si_{0.7}Ge_{0.3}$ n-channel junctionless TFET (nJLTFET) is proposed in which a thin $Si_{0.7}Ge_{0.3}$ region is incorporated with Ge as the source, drain, and channel material replacing Si. Ge is used in place of Si to exploit the advantage of a smaller bandgap (0.7 eV in comparison to 1.1 eV), which plays an important role in tunneling phenomena. The use of a thin pocket region toward the source side improves transfer characteristics by increasing the ON current. To increase channel mobility, the channel region is kept at a low doping profile in comparison to source-drain region doping. The doping of a thin pocket region is also kept high for nJLTFET. The proposed pocket $Ge\text{-}Si_{0.7}Ge_{0.3}$ nJLTFET has been optimized in terms of gate contact and oxide region materials to obtain better ON/OFF performance along with sufficient ON current. Similarly, ultra-small $Ge\text{-}Si_{0.7}Ge_{0.3}$ p-channel junctionless TFET (pJLTFET) is also designed and characterized for different gate contact and oxide material as well as doping profile. In $Ge\text{-}Si_{0.7}Ge_{0.3}$ pJLTFET, the thin pocket region is kept at a low doping profile in comparison to Ge-Si$_{0.7}$Ge$_{0.3}$ nJLTFET, to match their ON- and OFF-state characteristics. The transfer characteristics of nJLTFET and pJLTFET were compared and both found suitable for implementation in complementary MOS (CMOS) technology.

Trransistors designed with nJLTFET and pJLTFET can be further implemented in the future with low-power design DRAM [22] cell and SRAM [23, 24] cells. This chapter deals with the design and analysis of Ge-Si$_{0.7}$Ge$_{0.3}$ nJLTFETs and pJLTFETs for low power, low voltage applications. The analysis is performed for DC and AC parameters including the effect of temperature on device ON/OFF-state performance.

11.2 DEVICE STRUCTURES AND DIMENSIONS

The proposed Ge-Si$_{0.7}$Ge$_{0.3}$ nJLTFET has been implemented on the technology computer-aided design (TCAD) device simulator. Ge is used as a source, drain, and channel material replacing Si with the advantage of a lesser energy bandgap and higher tunneling rate. The three regions, source, drain, and channel, are doped with similar impurity atoms to make an n-type region, keeping source/drain doping at 1×10^{20} cm^{-3} and channel doping at 1×10^{16} cm^{-3}. A local BTBT Kane's model is followed along with the Shockley-Read-Hall (SRH) model for 2D/3D device simulation. The Lombardi model is used for mobility analysis in the presented simulation work. Kane's tunneling model is used in the simulator to calculate the carrier generation by the band-to-band tunneling phenomenon.

Kane's [25] model can be expressed as tunneling generation rate as follows:

$$G(E) = A \frac{|E|^2}{\sqrt{E_g}} e^{-B \frac{E_g^{3/2}}{|E|}} \qquad (11.2)$$

where E is the magnitude of the electric field and E$_g$ the energy bandgap. A and B are constant parameters that depend on device structures and materials. The values of these two parameters are adjusted to match the experimental I$_d$-V$_{gs}$ characteristics.

Figure 11.2 describes the 2D and 3D Ge-Si$_{0.7}$Ge$_{0.3}$ nJLTFET structure. A high workfunction Pt is used as gate contact material with a high value of work function 5.7 eV. High dielectric constant material HfO$_2$ (25) is used as an oxide region under the gate, replacing SiO$_2$ (dielectric constant = 3.9).

Figure 11.3 shows the energy-band diagram of Ge-Si$_{0.7}$Ge$_{0.3}$ nJLTFET obtained on visual TCAD simulation. The minimum of conduction and maximum of valance band are plotted against the changes in bias conditions. The changes observed in the conduction band are similar to those observed in the valance band, showing ideal behavior of the channel region in the proposed Ge-Si$_{0.7}$Ge$_{0.3}$ nJLTFET. The channel potential rises with an increase in gate bias voltage for a constant drain to source voltage, as shown in Figure 11.4.

Figure 11.5 indicates the electric field, with additional peaks in the source-pocket region. Figure 11.6 describes the variation of hole density in the pocket region, depending on the gate and drain bias conditions. High hole density in the OFF-state condition offers less effect of the parasitics. The proposed Ge-Si$_{0.7}$Ge$_{0.3}$ nJLTFET achieves better switching speed with a low value of hole density in the ON-state. Therefore, the proposed Ge-Si$_{0.7}$Ge$_{0.3}$ nJLTFET supports parasitic BJT triggering between the source-channel region in the ON-state.

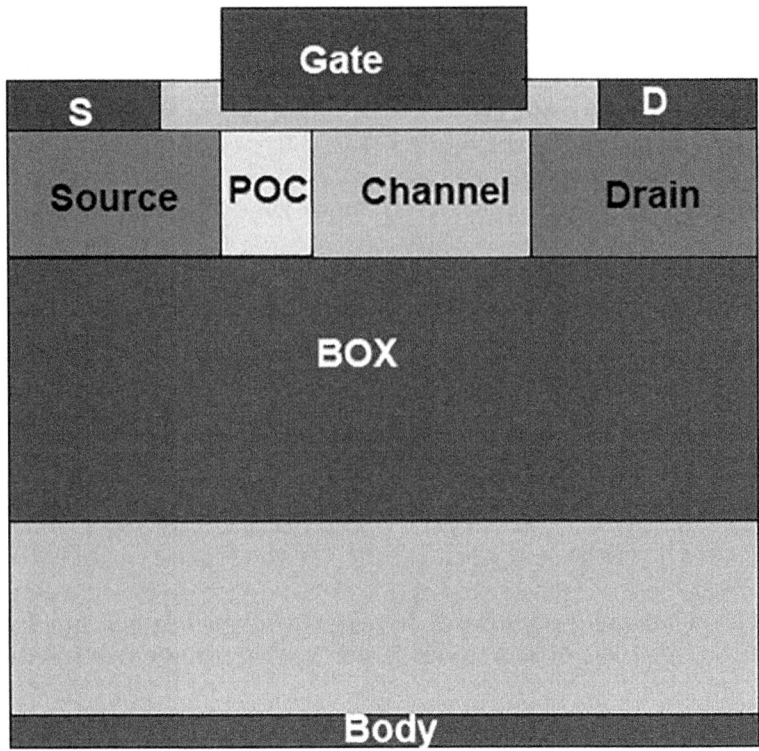

FIGURE 11.2   2D view of pocket $Si_{0.7}Ge_{0.3}$ JLTFET with Ge wafer

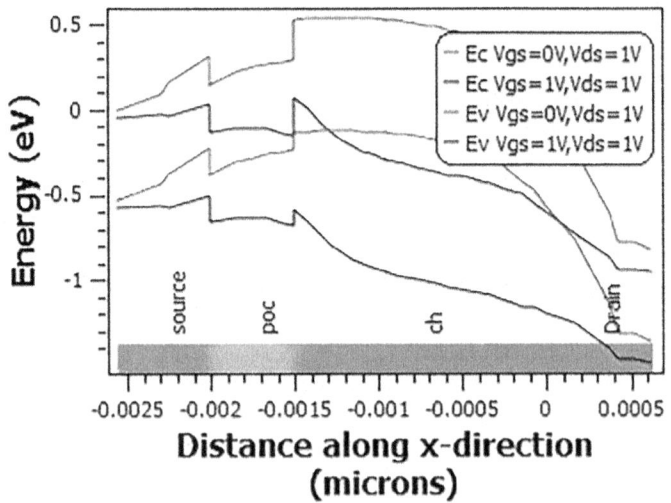

FIGURE 11.3   Energy-band diagram of Ge-$Si_{0.7}Ge_{0.3}$ nJLTFET along its x-axis

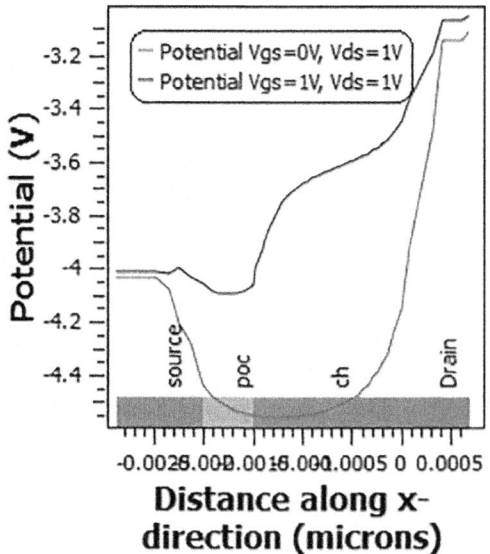

FIGURE 11.4   Channel potential of Ge-Si$_{0.7}$Ge$_{0.3}$ nJLTFET along its x-axis

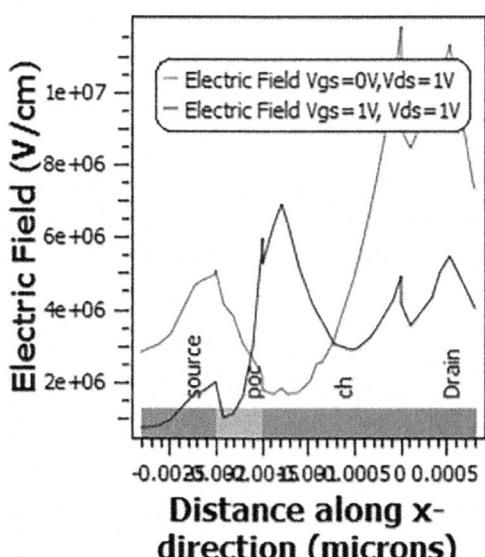

FIGURE 11.5   Electric field of Ge-Si$_{0.7}$Ge$_{0.3}$ nJLTFET along its x-axis

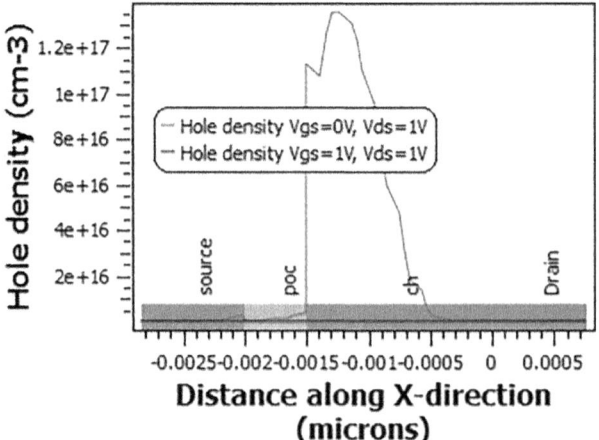

FIGURE 11.6   Hole density of Ge-Si$_{0.7}$Ge$_{0.3}$ nJLTFET along its x-axis

11.3   SUBTHRESHOLD PERFORMANCE PARAMETERS

A decrease in transistor size leads to various adverse effects like degradation in sub-threshold performance parameters.

The drain current should ideally be zero in the subthreshold region of operation when the applied gate voltage is less than the threshold voltage of the transistor. Practically, however, there is some leakage of current because of the transistor para-sitic effect that deteriorates the device subthreshold performance. The important sub-threshold performance parameters are:

i. *Threshold voltage* (V_{th}): the required gate to source voltage for channel enhancement in the transistor.

ii. *Subthreshold slope* (SS): a measure of the inverse of the slope of drain current to the gate voltage curve. The steepness of the subthreshold curve decides the ON/ OFF current ratio, which is an important measure for low-power applications.

$$SS = \frac{dVgs}{d\left(logI_D\right)}\left(mV / decade\right) \tag{11.3}$$

iii. *Drain-induced barrier lowering* (DIBL): defined as the ratio of the difference in gate voltage to the difference in drain voltage concerning the gate. This shows the effect of drain-induced electric field over the channel electric field. For proper operation of a device, this effect should be as low as possible.

$$DIBL = \frac{\Delta V_{th}}{\Delta V_{DS}} = \frac{V_{th1} - V_{th2}}{V_{DS1} - V_{DS2}} \tag{11.4}$$

iv. *OFF-state leakage current:* a measure of the drain current when the transistor is in OFF-condition ($V_{GS} = 0V$ when $V_{DS} = V_{DD}$). For low-power applications, the leakage current should be as low as possible.

These parameters are technology dependent. They must be kept under the limit for proper transistor operation and efficient use of the IC area. The transistor performance enhancement is mainly carried out by:

i. *Gate engineering:* in gate engineering, the gate contact material is varied depending on its work function. Oxide material can be used between gate contact and channel, according to their dielectric constant as per the requirements of a particular device.
ii. *Channel engineering:* a choice of suitable channel material is explored with different doping levels, as per the individual device's behavior.

11.4   RESULTS AND DISCUSSION

The 2D structure of novel Ge-Si$_x$Ge$_{1-x}$ nJLTFET is designed on a visual TCAD 2D device simulator. A good value of I$_{ON}$/I$_{OFF}$ is observed in Figure 11.7 for Pt/HfO. The use of the high-K dielectric material HfO$_2$ reduces the hot-electron effect, which reduces overall leakage current. A low value of leakage current and the steep subthreshold characteristics of proposed Ge-Ge-Si$_{0.7}$Ge$_{0.3}$ nJLTFET overcome the scaling limits with smaller dimensions. Figure 11.7 also compares the drain current of nJLTTFET with other TFET structures for different drain bias conditions. The proposed Ge-Si$_{0.7}$Ge$_{0.3}$ nJLTFET has a low value of leakage current and static power dissipation in comparison to its Si-Si$_{0.7}$Ge$_{0.3}$ hetero-junction structure, with a

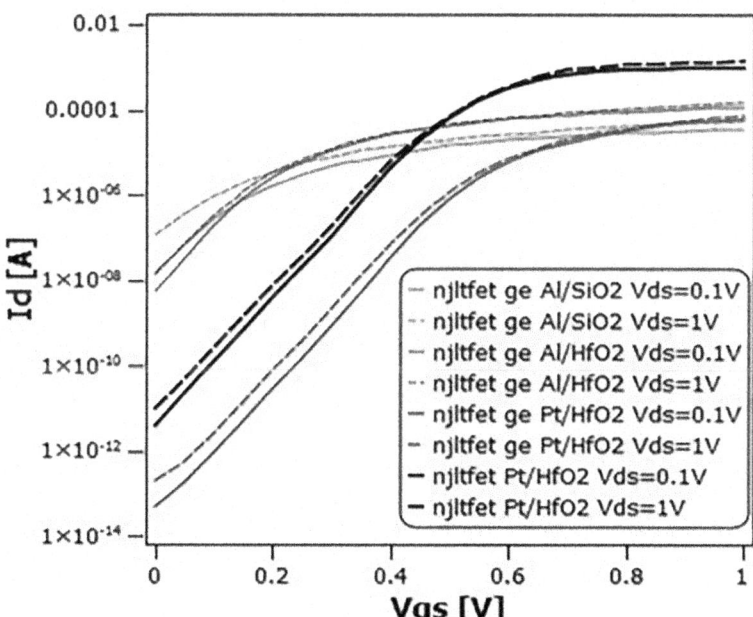

FIGURE 11.7   I$_d$ versus V$_{gs}$ of gate-engineered pocket Ge-Si$_{0.7}$Ge$_{0.3}$ nJLTFET in comparison to Si-Si$_{0.7}$Ge$_{0.3}$ nJLTFET

reduction in ON-state current but maintaining overall current gain. Kane's tunneling model is used in the simulator to calculate carrier generation by the band-to-band tunneling phenomenon. The tunneling phenomenon depends on the bandgap of the material used in the channel region. First, simulation is performed for a conventional junction-based tunnel FET then its performance compared with that of the pocket tunnel FET.

The low value of channel doping (1×10^{16} cm^{-3}) shows good ON- and OFF-state characteristics in comparison to high-channel doping (1×10^{20} cm^{-3}), as depicted in Figure 11.8. The proposed nJLTFET shows ideal I_d versus V_{ds} characteristics with variations in V_{gs} ranging from 0 to 1 V, as depicted in Figure 11.9. Figure 11.10 presents drain current variations with gate voltage and shows that the value of drain current decreases with increasing gate voltage.

The transfer characteristics of nJLTFET and pJLTFET are compared in Figure 11.11, indicating nearly ideal matching characteristics. This shows their compatibility toward CMOS-based circuit implementations.

Table 11.1 shows the advantages of the proposed Ge-Si$_x$Ge$_{1-x}$ nJLTFET compared with similar species in the available literature in terms of ON- and OFF-state performance: it has an I_{on}/I_{off} ratio of up to ~7 × 10^7, which is good compared with TDJLT [18]. Table 11.2 shows the subthreshold performance of Ge-Si$_x$Ge$_{1-x}$ nJLTFET in comparison to other TFET structures. The subthreshold slope and DIBL of the

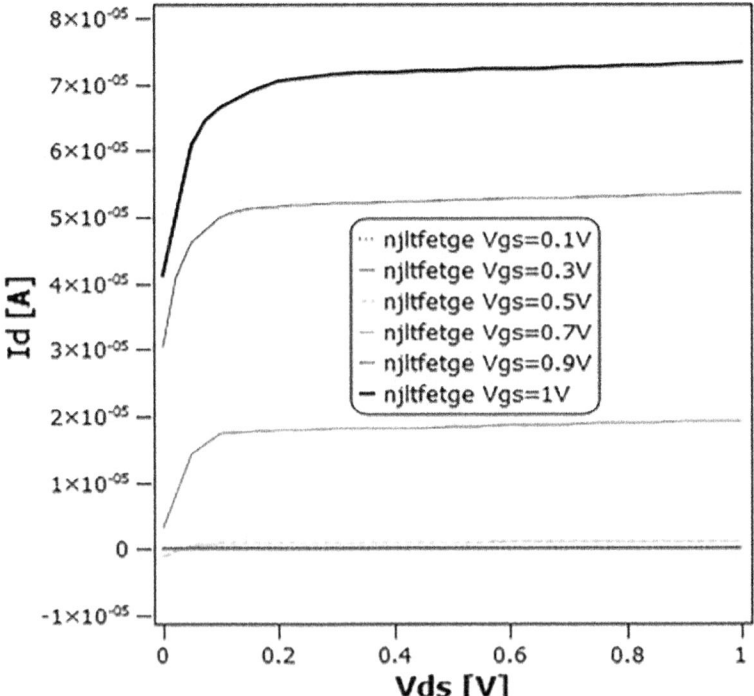

FIGURE 11.8 I_d versus V_{gs} of pocket Ge-Si$_{0.7}$Ge$_{0.3}$ nJLTFET with different pocket region doping

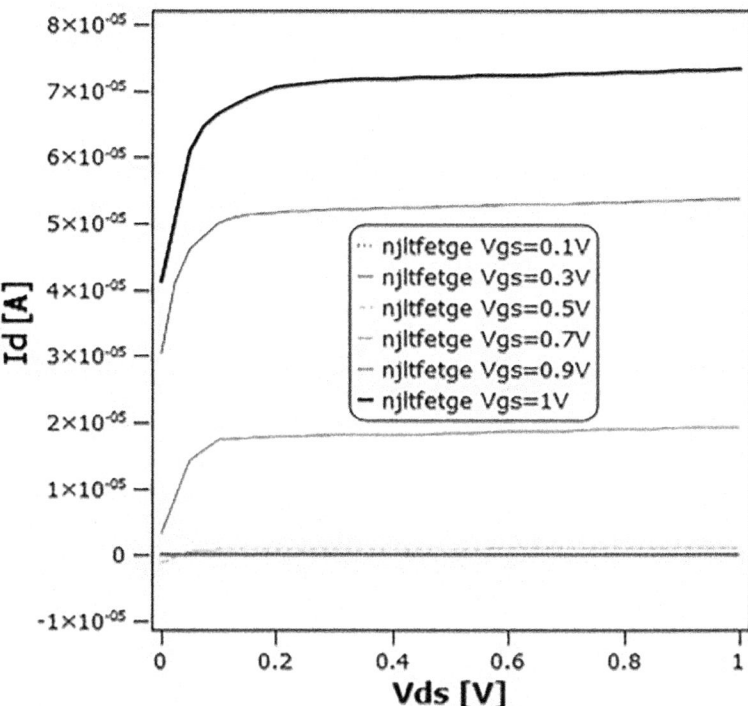

FIGURE 11.9 I$_d$ vs V$_{ds}$ of pocket Ge-Si$_{0.7}$Ge$_{0.3}$ nJLTFET with variation in gate voltage (V$_{gs}$)

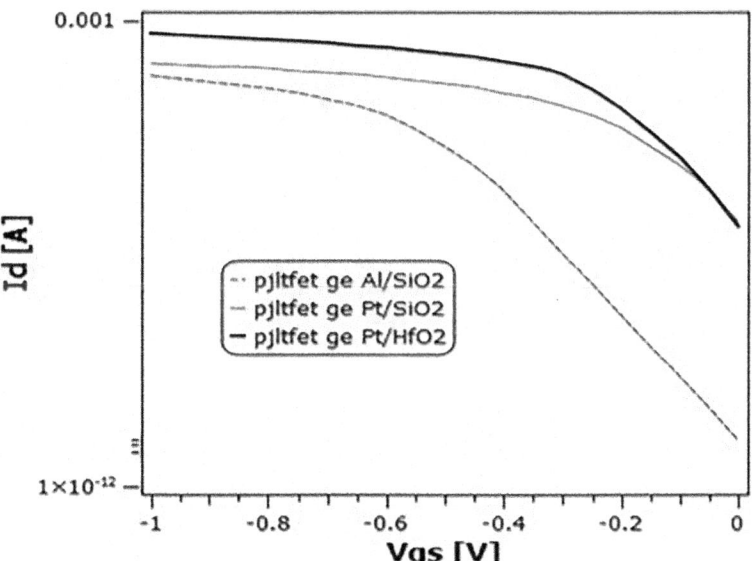

FIGURE 11.10 I$_d$ vs V$_{gs}$ of gate-engineered pocket Ge-Si$_{0.7}$Ge$_{0.3}$ pJLTFET

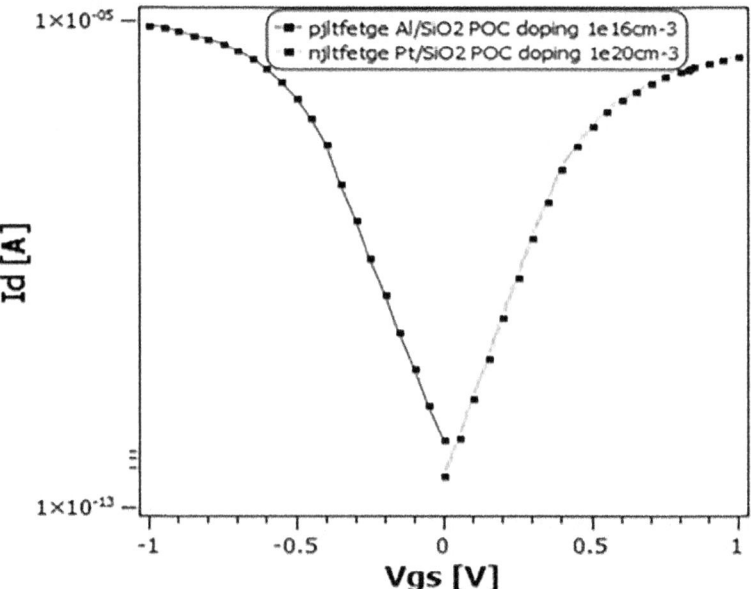

FIGURE 11.11 Comparison of I_d vs V_{gs} of ultra-small pocket Ge-Si$_{0.7}$Ge$_{0.3}$ nJLTFET and pJLTFET

TABLE 11.1
ON- and OFF-state performance comparison of Ge-Si$_{0.7}$Ge$_{0.3}$ nJLTFET with available structures

Device type (Lg = 2 nm)	I_{OFF} (A)	I_{ON} (A)	I_{ON}/I_{OFF}
Ge-Si$_{0.7}$Ge$_{0.3}$ nJLTFET with Al/SiO$_2$	1.73×10^{-8}	4.48×10^{-6}	258.4
Ge-Si$_{0.7}$Ge$_{0.3}$ nJLTFET with Al/HfO$_2$	5.20×10^{-9}	1.16×10^{-5}	2231
Ge-Si$_{0.7}$Ge$_{0.3}$ nJLTFET with Al/HfO$_2$	5.20×10^{-9}	1.16×10^{-5}	2231
Ge-Si$_{0.7}$Ge$_{0.3}$ nJLTFET with Pt/HfO$_2$	5.48×10^{-14}	3.86×10^{-6}	7×10^7

---

proposed Ge-Si$_x$Ge$_{1-x}$ nJLTFET are 73.4 mV/decade and 18.5 mV/V, respectively. This shows the ideal nature of the subthreshold performance for Ge-Si$_{0.7}$Ge$_{0.3}$ nJLT-FET in comparison to Si-Si$_{0.7}$Ge$_{0.3}$ nJLTFET.

11.4.1 TEMPERATURE ANALYSIS

Temperature analysis plays an important role in reliable device operation under harsh temperature conditions. Device performance normally deteriorates with very high and very low temperatures, so design and analysis based on varying temperatures are crucial to check device sensitivity with different temperatures (from 250–400 K). Figure 11.12 shows that the value of the OFF-state current varies with temperature

TABLE 11.2

Subthreshold performance comparison of gate-engineered Ge-Si$_{0.7}$Ge$_{0.3}$ nJLTFET

Device type	SS (mV/decade)	DIBL (mV/V) (Id =1 × 10-9 A)
Ge-Si$_{0.7}$Ge$_{0.3}$ nJLTFET with Al/SiO$_2$	15.49	78.1
Ge-Si$_{0.7}$Ge$_{0.3}$ nJLTFET with Al/HfO$_2$	35.47	19.5
Ge-Si$_{0.7}$Ge$_{0.3}$ nJLTFET with Pt/HfO$_2$	73.40	18.5
Si-Si$_{0.7}$Ge$_{0.3}$ nJLTFET with Al/HfO$_2$	95.77	19.1
Si- Si$_{0.7}$Ge$_{0.3}$ nJLTFET with Pt/HfO$_2$	124.25	26.55
Ge-Si$_{0.7}$Ge$_{0.3}$nJLTFET with Al/SiO2	15.49	78.1
Ge-Si$_{0.7}$Ge$_{0.3}$nJLTFET with Al/HfO2	35.47	19.5

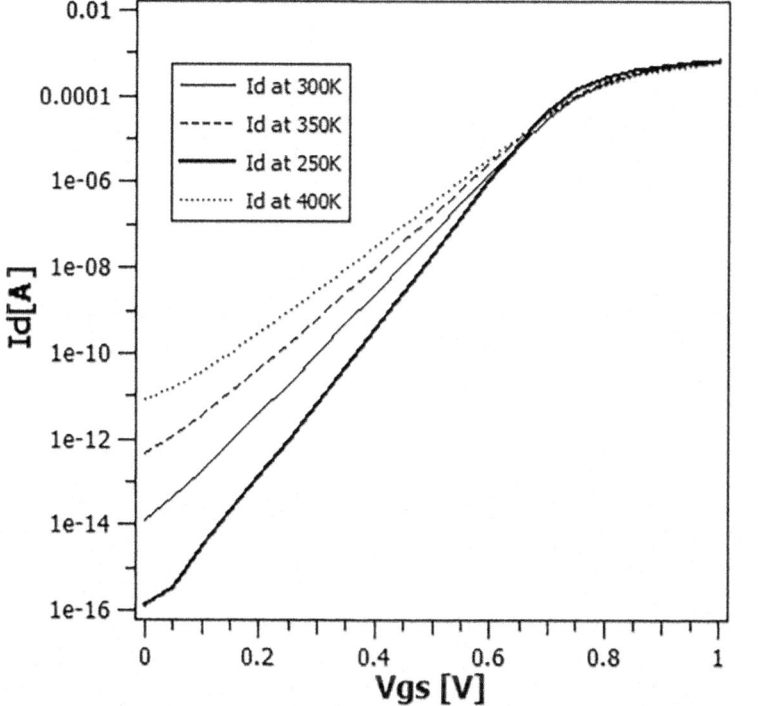

FIGURE 11.12 Variation of drain current with gate voltage with varying temperature

but the ON-state current is less affected by temperature variations. Figure 11.13 shows drain current versus drain voltage variations. As the temperature increases, the ON-state current increases but the nature of the curve remains ideal. It shows that the gate has good control over the channel even at varying temperatures. Also, it shows a high value of output resistance, with a slight tilt in the slope toward the drain voltage axis. Temperature analysis shows that there is a small variation in OFF-state current (<10%) with a constant value of ON-state current for nJLTFET.

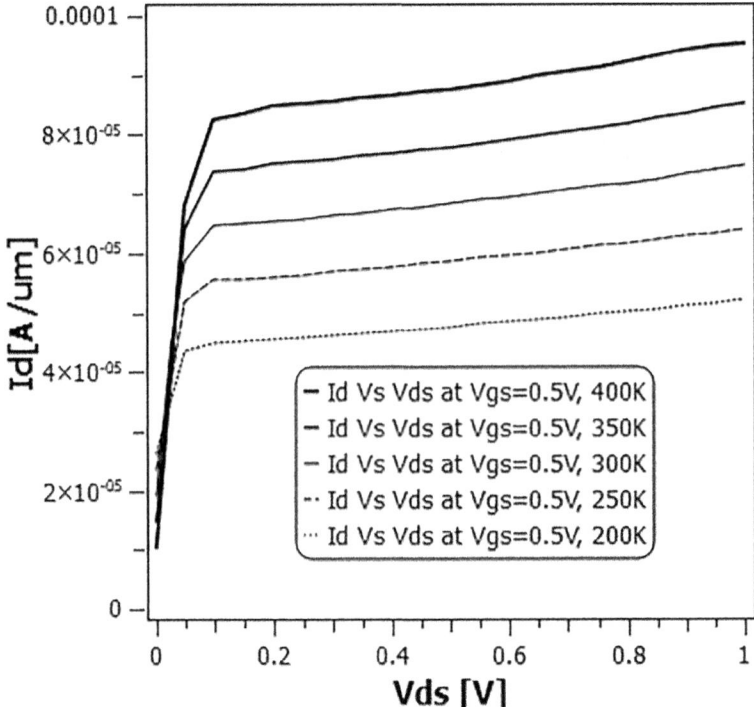

FIGURE 11.13   Variation of drain current with drain voltage with varying temperature

11.5   CONCLUSION

This chapter covers the design and analysis of an efficient Ge-Si$_x$Ge$_{1-x}$ nJLTFET for low power, low voltage operations. nJLTFET and pJLTFET have been compared for similar operation conditions to check their suitability for a CMOS-based digital circuit. The proposed Ge-Si$_x$Ge$_{1-x}$ nJLTFET has an I_{ON}/I_{OFF} ratio of more than ~7 × 10^7, a subthreshold slope of 73.4 mV/decade, and a DIBL of 18.5 mV/V. The ON/OFF characteristics of the proposed pJLTFET and nJLTFET match, which shows their suitability for CMOS-based circuit and memory applications. The proposed Ge-Si$_x$Ge$_{1-x}$ nJLTFET can be utilized for low-power circuit design and play an important role in large capacity memory design for Internet of Things systems. The device performs well under harsh temperature conditions (250–400 K), with a small variation in OFF-state current.

REFERENCES

1. Colinge JP (2008) The new generation of SOI MOSFETs. *Rom. J. Inf. Sci. Technol,* 11:3–15.
2. Colinge J, Gao M, Romano-Rodriguez A, Maes H, and Claeys C (1990) Silicon-on-insulator gate-all-around device. *Tech. Dig. IEDM,* pp. 595–598.

3. Manoj CR, Meenakshi N, Dhanya V, and Ramgopal RV (2008) Device design and optimization considerations for bulk FinFETs. *IEEE Transactions on Electron Devices*, 55:609–615.

4. Tripathi SL, Mishra R, and Mishra RA (2013) *"High performance Bulk FinFET with Bottom Spacer." IEEE CONNECT.*

5. Hui Z, Yeo Y-C, Rustagi Subhash C, and Samudra Ganesh Shankar (2008) analysis of the effects of fringing electric field on FinFET device performance and structural optimization using 3-D simulation. *IEEE Transactions on Electronics Devices*, 55:1177–1184.

6. Nirschl Thomas, Wang Peng-Fei, Hansch Walter, et al (2004) *The tunneling field effect transistor (TFET): the temperature dependence, the simulation model, and its application .IEEE*, pp.713–716.

7. Khatami Yasin, and Banerjee Kaustav (2009) Steep subthreshold slope n- and p-type tunnel-FET devices for low-power and energy-efficient digital circuits *IEEE Transactions on Electron Devices*, 56:2752–2761.

8. Avci U E et al.(2012) *Understanding the feasibility of scaled III–V TFET for logic by bridging atomistic simulations and experimental results, Proceedings on VLSI Technology and (VLSIT) Symposium*, Honolulu, HI, USA, pp. 183–184.

9. Avci Uygar E., Morris Daniel H., and Young Ian A. (2014) Tunnel field-effect transistors: prospects and challenges. *IEEE Journal of Electron Device society*, 2:44–49.

10. Sharma Ankit, Goud A. Arun, and Roy Kaushik (2015) *P-channel tunneling field effect transistor (TFET): sub 10nm technology enablement by GaSb-InAs with doped source underlap.* in Device research conference (DRC) IEEE***.

11. Tang, Wei-Bin, Song, Ya-Feng, and Xiangyu, Liu (2018) Study and theoretical calculation of germanium-tin n-tunneling FET for low off-state current,*Journal of Nanoelectronics and Optoelectronics*, 13: 965–970.

12. Ahmad Syed, Afzal Alam, Naushad, and Amin, S. Intekhab (2018) Impact of pocket-size variation on the performance of GaAs$_{0.1}$Sb$_{0.9}$/InAs based heterojunction double gate tunnel field effect transistor. *Journal of Nanoelectronics and Optoelectronics*, 13: 1009–1018.

13. Yang Zhaonian (2016) Tunnel field-effect transistor with and L-shaped gate. *IEEE Electron Device Letters*, 37:839–842.

14. Li Wei, Liu Hongxia, Wang Shulong, Chen Shupeng, and Yang Zhaonian (2017) Design of high performance Si/SiGe hetero junction tunneling FETs with a T-shaped gate. *Nanoscale Research Letters*, 12:198.

15. Li Wei, Liu Hongxia, Wang Shulong, and Chen Shupeng (2017) Reduced miller capacitance in U-shaped channel tunneling FET by introducing heterogeneous gate dielectric *IEEE Electron Device Letters*, 38:403–406.

16. Wang Wei, and Wang Peng-Feietal (2014) Design of U-shape channel tunnel FETs with SiGe source regions *IEEE Transactions on Electron Devices*, 61:193–197.

17. Gundapaneni S, Bajaj M, Pandey RKK Murali V R, Ganguly S, and Kottantharayil A (2012) Effect of band-to-band tunneling on junction less transistors. *IEEE Transactions on Electron Devices*, 59:1023–1029.

18. Avinash Lahgere, and Mamidala Jagadesh Kumar (2017) A tunnel dielectric-based junction less transistor with reduced parasitic BJT action, *IEEE Transactions on Electron Devices*, 64:3470–3475.

19. Ahn TJ, and Yu YS (2018) Electrical characteristics of Ge/Si-based source pocket tunnel field-effect transistors. *Journal of Nanoscience and Nanotechnology* 18:5887–5892.

20. Seo JH, Yoon YJ, Kang IM (2018) Design optimization of Ge/GaAs-based hetero junction gate-all-around (GAA) arch-shaped tunneling field-effect transistor (A-TFET), *Journal of Nanoscience and Nanotechnology*, 18:6602–6605.
21. Li W, Liu H, Wang S, Chen S, and Yang Zn (2017)Design of high performance Si/SiGe hetero junction tunneling FETs with a T-shaped gate,*Nanoscale Research Letters*, 12:198
22. Li et al (2017) The programming optimization of capacitorless 1T DRAM based on the dual-gate TFET, *Nanoscale Research Letter* 12:524.
23. Makosiej et al (2012) A 32nm tunnel FET SRAM for ultra low leakage, *IEEE International Symposium on Circuits and Systems*, 2517-2520.
24. Ahmad et al (2018) Robust TFET SRAM cell for ultra-low power IOT applications, *AEU - International Journal of Electronics and Communications*, Elsevier, 89:70–76.
25. K-H. Kao, S. V. Anne, G. V. William, S. Bart, G. Guido, and D. M. Kristin "Direct and indirect band-to-band tunneling in germanium-based TFETs," *IEEE Transactions on Electron Devices*, 59(2), pp. 292–301, Feb. 2012.

12 Influence of Moisture Uptake on the Mechanical Properties of Natural Fiber-Reinforced Polymer Composites

Partha Pratim Das, Aseem Acharya and Vijay Chaudhary

CONTENTS

12.1 INTRODUCTION

Fiber is used to reinforce polymer composites for low-cost applications that should be sustainable in different environments. The level of moisture absorption decreases as the fraction of fiber volume decreases, due to the high content of cellulose. The water absorption pattern of these composites at room temperature was found to adopt Fickian behavior, although it exhibited non-Fickian behavior at elevated temperatures [1]. The tensile properties of composite specimens submerged in water were measured and contrasted with dry specimens. There was evidence of a decrease in the mechanical properties of the immersed composites in water as compared with dry samples. The percentage of fiber weight increased due to the high cellulose content, the percentage of moisture uptake also increased [2].

Wood flour composites/polyvinylchloride were prepared by compression molding using Angelim Pedra sapwood and heartwood as filler, according to Singh et al. Specimens of the composites were subjected to water immersion and impact testing. The findings showed that moisture uptake of the composites increased marginally with increased immersion time and wood content [3]. In a humid atmosphere and when immersed in water, all polymer composites absorb moisture. The effect of

TABLE 12.1
Advantages and Disadvantages of Natural Fiber

Advantages	Disadvantages
Low cost	Hydrophilic behavior
Recyclable	Dimensional instability
Zero fingerprint CO_2	Low thermal resistance
Biodegradability	Anisotropic behavior
Low density and high specific mechanical properties	Discontinuous

moisture absorption contributes to the deterioration of the fiber-matrix interface area, which causes low stress transfer efficiencies, resulting in deterioration of mechanical properties [4]. The key concerns about the use of natural fiber-reinforced composite materials are their susceptibility to moisture absorption and its impact on physical, mechanical, and thermal properties. Therefore, it is important that this problem is solved so that natural fiber can be considered as a viable reinforcement in composite materials. Several studies in the use of natural fiber-reinforced polymeric composite have shown that the sensitivity of certain mechanical and thermal properties to moisture uptake can be reduced using coupling agents and fiber surface treatments. Table 12.1 shows the advantages and disadvantages of natural fibers.

12.2   MOISTURE UPTAKE BEHAVIOR OF NATURAL FIBERS

The hydrophilic behavior of natural fibers is greater than that of synthetic fiber. Natural fibers' hydrophilicity and high capillarity characteristics, and their use in mainly hydrophobic polymer matrix composites, can cause low adhesion at the matrix interface [5]. Low interfacial adhesion is associated with a low polarity and chemical affinity between the matrix and the fiber, which causes the formation of voids at the interface, and the initiation of failures that compromise the composites' mechanical performance. The design of the fibers and matrix material, relative humidity (RH), and processing methods affect the characteristics of polymers reinforced with natural fibers immersed in humid environments.

The water absorption rate in composite materials (in liquid or vapor phases) depends on a variety of factors, such as temperature, volumetric fiber fraction, reinforcement orientation, natural fiber permeability, porosity, exposed surface area, diffusiveness, and surface protection [6]. Figure 12.1 shows the mechanism of moisture absorption in fibers.

The absorption of moisture by polymers reinforced by cellulosic fibers occurs because of the dissolution of water in the polymer structure, because of the hydrogen bonds between the water and the hydrophilic groups present in the composite components, and because of the micro-cracks on the composite surface that are responsible for the transport and deposition of water [7].

Polymer-absorbed water consists of free water as well as bound water [8]. Free water molecules are able to travel independently across the void spaces, whereas bound water is delimited into polar polymer groups [9]. Once a polymer composite reinforced with natural fibers is exposed to moisture, free water penetrates and binds with hydrophilic fiber groups, forming intermolecular hydrogen bonds with the

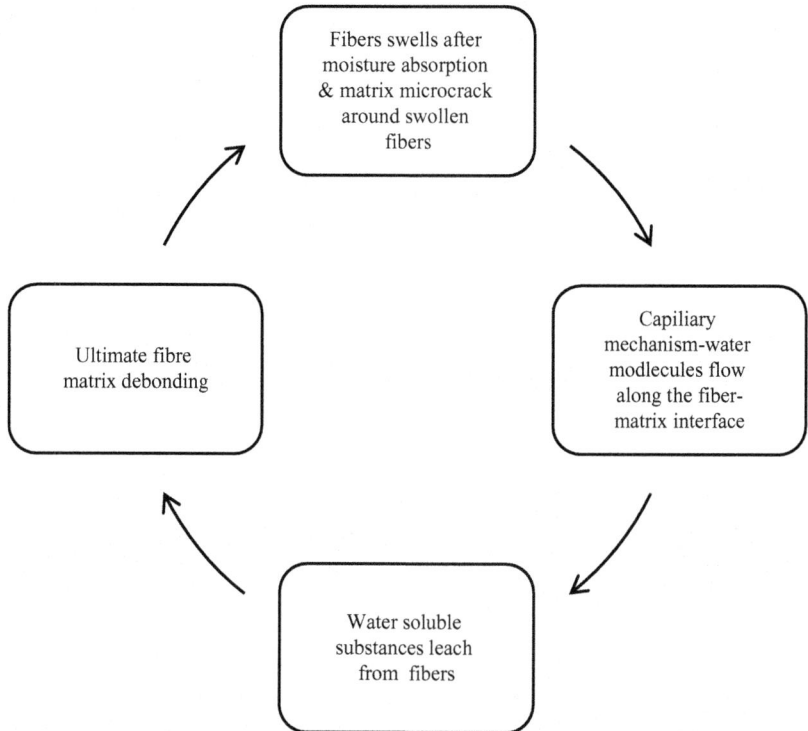

FIGURE 12.1   Mechanism of moisture uptake in natural fibers

fiber and reducing interfacial adhesion between fiber and matrix. The degradation process occurs with the swelling of cellulose fibers, causing an increase in tension in the interface regions that result in its becoming fragile and thereby contributing to the development of micro-cracks in the matrix around the swollen fibers [10]. This promotes capillarity and transport of moisture via micro-cracks, causing deterioration of the fibers, which eventually leads to debonding of fiber and the matrix.

Dhakal et al. [11] demonstrated an improvement in fiber volume uptake for composite polyester/hemp in water at 25°C. They reported that the absorption of moisture contributes to fiber swelling, resulting in micro-cracks forming in the matrix. As the composite cracks and weakens, capillarity becomes active and water is transported through micro-cracks. The capillarity mechanism could involve the flow of water molecules along fiber/matrix interfaces as well as a process of diffusion through the bulk matrix. This could result in debonding of the fiber and the matrix.

Polymer composite materials reinforced by natural fiber are often exposed to humid conditions. The water molecules can move within the polymer, altering its physical and mechanical properties. The key parameters defining the process of moisture sorption are the chemical composition and microstructure of the polymers.

Three distinct mechanisms govern the diffusion of moisture in polymeric composites [12]. The first involves the diffusion of water molecules within micro-gaps between polymer chains. The second involves moisture transfer through gaps and faults at the interface between fiber and matrix. This is a result of poor wetting and

TABLE 12.2
The Equilibrium Moisture Content of Different
Natural Fibers at 65% Relative Humidity and 21°C

Fiber	Equilibrium Moisture Content (%)
Sisal	11
Hemp	9
Jute	12
Abaca	15
Ramie	9
Pineapple	13
Coir	10
Bagasse	9.8
Bamboo	9.9
Flax	7

impregnation during initial fabrication. The third involves the transport of water molecules through micro-cracks in the matrix that originated from the manufacturing process [13]. Table 12.2 shows the equilibrium moisture content of different natural fibers at 65% RH and 21°C.

12.3   MODELS USED TO STUDY THE MOISTURE UPTAKE BEHAVIOR OF NATURAL FIBER-REINFORCED POLYMER COMPOSITES

Different models and their analytical and numerical solutions have been suggested to describe the kinetics of moisture absorption in polymer composites [14]. However, these works consider diffusion through the solid to be unidirectional [15, 16]. Among the different models proposed in the literature to describe the water sorption kinetics in polymer composites, we can cite the Jacob-Jones's model, Fick's model, model with variable diffusivity, and the Langmuir's model. Figure 12.2 shows the various models used to study the mechanism of moisture uptake in natural fibers.

The most used is Fick's model, which states that water migrates inside the solid purely by diffusion [17, 18]. Other models also consider chemical reactions and leaching of low molecular weight components that affect the kinetics of water sorption. This chapter basically explains Fick's model.

In this model, it is assumed that the moisture absorption occurs only by diffusion. According to Fick's first law, the diffusive flux (J) is directly proportional to the concentration gradient.

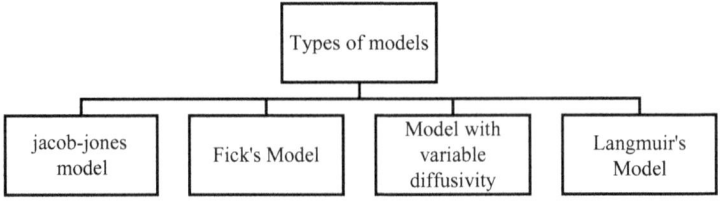

FIGURE 12.2   Models used to study the mechanism of moisture uptake in natural fibers

Mathematically, it can be expressed as

$$J = -D\frac{\delta M}{\delta y} \tag{12.1}$$

Here, J is the diffusive flux in mol m^{-2} s^{-1}; C is the diffusion coefficient; δM change in concentration of the material in g/m^3; δy change in distance in m

Fick's second law has been widely used, since it establishes the moisture diffusion in terms of concentration gradient in the solid, as follows:

$$\frac{\delta M}{\delta t} = \Delta\left(D\ \Delta\ M\right) \tag{12.2}$$

Here, $\dfrac{\delta M}{\delta t}$ moisture transfer rate; D is the diffusion coefficient; and δM change in concentration of the material in g/m^3.

Generally, the diffusion coefficient is constant, depending on the temperature and/ or the amount of moisture content in the fiber.

12.4   EFFECT OF MOISTURE UPTAKE ON MECHANICAL PROPERTIES

In general, fibers have a tendency to absorb water. Indeed, in advanced applications in medicine, building, etc., superabsorbent polymers are gaining popularity; at the same time, thermoplastic capabilities are leading to many changes in processing and properties. Moisture uptake is the ability of a plastic or a polymer to remove moisture from its environment. The absorption of moisture has been shown to act as a plasticizer, reducing the glass transition temperature and plastic strength – a reversible effect. Absorbed water can also contribute to irreversible deterioration of the structure of polymers. Exposure to humidity, immersion, and exposure to boiling water can result in distinctly different material responses. The equilibrium moisture content can be used to compare the amount of water absorbed by different types of natural fibers when they are exposed to moisture.

Moisture uptake is expressed as increase in weight percent or % weight gain of a specimen under the following testing procedures:

- Moisture uptake for 24 hours at 23°C: Immersion of a specimen in distilled water for 24 hours at 23°C
- Moisture uptake for 24 hours at 100°C: Immersion of a specimen in distilled boiling water for 24 hours
- Moisture uptake at saturation: Immersion of a specimen in distilled water at 23°C. Measurement occurs when the polymer does not absorb any more water and moisture uptake is at equilibrium.

The percentage increase is given as:

$$\%\ \text{increase in weight} = \frac{\text{wet weight} - \text{conditioned weight}}{\text{Conditioned weight}} \times 100 \tag{12.3}$$

Several authors showed a relationship between moisture and mechanical properties of plant fibers [19]. Moisture in any form is deleterious to composites of polymers, especially those reinforced with natural fibers [20, 21].

Natural fibers are hydrophilic, and fiber moisture not only serves as a plasticizer but also makes impregnation of polymer more difficult, causing poor adhesion to the polymer matrix-fiber interface, resulting in internal strain, porosity, and premature device failure. Davies et al. [22] also experimentally observed a propensity for the Young's module to decrease with increased RH for flax and nettle fibers. Symington et al. [23–25] also highlighted this pattern for flax fiber. Bio-composites typically exhibit worse mechanical properties than synthetic fiber-reinforced composites as moisture uptake adversely affects the composites' durability and physical and mechanical integrity. Knowing the influence of humidity on the composite material is therefore important for outdoor applications.

Surface alteration (chemical treatment) of the fiber or the matrix can minimize the incompatibilities between natural fiber-reinforced composites and water. At a high fiber content, the mechanical properties of vegetable fiber-reinforced composites improve significantly. Nonetheless, as fibrous polymer composites absorb moisture, they suffer from swelling, plasticizing, dissolving, leaching, and/or hydrolyzing, resulting in discoloration, fragility, lower heat and weather resistance, and poorer mechanical properties. The amount of water a sample absorbs varies depending on its size, length, void fraction (free volume available), temperature, surface area, surface safety, and exposure time. The effects of composite moisture and temperature on performance parameters such as tensile and shear strengths, elastic frames, fatigue behavior, cracking, break stress, dynamic impact response, and electrical resistance were investigated by Symington et al. [23]. UmitHuner et al. [24] investigated the effect of water absorption on the mechanical properties of epoxy composites enhanced by flax fiber. To research the effects of water absorption on mechanical properties, flax fiber-reinforced epoxy composites were subjected to water immersion tests. The percentage of absorption of moisture increased as the fraction of fiber volume increased due to the high cellulose content. The tensile and flexural properties of reinforced epoxy specimens have been found to deteriorate with increased moisture uptake percentage.

Faike et al. [26] documented the effect of water absorption on the hardness properties of epoxy reinforced with glass fibers. The epoxy resin matrix was reinforced by 0–90° fiber orientation with 25 percent volume fraction. The hardness value achieved at room temperature decreased with increasing immersion time in water. Chandramohan et al. [27] investigated the effect of dry and wet environments on the tensile properties and hardness of bio-epoxy composites. The amount of moisture absorbed decreased as the fraction of the fiber volume decreased because of the fiber's high cellulose content. The tensile and flexural properties of natural fiber-reinforced epoxy composite specimens have been found to decrease with an increase in moisture absorption percentages [28]. At elevated temperature, humidity-induced degradation of composite samples was observed. The water absorption pattern of these composites was found to obey Fickian behavior at room temperature, but not at higher temperatures. Table 12.3 displays previous research on the effects of moisture absorption on the behavior of natural fiber-reinforced polymer composite [29–38].

TABLE 12.3

Previous Research into Moisture Uptake Behavior of Natural Fiber-Reinforced Polymer Composites [29–38]

Authors	Fiber/polymer	Fabrication Techniques	Moisture Uptake	Time	Effect on Mechanical Properties	References
Priyanka et al.	Banana-fiber-reinforced functionalized polypropylene	Co-rotating twin screw extruder	11%	4680 h	Decrease in tensile strength was observed for wet sample	[30]
Anjali Singh et al.	Jute-fiber-reinforced modified polyethylene composites	Co-rotating twin screw extruder	10%–30%	6600 h	Decrease of 5%–9% was observed in tensile strength for wet sample	[31]
Chittaranjan et al.	*Lantana-camara*-fiber-reinforced epoxy composite	Hand lay-up technique	13.72%	80 h	At 10% fiber concentration, a decrease of 0.351 MPa was observed. Tensile Strength is spiked up at 30% fiber concentration	[32]
Garish et al.	Sisal/coconut coir natural fibers	Hand lay-up technique	For 20% wt of fiber composite, 3.71% For 30% wt of fiber composite, 5.31% For 40% wt of fiber composite, 8.65%	672 h	Decrease in tensile strength was observed: For 20% wt, decrease was 1 MPa For 30% wt, decrease was 22 MPa For 40% wt, decrease was 8 MPa	[33]
Gupta et al.	Hybrid sisal/jute fiber polymer composite	Hand lay-up technique	A decrease of about 50% was observed with respect to jute composite	24 h	Increase in tensile modulus and tensile strength is observed	[34]
Sideridis et al.	Low-content iron particles reinforced in epoxy resin composites		75%	280 days	Enhancement in mechanical properties is seen due to increased movement in polymer chains	[29]
Daramola et al.	Pineapple	Polyester	31%	336 h	Water absorption increases and rate of absorption decreases with increase in immersion time	[35]
Muñoz et al.	Flax	Epoxy	(6.23–6.56)%	768 h	Water absorption increases with increase in fiber weight fraction	[36]
Venkatesh et al.	Sisal and bamboo	Epoxy	19.6%	55 h	Water absorption increases with increase in fiber length	[37]
Girisha et al.	Sisal and coconut fiber	Epoxy	8.64%	672 h	Water absorption decreases after saturation point	[38]

Dynamic mechanical analysis also showed that the reinforcement of natural fibers has significantly enhanced the thermal stability of epoxy under dynamic loading conditions [39]. Dhakal et al. [11] investigated the effect of water absorption on the mechanical properties of composites of unsaturated polyester matrix reinforced with hemp fiber. Specimens of composites containing fractions of 0, 0.10, 0.15, 0.21, and 0.26 fiber volume were prepared. Water absorption experiments were carried out by immersing specimens for varying durations in a de-ionized water bath at 25°C and 100°C. The percentage of absorption of moisture increased as the fraction of the fiber volume increased, due to the high cellulose content. The researchers concluded that composites' tensile and flexural properties decrease with an increase in percentage of moisture uptake.

12.5 CONCLUSION

Several key points regarding the influence of moisture uptake on mechanical properties of natural fiber-reinforced polymer composites are as follows:

- Natural-fiber-enhanced composites are of interest due to their biodegradability, low cost, low relative density, high basic mechanical properties, and sustainable character. Such composites are likely in the near future to find more and more uses, as many studies lead us to understand and develop their properties.
- Understanding the hygroscopic nature of these materials for use under various weathering environments is a key problem.
- Many studies into the relation between the microstructure and the hydrophilic behavior of plant fibers are discussed and reviewed in this chapter.
- The effect of moisture on the fibers' properties, as well as the final properties of the composites that they reinforce, are also reviewed.
- The absorption of moisture by fibers and their composites has been found to significantly affect the mechanical and structural properties of the composites.
- One of the important aspects is that hybridization with synthetic fiber decreases the moisture uptake behavior in natural fiber-based polymer composites.
- Developed in recent years, nanotechnology is a new field of study that uses different nanofillers and ceramic coatings (nanocoatings) to enhance natural fiber moisture uptake properties.

REFERENCES

1. Rashdi, A. A. A., Salit, M. S., Abdan, K., and Megat, M. M. H. 2010. "Water absorption behaviour of kenaf reinforced unsaturated polyester composites and its influence on their mechanical properties". *Pertanika Journal of Science and Technology*, vol. 18, no. 2, pp. 433–440.
2. Iulianelli, G., Tavaresm, M. B., and Luetkmeyer, L. 2010. "Water absorption behavior and impact strength of PVC/Wood flour composites". *Chemistry & Chemical Technology*, vol. 4, no. 3, pp. 225– 229.
3. Alamri, H., and Low, I. M. 2012. "Mechanical properties and water absorption behaviour of recycled cellulose fibre reinforced epoxy composites". *Polymer Testing*, vol. 31, pp. 620–628. doi: 10.1016/j.polymertesting.2012.04.002.

4. Anbukarasi, K., and Kalaiselvam, S. 2015. "Study of effect of fibre volume and dimension on mechanical, thermal, and water absorption behaviour of luffa reinforced epoxy composites". *Materials and Design*, vol. 66, pp. 321–330. doi:10.1016/j.matdes.2014.10.078.

5. Espert, A., Vilaplana, F., and Karlsson, S. 2004. "Comparison of water absorption in natural cellulosic fiber from wood and one-year crops in polypropylene composites and its influence on their mechanical properties". *Composites Part A: Applied Science and Manufacturing*, vol. 35, no. 11, pp. 1267–1276.

6. Li, Y., Mai, Y. W., and Ye, L. 2000. "Sisal fibre and its composites: A review of recent developments". *Composites Science and Technology*, vol. 60. no. 11, pp. 2037–2055.

7. Azwa, Z. N., Yousif, B. F., Manalo, A. C., and Karunasena, W. 2013. "A review on the degradability of polymeric composites based on natural fibres". *Materials & Design*, vol. 47, pp. 424–442.

8. Chen, H., Miao, M., and Ding, X. 2009. "Influence of moisture absorption on the interfacial strength of bamboo/vinyl ester composites". *Composites Part A: Applied Science and Manufacturing*, vol. 40, no. 12, pp. 2013–2019.

9. Sreekala, M. S., Kumaran, M. G., and Thomas, S. S. 2002. "Water sorption in oil palm fiber reinforced phenol formaldehyde composites". *Composites Part A*, vol. 33, pp. 763–777.

10. Dhakal, H. N., Zhang, Z. Y., and Richardson, M. O. W. 2007. "Effect of water absorption on the mechanical properties of hemp fibre reinforced unsaturated polyester composites". *Composites Science and Technology*, vol. 67, no. 7–8, pp. 1674–1683.

11. Dhakal, H. N., Zhang, Z. Y., and Richardson, M. O. W. 2007. "Effect of water absorption on the mechanical properties of hemp fiber reinforced unsaturated polyester composites". *Composites Science and Technology*, vol. 67, pp. 1674–1683.

12. Joseph, P. V., Rabello, M. S., Mattoso, L. H., Joseph, K., and Thomas, S. 2002. "Environmental effects on the degradation behavior of sisal fiber reinforced polypropylene composites". *Composites Science and Technology*, vol. 62, no. 10–11, pp. 1357–1372.

13. Paul, S. A., Boudenne, A., Ibos, L., and Candau, Y. 2008. "Effect of fiber loading and chemical treatments on thermophysical properties of banana fiber/polypropylene commingled composite materials". *Composites Part A: Applied Science and Manufacturing*, vol. 39, no. 9, pp. 1582–1588.

14. Glaskova, T. I., Guedes, R. M., Morais, J. J., and Aniskevich, A. N. 2007. "A comparative analysis of moisture transport models as applied to an epoxy binder". *Mechanics of Composite Materials*, vol. 43, no. 4, pp. 377–388.

15. Najafi, S. K., Kiaefar, A., Hamidina, E., and Tajvidi, M. 2007. "Water absorption behavior of composites from sawdust and recycled plastics". *Journal of Reinforced Plastics and Composites*, vol. 26, no. 3, pp. 341–348.

16. Katzman, H. A., Castaneda, R. M., and Lee, H. S. 2008. "Moisture diffusion in composite sandwich structures". *Composites Part A: Applied Science and Manufacturing*, vol. 39, no. 5, pp. 887–892.

17. Xiao, G. Z., and Shanahan, M. E. R. 1998. "Swelling of DGEBA/DDA epoxy resin during hygrothermal ageing". *Polymer*, vol. 39, no. 14, pp. 3253–3260.

18. Maggana, C., and Pissis, P. 1999. "Water sorption and diffusion studies in an epoxy resin system". *Journal of Polymer Science Part B: Polymer Physics*, vol. 37, no. 11, pp. 1165–1182.

19. Chaudhary, V., Bajpai, P. K., and Maheshwari, S. 2018. "Effect of moisture absorption on the mechanical performance of natural fiber reinforced woven hybrid bio-composites". *Journal of Natural Fibers*. doi:10.1080/15440478.20 18.1469451.

20. Chaudhary, V., Bajpai, P. K., and Maheshwari, S. 2017a. "Studies on mechanical and morphological characterization of developed jute/hemp/flax reinforced hybrid

composites for structural applications". *Journal of Natural Fibers*, vol. 15, pp. 80–97. doi:10.1080/15440 478.2017.1320260.

21. Chaudhary, V., Bajpai, P. K., and Maheshwari, S. 2017b. "An investigation on wear and dynamic mechanical behavior of jute/hemp/flax reinforced composites and its hybrids for tribological applications". *Fibers and Polymers*, vol. 19, pp. 403–415. doi:10.1007/ s12221-018-7759-6.

22. Davies, G. C., and Bruce, D. M. 1998. "Effect of environmental relative humidity and damage on the tensile properties of flax and nettle fibers". *Textile Research journal*, 68, 623–629.

13 Exploring the Potential of Nanotechnology in Agriculture
Current Research and Future Prospects

Mitali Mishra, Ashutosh Kumar Pandey, Kritika Pandey, Saurabh Dixit, Fatima Zohra, Aparna Seth and Sanchita Singh

CONTENTS

13.1 INTRODUCTION

Nanotechnology is a multidisciplinary branch of science that is linked with many other areas of science like chemistry, physics, biology, engineering, electronics, and material sciences (Wang and Wang 2014). It can be defined as a field of applied sciences and engineering that focuses on controlling matter of size 1–100 nm and fabrication of tools, devices, and systems on this scale. Nanotechnology recasts the existing sciences, employing novel or advanced terms. Undoubtedly, exceptional features of materials at the nanometer scale make them suitable for the fabrication and design of new tools and systems in order to facilitate sustainable development in the agriculture sector. In this sector, nanotechnology interventions provide tools like nanoparticles (NPs), nanorods, nanocapsules, and quantum dots that help to increase yield, promote a targeted supply of nutrients, detect pathogens, absorb toxic compounds, and confer tolerance against biotic and abiotic stress. Some of the applications of nanotechnology are shown in Figure 13.1. NPs of silicon, carbon, titanium,

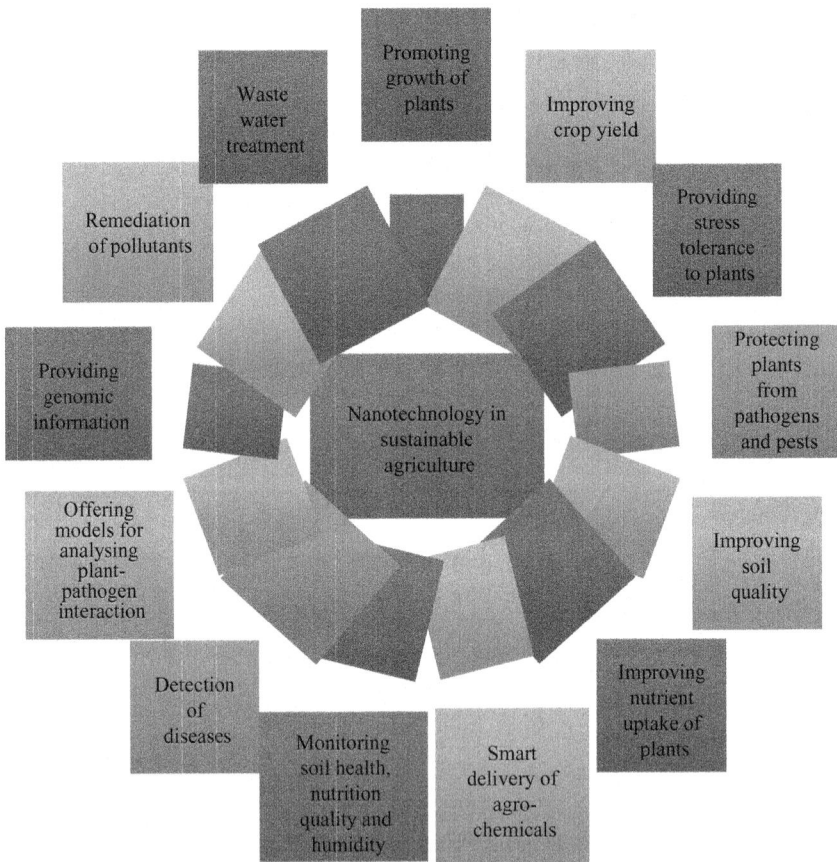

FIGURE 13.1 Applications of nanotechnology in sustainable agriculture

silver, copper, etc. have been reported to be good growth enhancers with positive effects on growth of seedlings, roots, and shoots. However, the effects of these NPs on plants depends on their composition, amount, size, charges, chemical properties, and the affinity of the plants to these NPs. Development of novel nanobased fertilizers and pesticides with modified carrier systems has further improved the productivity of crops by increasing nutrient absorption from the soil, protecting the plants from pest attacks. In addition, nanobiotechnology makes the understanding of plant-pest or plant-parasite interactions more feasible. Nanofabricated xylem vessels are found to be potent systems in the understanding of plant-pathogen interactions.

Practices that ensure the early detection of disease are of importance in agronomy. Nanobiosensors and nanobarcodes are the new vistas in this regard as they facilitate earlier and more precise detection of diseases. They can also be used in analyzing the health, moisture content, and nutrition contents of soil. In the light of the various potential applications of nanotechnology, there is a need to explore nanomaterials further. There is a need to find nanomaterials that ca show less toxicity towards the agro-ecosystem after release into the environment and, more importantly their relevance should be explored up to field conditions. This chapter envisages the applications of different types of nanobased miniature systems, tools, and techniques in agronomy, covering new advancements in nanotechnology in the field of agriculture as well as the scope of this multidisciplinary science in developing countries like India.

13.2 MULTIFACETED ROLE OF NANOTOOLS AND THEIR POTENTIAL APPLICATIONS

13.2.1 NANOPARTICLES

The role of different NPs in the field of agriculture is to monitor soil quality, maintain food quality, reduce the passage of chemicals and heavy metals into plant tissue, increase the absorption of nutrients from the soil, protect plants from abiotic and biotic stress, and increase the yield of a desired product. Their small size (1–100 nm) and greater surface-area-to-volume ratio make the properties of NPs different from those of the same material in bulk.

13.2.1.1 Silicon Nanoparticles

Silicon is considered an essential microelement. It is not required for growth but plays a vital role in plant adaptation during adverse conditions and influences the metabolic activities, growth, development, and productivity of a plant. Unpredictable changes in abiotic factors, i.e., wind, water, atmosphere, sunlight, and chemicals, causing abiotic stress, affects the physiological processes of plant that further damage its cell structure, functioning, and metabolic activities. Emergence of a nano-silicon film at the epidermis proves helpful in conditions such as dry spells and dampness and also protects vegetation from fungal, bacterial, and nematodal infection (Rastogi et al., 2019). Silicon NPs improve the antioxidant defense mechanism against heavy metal stress by acting against reactive oxygen species (ROS), a response generated during abiotic stress that is responsible for plant cell disruption.

At times of high chemical stress, silicon prevents the uptake of heavy metals from the soil thus regulating the ROS profile. The mesoporosity of silicon NPs makes them a good delivering agent or nanocarrier in the field of agriculture; they are also used as nanopesticides, nanoherbicides, and nanofertilizers to protect a plant from pests and unwanted vegetation. It is an agent for cell-wall strengthening, water retention in soil, and soil monitoring (Rastogi et al., 2019).

13.2.1.2 Carbon Nanoparticles

Composed of one or more layers of graphene sheets, these NPs can be metallic or semiconductors. They are strong and stable in high temperatures, acids, and changing environments. These NPs show conductive, optical, thermal, quantum, and surface plasmon resonance properties. Types of carbon NPs for the improvement of agricultural practices and regulating plant growth are carbon nanodots, carbon nanotubes (CNTs), nanocones, nanohorns, fullerols, and nano-onions. Carbon nanomaterials are toxic to soil microbes, maintain soil quality, improve nutrient cycling decomposition of organic matter, and help in establishing a symbiotic relationship of soil microbes with plant species as well as co-habitation of terrestrial plant species (Mukherjee et al., 2016).

Acting as nanocarrier or delivering agent for targeting agrochemicals to the host, a CNT releases only the amount of chemicals necessary for growth and development, maintaining the soil quality and environment, and protecting other plants from harmful chemicals. Exposure of CNTs to plants forming an aligned network facilitates the uptake of water, increases germination rate, and increases growth of both root and shoot when compared with the untreated control (Mukherjee et al., 2016). Single-walled CNTs are found to increase photosynthetic activity of plants. Their accumulation is observed in chloroplasts: three times higher photosynthetic activities with an enhanced electron transport rate have been reported in *Spinacia oleracea* with single-walled CNTs in comparison to control (Giraldo et al., 2014).

CNT sponge is a 3-D material that is elastic, compressible, flexible, lightweight, super hydrophobic, and oleophilic. The sponge is efficient in absorbing all the water contaminants caused by the presence of pesticides, fertilizers, oil, and pharmaceuticals in agricultural fields. Use of fullerol in plants (e.g., bitter melon) is responsible for increasing phytomedical components (like cucurbitacin, lycopene) that have anticancer and antidiabetic properties and can be used to treat diseases like AIDS (Ng et al., 1992).

Carbon nanohorn enhances the growth of terrestrial plants and the transport of organic co-contaminants. The study of carbon NPs other than CNTs is still at a preliminary stage.

13.2.1.3 Copper Nanoparticles

The synthesis of copper NPs takes place by physical, chemical, or biological methods through the process of reduction. Its cytotoxic nature makes it a good agent to alter microbial communities belonging to different genera and species. Agrochemicals are responsible for growth inhibition and DNA damage of pathogens such as *Staphylococcus aureus, Bacillus subtilis, Escherichia coli, Shigella dysenteriae, Salmonella typhi, Klebsiella pneumonia* in protecting the crops from deteriorating

(Kasana et al., 2017). Disease management in agriculture is sustained due to the antimicrobial activity of copper NPs. These NPs have an advantage in food packaging applications as they maintain the quality of food by increasing its shelf life and ensuring food safety.

13.2.1.4 Silver Nanoparticles

Silver NPs act as a delivery system in the agriculture sector for nutrients, growth hormones, DNA, and various chemicals to obtain better-quality crop production. These NPs have larvicidal, bactericidal, antifouling, antiviral, and anti-inflammatory properties. It takes part in disease management against diseases such as anthracnose disease, which is caused by *Colletotrichus* species (Babu et al., 2013).

13.2.1.5 Titanium Nanoparticles

Titanium oxide NPs are one of the best-explored NPs in agriculture and are widely used in products that can protect plants under different stress conditions. Several studies have discussed the role of titanium oxide NPs in controlling disease outbreaks in crops (Bowen et al., 1992). According to a study, treatment of rice, olive, and maize by TiO_2 NPs protects these crops from *Curvularia* leaf spot and bacteria leaf blight diseases (Chao and Choi, 2005; Glenn et al., 2000). Also, occurrence of rice blast and tomato spray mold was observed to be reduced after treatment with TiO_2 NPs. The combined activity of TiO_2, silica, and Al was found effective against downy and powdery mildew diseases (Bowen et al., 1992).

Titanium NPs instigate high photocatalytic activities and increase the light-absorbing capacities of leaves, so have the ability to increase photosynthetic activities of plants. In addition, these metallic NPs protect against chloroplast aging, trigger Rubisco carboxylation, and assist electron transport chain and chlorophyll phosphorylation activities (Gao et al., 2008, 2006; Hong, Yang, et al., 2005, Hong, Zhou, et al., 2005; Linglan et al., 2008; Qi et al., 2013). Furthermore, TiO_2 NPs also improve transpiration rates and water conductance in leaves (Lei et al., 2007).

13.2.2 Quantum Dots

Quantum dots are nanocrystals composed of semiconducting materials that have quantum mechanisms with improved photostability to track activities at the molecular or cellular level. CdSe, CdS, and CdTe are colloidal semiconductor quantum dots with a diameter of 210 nm that have high visual luminescence and are used as biological labels, tracking agents, and biosensors. Classification of biological labels according to their shape, size, and structure has greatly helped in tracing the residues of pesticides, fertilizers, herbicides, and growth-promoting hormones that are harmful and cause many side effects to the human body (Ung et al., 2012).

UbiGro film made up of UbiQDs (ubiquitous quantum dots) tremendously increases crop yield, crop quality, and faster growth cycle. The nanomaterial-based film converts ultraviolet, blue, and green sunlight into an orange glow, optimizing the light spectrum for a better yield. Quantum dots acting as biosensors can detect the presence of avian influenza viruses such as H5N1.

Recent studies have an interest in developing these quantum-confined nanocrystals as fluorescent probes for medical applications. They enable physicians to quickly detect the presence of a particular set of proteins that strongly indicates the onset of myocardial infarction. Semiconductor quantum dots provide various benefits, like size, composition tunable emission from visible to infrared wavelengths, large absorption coefficients across a wide spectral range, and really high levels of brightness and photostability. It has been estimated that quantum dots are 20 times brighter and 100 times more stable than traditional fluorescent reporters (Walling et al., 2009).

The use of quantum dots for highly sensitive cellular imaging has seen major advances. The improved photostability of quantum dots, for example, allows many consecutive focal-plane images to be recorded, which can be reconstructed into a high-resolution 3D image (Tokumasu et al., 2005). Another application that takes advantage of the extraordinary photostability of quantum dot probes is the real-time tracking of molecules and cells over extended periods of time (Dahan et al., 2003). Antibodies, streptavidin (Howarth et al., 2008), peptides (Åkerman et al., 2002), DNA (Farlow et al., 2013), nucleic acid aptamers (Dwarakanath et al., 2004), or small-molecule ligands (Zherebetskyy et al., 2014) can be used to target quantum dots to specific proteins on cells.

13.2.3 NANORODS

A nanomaterial is considered as ideal only when it can properly, specifically interact with plant systems as well as being tracked easily within the plant system so that understanding of these nanomaterials and plant systems can become more feasible. In this regard, nanorods, which have specific optoelectronic properties, have been gaining the attention of researchers (Bulovic et al., 2004). According to a study, gold nanorods have low phytotoxicity, enhanced uptake, the ability to transport 2,4-dichlorophenoxyacetic acid efficiently in tobacco cell cultures, and have specific *in vivo* detection potential. These are the critical features required for "smart delivery" systems (Nima et al., 2014). Nanorods as surface enhanced Raman spectroscopy (SERS) substrates can also be used to detect pesticides in fruits and vegetables (Zhang et al., 2015).

13.2.4 NANOCAPSULES

Encapsulation (micro-encapsulation/macro-encapsulation) is the process of coating a substance or implanting it in a matrix. Encapsulation protects a substance from degradation in unfavorable environments, facilitates controlled discharge, and ensures specific targeting (Ezhilarasi et al., 2013; Ozdemir and Kemerli, 2016). Nanocapsules are vesicle-like structures in which substances are restricted to the liquid cavity and covered by a membrane made up of polymers (Couvreur et al., 1995). Microcapsules and NPs are gaining attention in drug delivery and the delivery of other biological molecules to specific sites in plants (Beck et al., 2008). These nanocapsules can also be used in making nanorobots/nanobot systems that can be guided by magnetic resonance imaging (Vartholomeos et al., 2011).

13.3 NANOMATERIALS AND NANOSYSTEMS IN SUSTAINABLE AGRICULTURE

13.3.1 NANO PESTICIDES

The global population is increasing at an alarming rate nowadays, resulting in increasing food demand, which is not fulfilled due to damage of crops by pests and insects every year. Therefore, there is a need to protect crops from such pests; for this, pesticides are used. The use of synthetic pesticides like carbamates, organophosphates, etc. causes environmental pollution, and over time the pests become resistant to such pesticides. One of the commonly used traditional synthetic pesticides DDT (dichlorodiphenyltrichloroethane) is very harmful, not only for plants and soil but also for humans and animals.

Recent advancements in the field of nanotechnology have led to the formulation of nano pesticides (nano structures with pesticidal properties), which are often referred as nano-agrochemicals. These are eco-friendly and less toxic in nature (Chhipa, 2017). They play an important role in the field of agriculture and their use will definitely pave the way for better, advanced agricultural practices. However, public awareness is one of the major factors hindering their use. There is a fear that these nano-agrochemicals might be rejected since many consider them to be unsafe for consumption (Kah, 2015).

Nano pesticides exist in many forms like liquid, gels, etc. (Zaki et al., 2017). Different formulations of NPs have been developed to control the pests of attacks on crops.

Use of nano pesticides in agriculture leads to higher yield of crops and less damage to plants. These pesticides do not degrade as easily as conventional pesticides; therefore, even a small amount is very effective. Two basic types of nanomaterials are used in agricultural practices: CNTs and nanomaterials based on metals and metal oxides (Dubey and Mailapalli, 2016).

Nanopesticides help release the main active component at a particular targeted site and protect the component from environmental degradation. They also help increase the solubility of the component and do not cause any harm to the crops (Bhan et al., 2018). Cotton leafworm *Spodoptera littoralis* Boisd (Lepidoptera:Noctuidae) is a very destructive pest that causes damage to cotton crops. To protect crops from this pest, farmers use many traditional pesticides but unfortunately the insect has become resistant against such pesticides. *Bacillus thuringiensis* (Bt) is also used as a biopesticide against cotton leafworm but is not very effective. A nanocomposite made of sodium titanate nanotubes and Bt-based biopesticides has proven very effective and useful in controlling cotton leafworm (Zaki et al., 2017).

13.3.2 NANOFERTILIZERS

An important concept in terms of sustainable agriculture is smart fertilizers. These are nanofabricated systems that can act as carriers of plant nutrients and facilitate their controlled discharge: they can be applied to crops by irrigation or spraying. Nanobased fertilizers not only have the potential to provide an efficient nutrient supply but also minimize nutrient losses and hence improve productivity. The possible mechanisms of supplying nutrients are as follows:

a. In the form of emulsions or minute particles (CNTs, titanium oxide NPs, silicon NPs, fullerenes) in different growth stages of plants.
b. In the form of capsules that are designed to facilitate the slow and controlled release of nutrients. Discharge of nutrients is induced by environmental factors or artificial stimulations (Millán et al., 2008; Perrin et al., 1998).

Some mechanisms for the controlled release of nutrients through nanotechnology are as follows (Aouada and De Moura, 2015):

- Nanocapsules release nutrients slowly so that they are assimilated properly and leaching can be controlled.
- These nanocapsules are target specific. Shells break under the influence of water content or pH variations within cells as they approach a particular plant organ. The opening of shells can also be initiated artificially in response to magnetic or ultrasonic pulses.
- Recognition of specific targets is mediated by functional groups attached to the surface of shells/capsules.

Nutrients are trapped in a polymeric matrix in the capsules. These matrices are made up of biologically derived or synthetic polymers. Natural/biological polymers are preferred over synthetic ones as they are cost effective, easily available, and, most importantly, biodegradable (Corradini et al., 2010). Some of the polymers used for this purpose are chitosan, zeolites, and polyacrylic acid (Ditta and Arshad, 2016; Servin et al., 2015). According to a study, nanoporous zeolite is helpful in the slow release of fertilizer, by means of which a plant is able to absorb all the nutrients. It has a large carrying capacity due to its high surface area. Therefore, many molecules of fertilizers can be packed into it and be released according to the plant's need. The most important nutrient for plant growth is nitrogen but it leaches from the soil due to its high solubility. Therefore, nanoporous zeolite combined with urea helps in its increased uptake (Rameshaiah et al., 2015).

Cation exchangers can be used as a supplement with fertilizer to control nitrogen release. Clinoptilolite has percolative properties along with a high cation-exchange capacity. Therefore, it is used alongside fertilizer to provide many advantages such as controlled release of nitrogen and increased nitrogen absorption by plants. It reduces nitrogen leaching by hindering the process of nitrification (Preetha and Balakrishnan, 2017).

Chitosan obtained from chitin is a biopolymer of N-acetyl glucosamine and glucosamine. NPs based on chitosan have a high surface:volume ratio compared with bulk chitosan; therefore they are widely used in fertilizers.

Nanofertilizers help in plant growth up to a certain concentration limit, beyond which they start inhibiting instead of promoting growth. Nanomaterials possess different physical and chemical properties compared with their bulk form, due to change in intermolecular interactions and reduced particle size. An example of this is rock phosphate. When it is applied to plants as a nanomaterial, more phosphorus is available due to inhibition of the fixation process. Similarly, compared with bulk $ZnSO_4$, nano ZnO is seen to have a more positive effect on root growth and seed germination in peanut plants (Qureshi et al., 2018).

Nanofertilizers are far better than conventional fertilizers and are definitely going to bring a big change in the field of agriculture. As stated earlier, nanofertilizers have high surface area, which results in many reactions and thereby increases the rate of photosynthesis (AL-Tameemi et al., 2019).

The advantages of nanofertilizers over synthetic fertilizers are as follows:

• Nanofertilizers have a large surface area, which results in high reactivity and provides more nutrients to the plant.
• They are easily absorbed by the plant from the soil due to their small particle size. Their particle size being even smaller than the root pores proves to be a major factor in fertilizer absorption.
• They help improve the quality and yield of crops.
• Nanofertilizers are non-poisonous for the environment as well as for human health.
• They are also helpful in seed germination because they easily penetrate and provide nutrients to the germinating seed (Qureshi et al., 2018).

In spite of all these applications and effectiveness, more nanofertilizers are still to be explored. The main challenge for researchers is to develop and explore carriers that can supply nutrients in better controlled, synchronized way to the plants. The interaction mechanisms between nanofertilizers and plants need to be studied more precisely.

13.3.3 NANO BIOSENSORS

A biosensor is a device incorporating a biological sensing element integrated within a transducer; a nano biosensor is a biosensor on the nanoscale size (Malik et al., 2013). Nano biosensors are emerging as a great asset in various fields such as the medical field (for early diagnosis of diseases), food industries, defense, and agriculture. The extreme success of nano biosensor lies in NPs, nanocomposites, nanofilms, etc. Their high surface-to-volume ratio and small size, and the extreme sensitivity of nanocomposites and nanofilms makes a nano biosensor target-specific and provides it with an outstanding limit of detection.

Pesticides can be harmful with respect to public health. Several nano biosensors have made it possible to detect pests and contaminants, making their early detection less tedious. Organophosphorus pesticides such as dichlorvos and paraxon can be measured and monitored at very low levels by liposome-based biosensors (Vamvakaki and Chaniotakis, 2007). Biosensors developed using PS II (photosystem II) are known to bind several groups of herbicides isolated from photosynthetic organisms, so may have potential to monitor polluting chemicals, leading to the development of a low-cost, easy-to-use apparatus able to detect specific herbicides and eventually a wide range of organic compounds present in industrial and urban effluents, sewage, sludge, landfill, ground water, and irrigation water (Giardi and Piletska, 2006). A hollow gold NP deposited on a glassy carbon electrode, functionalized by the L-cysteine-chitosan-acetylcholinesterase in its hollow surface, has been developed. Nano biosensors (constructed by this assembly of working electrode and immobilized enzyme) can detect insecticides like carbofuran and chlorpyrifos in the range of 0.06–0.08 micrograms per decimeter (Srivastava et al., 2018).

An immunosensor conjugated with multiwall carbon nanotube (MWCNT)– graphene sheets-ethyleneimine polymer (GS-PEI)–gold NPs antibody conjugate is used to detect or quantify carbofuran (a widely used insecticide). This immunosensor is very specifically sensitive to carbofuran because of its anti-carbofuran monoclonal antibody (Zhu et al., 2013).

Plant diseases caused by infecting microorganisms are diagnosed by nano biosensors by measurement of relative activity (differential oxygen consumption in respiration) of "good microbes" and "bad microbes" in the soil. Two sensors are impregnated and made sensitive to good and bad microbes and the relative activity of both is measured. Specific kinds of DNA oligonucleotides can be detected by several optical nano biosensors using a single-stranded DNA conjugated with carbon nanotube probe (Cao et al., 2008). A biosensor probe using MWCNT along with zinc oxide and chitosan nanocomposites is able to identify and detect different DNA sequences (Zhang et al., 2008). To detect deep DNA damage, an electro-chemical biosensor was developed in which a screen printed carbon electrode with MWCNT chitosan conjugated with MWCNT is deposited (Galandova et al., 2008). Fertilizer consumption is increasing worldwide so as to enhance crop productivity. Although fertilizers promote plant growth and stimulate soil life, they are not environmentally friendly. It is necessary to measure components of fertilizers to understand the threshold value that would provide better productivity of plants as well as minimizing damage to the environment. Conventional analysis of soil sample is very complex and expensive and nano biosensors have proved a promising substitute. They can potentially evaluate the absolute concentration of minerals easily and quickly, thus promoting sustainable agriculture. A nano biosensor for detection of solid nitrate was made using polypyrrole film with nitrate. It is measured in the range of 10^{-5}–10^{-1} M and response time was less than 15 s (Pan et al., 2016).

Urea hydrolyzes and forms ammonium carbonate, which is toxic for germinating plants creating a necessity to detect its presence in soil. For this reason a calorimetric nano biosensor was fabricated using gold NPs, tetramethylbenzidine, and H_2O_2. It is ultrasensitive for urea, urease, and urease inhibitor (Deng et al., 2016). Zeolite-linked nano biosensors can modernize agriculture in the sense that they can sense deficiencies in either plant or soil and control the release of water, nutrients, and soil (OECD International Futures Programme, 2007).

Relative humidity (RH) is a term generally used to measure the amount of water vapor present in a gas mixture such as air. Humidity nanosensors are made of materials whose dielectric properties alter upon the absorption of water molecules, i.e., hygroscopic materials (Antonacci et al., 2018). Ag-Pd as interdigitated electrodes were used to fabricate an impedence nanosensor for humidity in which $Na_2Ti_3O_7$ (a hygroscopic material) nanotubes were coated on Al_2O_3 substrate. Its sensitivity ranges from 11 to 95% RH, response time is 2 s, and recovery time is 4 s, which is very quick with less than 3% variation of the impedance (Liu et al., 2011). A humidity sensor was fabricated by Zhao and colleagues using graphene oxide films. These films significantly improve the sensitivity and response time of sensors. Its efficient repeatability makes it unique in terms of stability (Zhao et al., 2011). $Ba_{0.7}Sr_{0.3}TiO_3$, $Na_2Ti_3O_7$, doped nanoporous $Ti_{0.9}Sn_{0.1}O_2$ thin films, $Bi_{0.5}Na_{0.5}TiO_3$–$Bi_{0.5}K_{0.5}TiO_3$, and $BaAl_2O_4$ composites are used as core materials for nano biosensors. These novel materials have great advantages in terms of sensitivity, repeatability, response time (Antonacci et al., 2018).

13.3.3.1   Nano Barcodes

A barcode is an optical machine-readable representation of data. Originally, barcodes represented data by varying the width and spacing of parallel lines or rectangular or geometrical shapes. Barcodes that use NPs and are of nano scale are called nano barcode. Nano bio-barcodes have been fabricated in recent years and are also known as DNA barcodes.

DNA barcodes have an inbuilt code and a probe for molecular detection of every single molecule where multiple detection is required (Li et al., 2005). DNA barcoding is used to store genome information of a particular species by using genetic markers of that species (plants or animals); ultrasensitive detection of proteins can be performed by these nano bio-barcodes and single-cell protein-expression experiments can also be performed easily and quickly. When it comes to diagnosis and analysis of a new disease, sensitivity and specificity is a challenge. A first-generation DNA barcode, fabricated using fluorescence-labeled Y DNA, has been designed by Soong and colleagues: it was able to detect eight different pathogens simultaneously (Um et al., 2006). Some nanomaterials with their functions are shown in Figure 13.2.

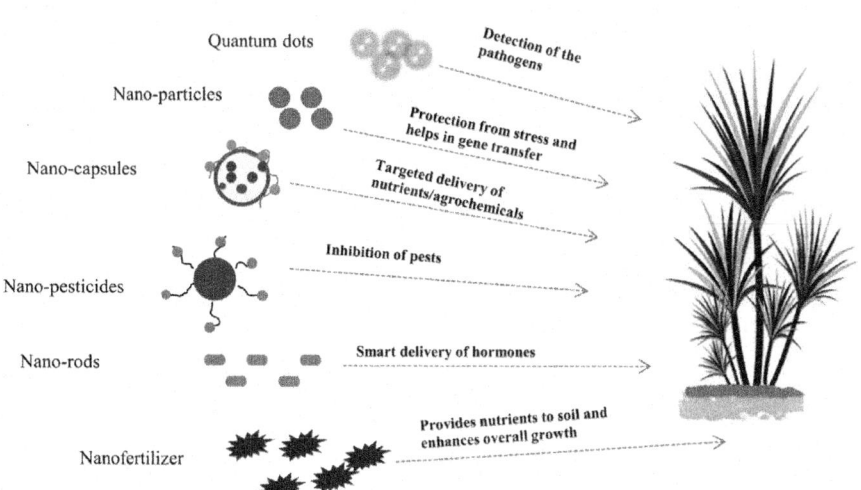

FIGURE 13.2   Nanomaterials used in agriculture and their various functions

13.4 NEW VISTAS OF NANOTECHNOLOGY

13.4.1 CELLULOSE NANOFIBERS

In tissues of plants, micro/microfibrils form cellulose fibers and nanofibers form microfibrils. The size of the nanofibers depends on the source from which they are obtained and generally ranges from 3 to 35 nm (Athinarayanan et al., 2015). These nanofibers are gaining attention as novel nanomaterials due to certain beneficial properties including extraordinary optical properties, high surface area, and excellent mechanical strength (Anastas and Warner, 1998; Wanyika et al., 2012). Chemically modified nanocellulose can be used in nanocomposite-based applications. These nanocomposites can be used further for water treatment, disease diagnosis, drug delivery systems, and detection of pathogens and pests.

13.4.2 NANOFABRICATED XYLEM VESSELS

Recently, nanofabricated xylem vessels have been developed in order to mimic natural xylem vessels. These xylem vessels can be used to study the mechanism of pathogenesis so that treatment strategies can be prescribed accordingly; this would not be possible using traditional processes (Cursino et al., 2009). A study been carried out to instigate nanobased fabrication of xylem vessels (Bandyopadhyay et al., 2013). By using these bio-mimicking xylem vessels, researchers have investigated the mechanisms of attachment for *Xylella fastidiosa* and *Escherichia coli* to xylem vessels (Cursino et al., 2009).

13.4.3 NANO-PHOTOCATALYSTS

Photocatalysis is a reaction in which a chemical reaction is accelerated by direct or indirect irradiation of a catalyst. Some NPs have a tendency to behave as a photocatalyst as, in the presence of light, valence electrons get excited and generate electron pairs (Ji et al., 2011). As a result of this, these irradiated NPs can behave as strong oxidizing agents to degrade toxic compounds. NPs that have been explored for this purpose are TiO_2 (Bhatkhande et al., 2002; Khataee et al., 2013), Ag (Ji et al., 2011), Au (Chauke et al., 2011), CdS (Guo et al., 2014), ZnS (Feigl et al., 2010), and some nanocomposites composed of silver and iron (III) oxide (Zhang et al., 2014). This strategy has been used for degrading pesticides that take a comparatively long time to decompose.

The unique properties of NPs make them suitable for imaging, therapy, and drug delivery. These NPs can be surface modified or incorporated into polymer- or lipid-based systems to perform multiple functions simultaneously.

It has been shown previously that NP can be loaded with a number of different drugs both via encapsulation and by covalent linkage. This section attempts to outline the vast array of NPs carrying an even wider range of drugs for disease treatment. The remarkable arrival of a targeted, physiologically shielded, deliver-on-demand tool has led to a huge advancement in scientific research. Furthermore, the variety of drugs being carried is not just limited to cancer therapy (Chen et al., 2010; Gvili and

Machluf, 2006; Lee et al., 2014): Lee et al. in (2014) used small interfering RNA (siRNA) / thiolated glycol chitosan NP to silence tumor necrosis factor genes for the treatment of rheumatoid arthritis (Lee et al., 2014). Polylactic-co-glycolic acid (PLGA) NPs were used in 2011 to deliver triplex-forming peptide nucleic acids, a powerful gene therapy agent for site-specific genomic recombination. These PLGA NPs delivered site-targeted, specific correction of disease-causing genetic mutations (McNeer et al., 2011). Other PLGA NPs have also been proposed as DNA vaccination tools (Gvili and Machluf, 2006).

13.5   CURRENT SCENARIO OF NANOTECHNOLOGY IN INDIA

Because the Indian economy is mainly dependent on agriculture, it is crucial to adopt practices that ensure better productivity and hence food security. In addition, soil diversity, climate change, and population growth offer more challenges in the agri sector. In this regard, integrated practices of nanotechnology in agriculture may offer great promise for sustainable agriculture.

In view of this, the Nanoscience and Technology Initiative (NSTI) was initiated in 2001. NSTI is focused on bridging the gaps between nanotechnological interventions and agriculture by initiating several nanoscience-related R&D projects, establishing different centers and facilities, developing human resources through different skill-development programs, and instigating collaborative programs and joint institution-industry projects.

Over recent decades nanotechnology-based tools and techniques have been applied in agriculture practices like nano fertilizers, nano pesticides, nano-agro-chemicals, waste water management, drug delivery, DNA/RNA delivery, plant disease diagnosis and treatment, plant genetics, etc.; applications of nanotechnology for sustainable agriculture are still in the primary stage. In terms of India, nanotechnology can pave pathways for sustainable agriculture by utilizing inputs more effectively and efficiently, reducing the production of by-products, or converting agricultural waste into valuable nanomaterials. Efficient utility of nanotechnology-based tools and techniques in basic practices of agriculture, production of value-added products, crops and food preservation, diagnosis of diseases, and genetic modifications of plants can bring vast advancements in agriculture sectors in India and hence can be a boon for the Indian economy. In the agriculture sector, there are still various possibilities to be investigated; therefore, there is a urgent need for better debates and communications related to the advantages of nanotechnology along with training for applying nanotechnology principles to this field (Bhagat et al., 2015).

13.6   FUTURE PROSPECTS OF NANOTECHNOLOGY IN AGRICULTURE

Innovative application of nanotechnology in the field of agriculture assists in handling challenges due to the insufficiency of land, water, resources, nutrients, and labor. Chemicals, in the form of pesticide, insecticide, and herbicide, while obtaining high mass product for meeting the needs of around 1.25 billion people, also create ecological imbalance, environmental pollution, and deterioration of land, soil, and water. Replacing these with nanopesticides, nanoinsecticides, nanoherbicides, and nanofertilizers enhances the reach of nutrients to root and shoot, thus promoting its growth and improving vegetation and pest control (Pandey, 2018). Nanotechnology is a field that will be highly appreciable in the near future due to its ability to cross all barriers and bring out the finest consequences.

NPs such as quantum dots, CNTs, metal NPs, nanobiosensors, nanobarcodes, and nanorobots are being utilized for sensing, monitoring, and analyzing the soil quality, for nutrient uptake, target drug delivery, dosage management, and plant health. Their potential applications would aid in conserving land, water, and resources. Demolishing poverty, malnutrition, and scarcity must be the principle aim of this forthcoming technology as it is supposed to provide stuff of righteous quality and quantity. Its implementation in the food sector for production, processing, protection, packaging, and preservation will be profitable for one and all. These nano-sized particles are also proficient in gene transfer to construct genetically modified plants and enhance the property of insect resistance like in Bt cotton, Bt maize, etc. (Vishwakarma et al., 2018). Some researchers have also been successful in embedding NPs that along with luciferase and luciferin interaction give off bright light in leaves of plants such as watercress, kale, spinach, and arugula. This led to the development of bioluminescent plants that are capable of providing light for about 4 hours and with further optimization will one day be able to illuminate a workspace.

The planning commission of India has endorsed research and development in nanotechnology and established the National Institute of Nanotechnology to rehabilitate Indian agriculture and boost the productivity rate. Having both on- and off-site effects in agriculture, the risks and safety measures of nanotechnology need to be thoroughly studied (Pandey, 2018).

13.7 CONCLUSION

Applications of nanomaterials in the area of agriculture have proved to be a boon in crop protection. Nanobased delivery of agrochemicals in the form of nanocapsules, nanospheres, micelles, nanocomposites, etc. is now possible with several advantages over the conventional methods of crop protection. The effects of encapsulated pesticides last for a long time because of their controlled release and protection against photocatalysis, hydrolysis, and/or other degrading mechanisms.

Considering the benefits of NPs over bactericides and fungicides, their potential as nanopesticides is remarkable. Real-time monitoring of bio-physicochemical properties related to crop improvement can now be achieved with high sensitivity through nanobased sensors. Gene transfer using nanotechnology is also attracting many researchers to efficient plant transformation. However, nanomaterials can be phytotoxic at certain concentrations, which should be assessed prior to their field application. Evaluation of the toxic effects of these nanomaterials on animals and the environment with their bioavailability is thus essential. Before releasing nanobased plant protectants to the market, they should undergo proper regulatory assessment to measure their efficiency and toxicity. Furthermore, it is essential to involve all social and economic personnel in open conversation to gain support and acceptance for this interdisciplinary technology.

REFERENCES

Åkerman, M.E., Chan, W.C.W., Laakkonen, P., Bhatia, S.N., Ruoslahti, E., 2002. Nanocrystal targeting in vivo. *Proc. Natl. Acad. Sci. U. S. A.* https://doi.org/10.1073/pnas.152463399

AL-Tameemi, A.J., AL-Aloosy, Y.A.M., Jumaa, S.S., 2019. Nano fertilizers and optimum crop productivity: a review.

Anastas, P.T., Warner, J.C., 1998. Green chemistry. *Frontiers (Boulder)*. 640.

Antonacci, A., Arduini, F., Moscone, D., Palleschi, G., Scognamiglio, V., 2018. Nanostructured (Bio)sensors for smart agriculture. *TrAC - Trends Anal. Chem.* https://doi.org/10.1016/j.trac.2017.10.022

Aouada, F.A., De Moura, M.R., 2015. Nanotechnology applied in agriculture: Controlled release of agrochemicals, in: *Nanotechnologies in Food and Agriculture*. Springer, pp. 103–118.

Athinarayanan, J., Periasamy, V.S., Alhazmi, M., Alatiah, K.A., Alshatwi, A.A., 2015. Synthesis of biogenic silica nanoparticles from rice husks for biomedical applications. *Ceram. Int.* https://doi.org/10.1016/j.ceramint.2014.08.069

Babu, M., Devi, V., Ramakritinan, C.M., Umarani, R., Taredahalli, N., Kumaraguru, A.K., 2013. Application of Biosynthesized Silver Nanoparticles in Agricultural and Marine Pest Control. *Curr. Nanosci.* https://doi.org/10.2174/15734137113096660103

Bandyopadhyay, S., Peralta-Videa, J.R., Gardea-Torresdey, J.L., 2013. Advanced analytical techniques for the measurement of nanomaterials in food and agricultural samples: a review. *Environ. Eng. Sci.* 30, 118–125.

Beck, R.C.R., Lionzo, M.I.Z., Costa, T.M.H., Benvenutti, E. V., Ré, M.I., Gallas, M.R., Pohlmann, A.R., Guterres, S.S., 2008. Surface morphology of spray-dried nanoparticle-coated microparticles designed as an oral drug delivery system. *Brazilian J. Chem. Eng.* https://doi.org/10.1590/S0104-66322008000200016

Bhagat, Y., Gangadhara, K., Rabinal, C., Chaudhari, G., Ugale, P., 2015. Nanotechnology in agriculture: a review. *J Pure App Microbiol* 9, 737–747.

Bhan, S., Mohan, L., Srivastava, C.N., 2018. Nanopesticides: A recent novel ecofriendly approach in insect pest management. *J. Entomol. Res.* 42, 263–270.

Bhatkhande, D.S., Pangarkar, V.G., Beenackers, A.A.C.M., 2002. Photocatalytic degradation for environmental applications - A review. *J. Chem. Technol. Biotechnol.* https://doi.org/10.1002/jctb.532

Bowen, P. a., Menzies, J.G., Ehret, D.L., 1992. Soluble Silicon Sprays Inhibit Powdery Mildew on grape leaves. *J. Am. Soc. Hortic. Sci.*

Bulovic, V., Mandell, A., Perlman, A., 2004. *Molecular memory device*. US 20050116256 A 1.

Cao, C., Kim, J.H., Yoon, D., Hwang, E.-S., Kim, Y.-J., Baik, S., 2008. Optical detection of DNA hybridization using absorption spectra of single-walled carbon nanotubes. *Mater. Chem. Phys.* 112, 738–741.

Chao, S.H.L., Choi, H.S., 2005. *Method for providing enhanced photosynthesis. Korea Research Institute of Chemical Technology*. Bull. South Korea Press 10.

Chauke, V.P., Antunes, E., Chidawanyika, W., Nyokong, T., 2011. Photocatalytic behaviour of tantalum (V) phthalocyanines in the presence of gold nanoparticles towards the oxidation of cyclohexene. *J. Mol. Catal. A Chem.* https://doi.org/10.1016/j.molcata.2010.11.023

Chen, Y., Zhu, X., Zhang, X., Liu, B., Huang, L., 2010. Nanoparticles modified with tumor-targeting scFv deliver siRNA and miRNA for cancer therapy. *Mol. Ther.* https://doi.org/10.1038/mt.2010.136

Chhipa, H., 2017. Nanopesticide: current status and future possibilities. *Agric Res Technol* 5, 1–4.

Corradini, E., de Moura, M.R., Mattoso, L.H.C., 2010. A preliminary study of the incorpora-tion of NPK fertilizer into chitosan nanoparticles. *Express Polym. Lett.* https://doi.org/10.3144/expresspolymlett.2010.64

Couvreur, P., Dubernet, C., Puisieux, F., 1995. Controlled drug delivery with nanoparticles: Current possibilities and future trends. *Eur. J. Pharm. Biopharm.*

Cursino, L., Li, Y., Zaini, P.A., De La Fuente, L., Hoch, H.C., Burr, T.J., 2009. Twitching motility and biofilm formation are associated with tonB1 in Xylella fastidiosa. *FEMS Microbiol. Lett.* https://doi.org/10.1111/j.1574-6968.2009.01747.x

Dahan, M., Lévi, S., Luccardini, C., Rostaing, P., Riveau, B., Triller, A., 2003. Diffusion Dynamics of Glycine Receptors Revealed by Single-Quantum Dot Tracking. *Science* (80). https://doi.org/10.1126/science.1088525

Deng, H.H., Hong, G.L., Lin, F.L., Liu, A.L., Xia, X.H., Chen, W., 2016. Colorimetric detec-tion of urea, urease, and urease inhibitor based on the peroxidase-like activity of gold nanoparticles. *Anal. Chim. Acta.* https://doi.org/10.1016/j.aca.2016.02.008

Ditta, A., Arshad, M., 2016. Applications and perspectives of using nanomaterials for sustain-able plant nutrition. *Nanotechnol. Rev.* https://doi.org/10.1515/ntrev-2015-0060

Dubey, A., Mailapalli, D.R., 2016. Nanofertilisers, nanopesticides, nanosensors of pest and nanotoxicity in agriculture, in: *Sustainable Agriculture Reviews.* Springer, pp. 307–330.

Dwarakanath, S., Bruno, J.G., Shastry, A., Phillips, T., John, A., Kumar, A., Stephenson, L.D., 2004. Quantum dot-antibody and aptamer conjugates shift fluorescence upon binding bacteria. *Biochem. Biophys. Res. Commun.* https://doi.org/10.1016/j.bbrc.2004.10.099

Ezhilarasi, P.N., Karthik, P., Chhanwal, N., Anandharamakrishnan, C., 2013. Nanoencapsulation Techniques for Food Bioactive Components: A Review. *Food Bioprocess Technol.* https://doi.org/10.1007/s11947-012-0944-0

Farlow, J., Seo, D., Broaders, K.E., Taylor, M.J., Gartner, Z.J., Jun, Y.W., 2013. Formation of targeted monovalent quantum dots by steric exclusion. *Nat. Methods.* https://doi.org/10.1038/nmeth.2682

Feigl, C., Russo, S.P., Barnard, A.S., 2010. Safe, stable and effective nanotechnology: Phase mapping of ZnS nanoparticles. *J. Mater. Chem.* https://doi.org/10.1039/b924697e

Galandova, J., Ziyatdinova, G., Labuda, J., 2008. Disposable electrochemical biosensor with multiwalled carbon nanotubes - Chitosan composite layer for the detection of deep DNA damage. *Anal. Sci.* https://doi.org/10.2116/analsci.24.711

Gao, F., Hong, F., Liu, C., Zheng, L., Su, M., Wu, X., Yang, F., Wu, C., Yang, P., 2006. Mechanism of nano-anatase TiO_2 on promoting photosynthetic carbon reaction of spin-ach. *Biol. Trace Elem. Res.* 111, 239–253.

Gao, F., Liu, C., Qu, C., Zheng, L., Yang, F., Su, M., Hong, F., 2008. Was improvement of spinach growth by nano-TiO2 treatment related to the changes of Rubisco activase? *BioMetals.* https://doi.org/10.1007/s10534-007-9110-y

Giardi, M.T., Piletska, E.V., 2006. *Biotechnological applications of photosynthetic proteins: biochips, biosensors and biodevices.* Springer.

Giraldo, J.P., Landry, M.P., Faltermeier, S.M., McNicholas, T.P., Iverson, N.M., Boghossian, A.A., Reuel, N.F., Hilmer, A.J., Sen, F., Brew, J.A., Strano, M.S., 2014. Erratum: Plant nanobionics approach to augment photosynthesis and biochemical sensing. *Nat. Mater.* 13 400–408 https://doi.org/10.1038/nmat3947

Glenn, D.M., Sekutowski, D.G., Puterka, G.J., 2000. *Method for providing enhanced photosynthesis.*

Guo, X., Chen, C., Song, W., Wang, X., Di, W., Qin, W., 2014. CdS embedded TiO2 hybrid nanospheres for visible light photocatalysis. *J. Mol. Catal. A Chem.* 387, 1–6.

Gvili, J., Machluf, M., 2006. 544. PLGA Nanoparticles for DNA Vaccination–Waiving Complexity and Increasing Efficiency. *Mol. Ther.* https://doi.org/10.1016/j.ymthe.2006.08.616

Hong, F., Yang, F., Liu, C., Gao, Q., Wan, Z., Gu, F., Wu, C., Ma, Z., Zhou, J., Yang, P., 2005. Influences of nano-TiO$_2$ on the chloroplast aging of spinach under light. *Biol. Trace Elem. Res.* https://doi.org/10.1385/BTER:104:3:249

Hong, F., Zhou, J., Liu, C., Yang, F., Wu, C., Zheng, L., Yang, P., 2005. Effect of Nano-TiO$_2$ on photochemical reaction of chloroplasts of spinach. *Biol. Trace Elem. Res.* https://doi.org/10.1385/BTER:105:1-3:269

Howarth, M., Liu, W., Puthenveetil, S., Zheng, Y., Marshall, L.F., Schmidt, M.M., Wittrup, K.D., Bawendi, M.G., Ting, A.Y., 2008. Monovalent, reduced-size quantum dots for imaging receptors on living cells. *Nat. Methods.* https://doi.org/10.1038/nmeth.1206

Ji, Z., Ismail, M.N., Callahan, D.M., Pandowo, E., Cai, Z., Goodrich, T.L., Ziemer, K.S., Warzywoda, J., Sacco, A., 2011. The role of silver nanoparticles on silver modified titanosilicate ETS-10 in visible light photocatalysis. *Appl. Catal. B Environ.* https://doi.org/10.1016/j.apcatb.2010.12.021

Kah, M., 2015. Nanopesticides and nanofertilizers: emerging contaminants or opportunities for risk mitigation? *Front. Chem.* 3, 64.

Kasana, R.C., Panwar, N.R., Kaul, R.K., Kumar, P., 2017. Biosynthesis and effects of copper nanoparticles on plants. *Environ. Chem. Lett.* https://doi.org/10.1007/s10311-017-0615-5

Khataee, A.R., Fathinia, M., Joo, S.W., 2013. Simultaneous monitoring of photocatalysis of three pharmaceuticals by immobilized TiO$_2$ nanoparticles: Chemometric assessment, intermediates identification and ecotoxicological evaluation. *Spectrochim. Acta - Part A Mol. Biomol. Spectrosc.* https://doi.org/10.1016/j.saa.2013.04.028

Lee, S.J., Lee, A., Hwang, S.R., Park, J.-S., Jang, J., Huh, M.S., Jo, D.-G., Yoon, S.-Y., Byun, Y., Kim, S.H., 2014. TNF-α gene silencing using polymerized siRNA/thiolated glycol chitosan nanoparticles for rheumatoid arthritis. *Mol. Ther.* 22, 397–408.

Lei, Z., Mingyu, S., Chao, L., Liang, C., Hao, H., Xiao, W., Xiaoqing, L., Fan, Y., Fengqing, G., Fashui, H., 2007. Effects of nanoanatase TiO$_2$ on photosynthesis of spinach chloroplasts under different light illumination. *Biol. Trace Elem. Res.* https://doi.org/10.1007/s12011-007-0047-3

Li, Y., Cu, Y.T.H., Luo, D., 2005. Multiplexed detection of pathogen DNA with DNA-based fluorescence nanobarcodes. *Nat. Biotechnol.* 23, 885–889.

Linglan, M., Chao, L., Chunxiang, Q., Sitao, Y., Jie, L., Fengqing, G., Fashui, H., 2008. Rubisco activase mRNA expression in spinach: Modulation by nanoanatase treatment. *Biol. Trace Elem. Res.* https://doi.org/10.1007/s12011-007-8069-4

Liu, X., Wang, R., Xia, Y., He, Y., Zhang, T., 2011. LiCl-modified mesoporous silica SBA-16 thick film resistors as humidity sensor. *Sens. Lett.* 9, 698–702.

Malik, P., Katyal, V., Malik, V., Asatkar, A., Inwati, G., Mukherjee, T.K., 2013. Nanobiosensors: concepts and variations. *ISRN Nanomater.* 2013.

McNeer, N.A., Chin, J.Y., Schleifman, E.B., Fields, R.J., Glazer, P.M., Saltzman, W.M., 2011. Nanoparticles deliver triplex-forming PNAs for site-specific genomic recombination in CD34+ human hematopoietic progenitors. *Mol. Ther.* https://doi.org/10.1038/mt.2010.200

Millán, G., Agosto, F., Vázquez, M., Botto, L., Lombardi, L., Juan, L., 2008. Use of clinoptilolite as a carrier for nitrogen fertilizers in soils of the Pampean regions of Argentina. *Cienc. e Investig. Agrar.* https://doi.org/10.4067/S0718-16202008000300007

Mukherjee, A., Majumdar, S., Servin, A.D., Pagano, L., Dhankher, O.P., White, J.C., 2016. Carbon nanomaterials in agriculture: A critical review. *Front. Plant Sci.* https://doi.org/10.3389/fpls.2016.00172

Ng, T.B., Chan, W.Y., Yeung, H.W., 1992. Proteins with abortifacient, ribosome inactivating, immunomodulatory, antitumor and anti-AIDS activities from Cucurbitaceae plants. *Gen. Pharmacol.* https://doi.or/g/10.1016/0306-3623(92)90131-3

Nima, Z.A., Lahiani, M.H., Watanabe, F., Xu, Y., Khodakovskaya, M. V., Biris, A.S., 2014. Plasmonically active nanorods for delivery of bio-active agents and high-sensitivity SERS detection in planta. *RSC Adv.* https://doi.org/10.1039/c4ra10358k

OECD International Futures Programme, 2007. *Small sizes that matter: Opportunities and risks of Nanotechnologies. Allianz.*

Ozdemir, M., Kemerli, T., 2016. Innovative applications of micro and nanoencapsulation in food packaging, in: *Encapsulation and Controlled Release Technologies in Food Systems*: Second Edition. https://doi.org/10.1002/9781118946893.ch12

Pan, P., Miao, Z., Yanhua, L., Linan, Z., Haiyan, R., Pan, K., Linpei, P., 2016. Preparation and evaluation of a stable solid state ion selective electrode of polypyrrole/electrochemically reduced graphene/glassy carbon substrate for soil nitrate sensing. *Int. J. Electrochem. Sci.* https://doi.org/10.20964/2016.06.7

Pandey, G., 2018. Challenges and future prospects of agri-nanotechnology for sustainable agriculture in India. *Environ. Technol. Innov.* https://doi.org/10.1016/j.eti.2018.06.012

Perrin, T.S., Drost, D.T., Boettinger, J.L., Norton, J.M., 1998. Ammonium-loaded clinoptilolite: A slow-release nitrogen fertilizer for sweet corn. *J. Plant Nutr.* https://doi.org/10.1080/01904169809365421

Preetha, P.S., Balakrishnan, N., 2017. A review of nano fertilizers and their use and functions in soil. *Int. J. Curr. Microbiol. App. Sci* 6, 3117–3133.

Qi, M., Liu, Y., Li, T., 2013. Nano-TiO2 improve the photosynthesis of tomato leaves under mild heat stress. *Biol. Trace Elem. Res.* https://doi.org/10.1007/s12011-013-9833-2

Qureshi, A., Singh, D.K., Dwivedi, S., 2018. Nano-fertilizers: a novel way for enhancing nutrient use efficiency and crop productivity. *Int. J. Curr. Microbiol. App. Sci* 7, 3325–3335.

Rameshaiah, G.N., Pallavi, J., Shabnam, S., 2015. Nano fertilizers and nano sensors–an attempt for developing smart agriculture. *Int. J. Eng. Res. Gen. Sci.* 3, 314–320.

Rastogi, A., Tripathi, D.K., Yadav, S., Chauhan, D.K., Živčák, M., Ghorbanpour, M., El-Sheery, N.I., Brestic, M., 2019. Application of silicon nanoparticles in agriculture. *3 Biotech.* https://doi.org/10.1007/s13205-019-1626-7

Servin, A., Elmer, W., Mukherjee, A., De la Torre-Roche, R., Hamdi, H., White, J.C., Bindraban, P., Dimkpa, C., 2015. A review of the use of engineered nanomaterials to suppress plant disease and enhance crop yield. *J. Nanoparticle Res.* https://doi.org/10.1007/s11051-015-2907-7

Srivastava, A.K., Dev, A., Karmakar, S., 2018. Nanosensors and nanobiosensors in food and agriculture. *Environ. Chem. Lett.* 16, 161–182.

Tokumasu, F., Fairhurst, R.M., Ostera, G.R., Brittain, N.J., Hwang, J., Wellems, T.E., Dvorak, J.A., 2005. Band 3 modifications in Plasmodium falciparum-infected AA and CC erythrocytes assayed by autocorrelation analysis using quantum dots. *J. Cell Sci.* https://doi.org/10.1242/jcs.01662

Um, S.H., Lee, J.B., Kwon, S.Y., Li, Y., Luo, D., 2006. Dendrimer-like DNA-based fluorescence nanobarcodes. *Nat. Protoc.* https://doi.org/10.1038/nprot.2006.141

Ung, T.D.T., Tran, T.K.C., Pham, T.N., Nguyen, D.N., Dinh, D.K., Nguyen, Q.L., 2012. CdTe and CdSe quantum dots: Synthesis, characterizations and applications in agriculture. *Adv. Nat. Sci. Nanosci. Nanotechnol.* https://doi.org/10.1088/2043-6262/3/4/043001

Vamvakaki, V., Chaniotakis, N.A., 2007. Pesticide detection with a liposome-based nano-biosensor. *Biosens. Bioelectron.* 22, 2848–2853.

Vartholomeos, P., Fruchard, M., Ferreira, A., Mavroidis, C., 2011. MRI-Guided Nanorobotic Systems for Therapeutic and Diagnostic Applications. *Annu. Rev. Biomed. Eng.* https://doi.org/10.1146/annurev-bioeng-071910-124724

Vishwakarma, K., Upadhyay, N., Kumar, N., Tripathi, D.K., Chauhan, D.K., Sharma, S., Sahi, S., 2018. Potential Applications and Avenues of Nanotechnology in Sustainable Agriculture, in: *Nanomaterials in Plants, Algae, and Microorganisms*. https://doi.org/10.1016/B978-0-12-811487-2.00021-9

Walling, M.A., Novak, J.A., Shepard, J.R.E., 2009. Quantum dots for live cell and in vivo imaging. *Int. J. Mol. Sci.* https://doi.org/10.3390/ijms10020441

Wang, Edina C., and Andrew Z. Wang, 2014. Nanoparticles and their applications in cell and molecular biology. *Integrative Biol* 6(1): 9–26.

Wanyika, H., Gatebe, E., Kioni, P., Tang, Z., Gao, Y., 2012. Mesoporous silica nanoparticles carrier for urea: Potential applications in agrochemical delivery systems. *J. Nanosci. Nanotechnol.* https://doi.org/10.1166/jnn.2012.5801

Zaki, A.M., Zaki, A.H., Farghali, A.A., Abdel-Rahim, E.F., 2017. Sodium titanate-bacillus as a new nanopesticide for cotton leaf-worm. *J Pure Appl Microbiol* 11, 7.

Zhang, S., Ren, F., Wu, W., Zhou, J., Sun, L., Xiao, X., Jiang, C., 2014. Size effects of Ag nanoparticles on plasmon-induced enhancement of photocatalysis of Ag-α-Fe2O3 nano-composites. *J. Colloid Interface Sci.* https://doi.org/10.1016/j.jcis.2013.12.012

Zhang, W., Yang, T., Huang, D., Jiao, K., Li, G., 2008. Synergistic effects of nano-ZnO/multi-walled carbon nanotubes/chitosan nanocomposite membrane for the sensitive detection of sequence-specific of PAT gene and PCR amplification of NOS gene. *J. Memb. Sci.* 325, 245–251.

Zhang, Z., Yu, Q., Li, H., Mustapha, A., Lin, M., 2015. Standing Gold Nanorod Arrays as Reproducible SERS Substrates for Measurement of Pesticides in Apple Juice and Vegetables. *J. Food Sci.* https://doi.org/10.1111/1750-3841.12759

Zhao, C.-L., Qin, M., Huang, Q.-A., 2011. Humidity sensing properties of the sensor based on graphene oxide films with different dispersion concentrations, in: *SENSORS*, 2011 IEEE. IEEE, pp. 129–132.

Zherebetskyy, D., Scheele, M., Zhang, Y., Bronstein, N., Thompson, C., Britt, D., Salmeron, M., Alivisatos, P., Wang, L.W., 2014. Hydroxylation of the surface of PbS nanocrystals passivated with oleic acid. *Science* (80). https://doi.org/10.1126/science.1252727

Zhu, Y., Cao, Y., Sun, X., Wang, X., 2013. Amperometric immunosensor for carbofuran detection based on MWCNTs/GS-PEI-Au and AuNPs-antibody conjugate. *Sensors* 13, 5286–5301.

14 Nanostructuring of Materials by Severe Deformation Processes

Aman J. Shukla, Devesh K. Chouhan and Somjeet Biswas

CONTENTS

14.1 INTRODUCTION

There is a keen interest in developing bulk nanostructured (NS) polycrystalline metallic materials by severe plastic deformation (SPD) methods [1]. Prof. Herbert Gleiter [2] was the first to indicate that NS materials contain a large grain boundary area fraction with an amorphous structure and consequently have exceptional properties. A bulk NS is defined as a solid with < 100 nm microstructural features in at least one dimension. There are two techniques to produce bulk nanostructured materials (BNM): (i) the bottom-up approach, by nanopowder compaction and (ii) the top-down approach, using SPD to obtain ultrafine/NS grains from a coarse-grain microstructure. SPD processes are inexpensive and industrially adaptable to produce BNMs and alloys, unlike the nanopowder compaction route. SPD can introduce large strains; with thermal aids and strain path changes, ultrafine/NS polycrystalline metallic materials can be obtained. There are various SPD processes, such as equal channel angular pressing (ECAP) [3], high-pressure torsion (HPT) [4], accumulative roll bonding (ARB) [5], multi-axial forging (MAF) [6, 7], and repetitive corrugation and straightening [8], etc. A large amount of strain can be introduced in these processes as the dimension of the specimen does not change after a deformation pass, and it can be introduced to further passes with strain path change. The bulk materials produced by various SPD processes led to the development of a wide range of NS and crystallographic textures for structural and functional applications.

Microstructural refinement during SPD is linked with introducing statistically stored dislocations and the evolution of geometrically necessary boundaries and grain boundaries in the material. The microstructural hierarchy is closely connected to the SPD methods and routes used. These materials possess remarkably high strength, ductility, toughness, and formability, properties that are of great importance in the automobile, aerospace, and biomedical implant industries as the materials can also be very lightweight. The first NS material using the SPD process was developed by Valiev and co-workers approximately 30 years ago [9]. These neoteric years show the sharp growth of research on this subject. Despite this, a considerable area of research remains arcanum and is waiting for us to unravel interesting and useful scientific developments. In this chapter, we will discuss the various SPD processes and will emphasize the principles of different SPD methods. The mechanism of the evolution of the hierarchical microstructure will be discussed. Special attention will be paid to the relationship between the microstructures and properties of BNMs.

14.2 WHAT ARE NANOSTRUCTURED MATERIALS?

The scale of these materials is in the nano range; thus, one can classify the types of NS as nanoparticles, nanocomposites, and BNMs, etc. They can also be defined with dimensionality. Quantum confinement is another parameter for distinguishing the NS materials. The nanoparticles have a "zero" dimension, which means that electrons are confined in all directions. There are a vast number of different applications for BNMs, such as ZnO nanoparticles used in cosmetic industries and various nanoparticles used in pharmaceutical industries. Quantum dots also have a broad use in the fields of optoelectronics and medicine. There are nanotubes, nanowires, nanorods, and nanoneedles, a lamellar structure in which electrons are allowed to move in only one direction; an image of nanoneedles is shown in Figure 14.1. It has only one micro/mesoscale dimension, and the other two are nanoscale. 1D NS materials have a huge impact on electronics, nanodevices, and provide alternative sources of energy, food, and security [10]. In 2D materials, in which two dimensions are outside the nanoscale, the electron has two directions for motion. This category includes thin-films, nano-coating, and graphene.

These 2D materials are used in the fabrication of nanodevices like sensors, nanocatalyst, etc. 2D structures like nanopores are also used in oil–water separation and water purification [11]. The action of NS materials depends on their morphology, size, shape, and surface-to-volume ratio, which are the main factors for desirable performance and activities.

3D nanostructures are free for electron motion in all three directions. With controlled morphology, 3D NS materials are widely used as electrode materials for batteries [12]. Owing to their large surface area, they provide a massive number of absorption

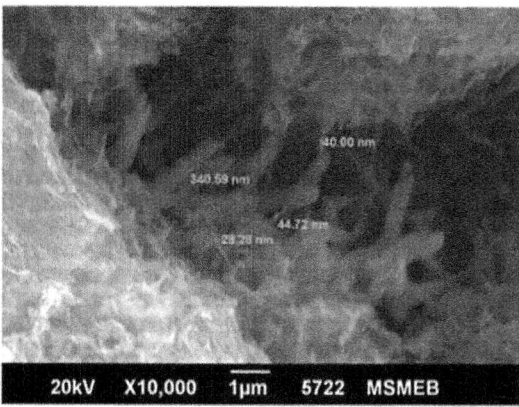

FIGURE 14.1   Growth of ZnO nanoneedles by an oxidation-assisted method

sites in very small space [13]; with porosity, in 3D it may help the better motion of molecules. Nanoflowers, fullerenes, and dendritic structures are in this category. Now taking into account BNMs, containing the two words "bulk" and "nanostructure", the bulk material contains nanostructured elements. The BNMs are closely related to the ultrafine-grained (UFG) structure, which has a <1 μm grain size, between submicron and nanosized grains. In UFG materials, high-angle grain boundaries (HAGB) are present in the majority, and the microstructures are usually equiaxed. The large fraction of HAGB is essential for the material to obtain the desired properties [14].

In order to be in the BNM category, an auxiliary requirement is the homogeneous distribution of nanoscale elements in the whole sample. NS materials exhibit other structural elements that are in the nanometer range, i.e., second phase particles, precipitates, nanosized twins, substructures, etc. These structural elements have a significant effect on the properties of the materials. For example, the microstructure of SPD-processed metals/alloys includes grain boundaries, which may form equilibrium, nonequilibrium, random, low-angle, and high-angle boundaries with special characteristics [15] depending on the processing techniques used. Moreover, the different characteristics of the boundary may have different mechanisms of transportation, i.e., diffusion, etc. These grain boundary properties lead to different mechanical, chemical, magnetic, and electrical properties. The various SPD techniques used in the formation of NS materials may give a platform for new advancement in the enhancement of mechanical properties such as strength and ductility, leading to new structural and functional applications of the materials.

14.3   METHODS OF SEVERE PLASTIC DEFORMATION

In SPD processes, there are some requirements that should be estimated while developing them for the formation of NS materials. First, the NS should be homogeneous within the entire volume of the grains. This uniformity is needed to ensure the properties of the processed material are stable. Second, when the sample experiences large strain, it should not have any surface/mechanical damage or cracks. And last, the formation of ultrafine grains with widespread HAGBs can show the relevant changes in the properties of materials.

Conventional deformation methods like rolling, extrusion, drawing, or forging alone are not very effective in the formation of NS. Therefore, it requires a special strategy (compound deformation) to impose large strains at ambient as well as high temperature.

Recently, researchers have combined conventional and unconventional techniques such as equal channel angular pressing (ECAP) + rolling, XYZ forging + rolling, ARB, and HPT followed by rolling to obtained nano crystalline metals/alloy [16–19]. Some results have shown the formation of sub- or nanocrystalline structures by XYZ forging or multi-axial/direction forging [17] and many passes of ARB [18, 19].

14.3.1 SEVERE PLASTIC DEFORMATION TECHNIQUES

14.3.1.1 Equal Channel Angular Pressing

ECAP is the process in which material deforms plastically via simple shear. Segal and co-workers established this method in the early 1980s. At the beginning of the 1990s, this process was widely used for the processing of UFG and nanostructured grains [20, 21, 22].

In the ECAP process, the lubricated billet is placed into a die that has two channels that intersect at an angle called the die channel angle [23]. The simple shear occurs along half of the die angle. During this process, a huge amount of plastic strain can be incorporated into each pass without altering the cross-sectional area of the sample, leading to desirable grain refinement and strengthening [24]. It is possible to repeat the process many times to get the NS grains. Figure 14.2 shows a schematic view of ECAP with Φ die angle and ψ arc angle.

The reference directions are important in the process: extrusion direction, transverse direction, normal direction, shear direction, and normal to shear direction. It can be seen that simple shear normally occurs at 45° towards the ECAP direction. The strain imposed in one pass can be calculated by the equation given below:

$$\varepsilon = \frac{1}{\sqrt{3}}\left\{2\cot\left(\frac{\Phi}{2}+\frac{\psi}{2}\right) + \psi\cosec\left(\frac{\Phi}{2}+\frac{\psi}{2}\right)\right\} \qquad (14.1)$$

Φ = die angle and ψ = arc of curvature

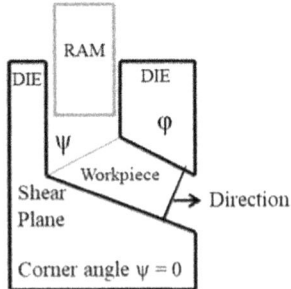

FIGURE 14.2   Schematic diagram of equal channel angular pressing

From this equation, we can calculate the von Mises strain for N number of passes, i.e., $N \times \varepsilon$.

With the calculation, Iwahashi et al. [25] showed the effect of the arc of curvature angle in the ECAP process. At $\psi = 0°$ the imposed strain is calculated as approximately 1.16.

ECAP has four routes categorized by the change in strain path: routes A, B_A, B_C, and C. These strain paths are related to the change in orientation of the billet before reinsertion into the die cavity during subsequent passes. In Figure 14.3, route A has no rotation along the extrusion direction during subsequent passes. For route B_A, the sample was subjected to 90° counterclockwise rotation along with clockwise rotation after each pass, or contrariwise; i.e. if the sample was rotated counterclockwise in the second pass then it will be rotated clockwise for the third pass, and vice versa. In route B_C, the sample rotation occurs clockwise or counterclockwise by 90°. In route C, the samples were subjected to 180° rotation along its longitudinal axis with every successive pass [26]. The changes in strain path are significant and relate to an increment in the number of passes as well as the grain refinement. The strain path change and shearing characteristics are shown in Figure 14.3.

It may be observed that the operating temperature is also a main factor in ECAP. If the ECAP operating temperature is lower than 0.4 of the melting point of the material it leads to loose ductility and the material becomes brittle for further

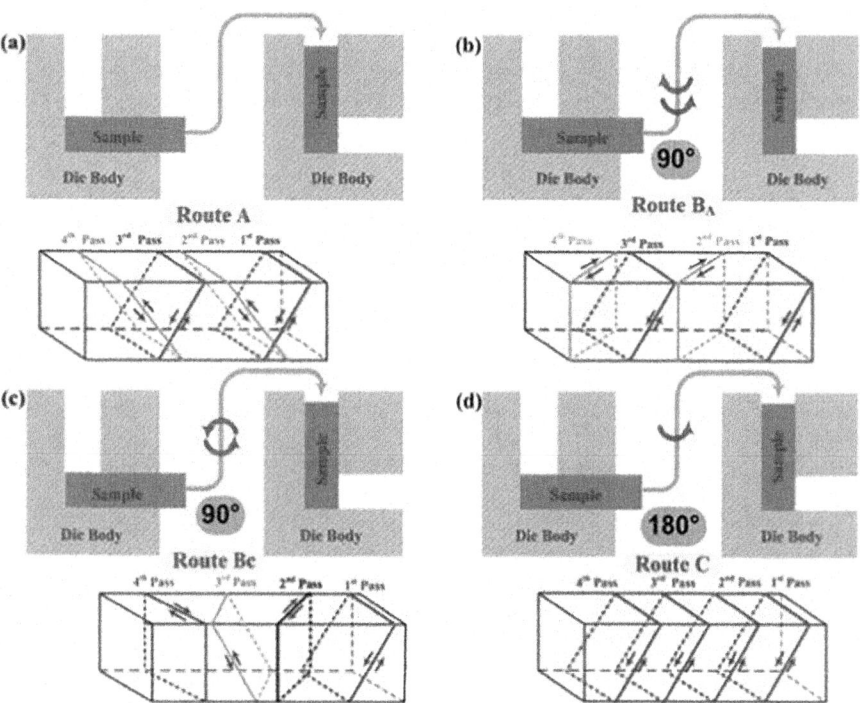

FIGURE 14.3 Illustrative diagram of equal channel angular pressing with routes and the change in strain path per pass

deformation, although this can be overcome by varying the temperature and applying back-pressure [27]. In the mechanism of grain refinement, a high operating temperature results in dynamic recrystallization and grain growth [28]. The true hydrostatic state of stress conditions can be achieved via application of back-pressure during ECAP processing to solve this issue. A compressive hydrostatic state of stress subdues the instability of material during the deformation of ECAP. It also improves the uniformity, increases the proportion of strain in each pass, and decrease defects and pores [29, 30].

14.3.1.2 High-Pressure Torsion

HPT is another SPD technique, in which the samples are subjected to torsional shear strain under high hydrostatic pressure. The basic concept of HPT was developed 70 years ago [4]. However, during the past 25 years HPT has been developed as an eminent scientific tool to achieve exceptional grain refinement and strength combination. In this process, a disk of the polycrystalline sample is placed in the middle of two anvils. The present die design, shown in Figure 14.4, is the evolution of the Bridgman [31] anvil-type instrument. A huge compressive load (normally at GPa) is applied and one of the anvils rotated to create a torsion force.

Valiev and co-workers investigated grain refinement of an Al alloy by the HPT process in 1988. They also reported the formation of UFG with augmented mechanical properties [16]. Apart from grain refinement [14], the HPT process leads to the development of characteristic texture [32], along with enhancement in plasticity [33], fatigue resistance [34], etc.

During the HPT process, the punch provides torque with contact friction at the interface in the middle of the punch and disk. The strain in torsion is calculated by:

$$\gamma(r) = \frac{2\pi n r}{l} \tag{14.2}$$

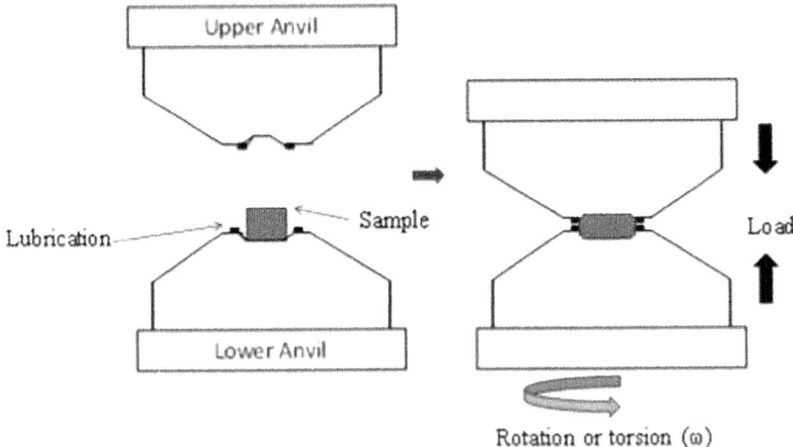

FIGURE 14.4   Schematic diagram of high-pressure torsion technique

where n is the number of rotations, r is the distance from the axis of the disk sample, and l the thickness of the sample. The von Mises yield equivalent strain could be given as:

$$\varepsilon(r) = \frac{\gamma(r)}{\sqrt{3}}$$
(14.3)

There are some disadvantages of the HPT process: first, it can only be used with a relatively small disk and is not suitable to process bulk samples, and second, microstructural inhomogeneity. Xu and co-workers [35] modified the process to carry out HPT of bulk samples and designated this as bulk-HPT.

14.3.2 ACCUMULATIVE ROLL BONDING

ARB is another important technique for severely plastically deforming metals and alloys [36]. This process is quite different from other SPD techniques and employs a rolling deformation technique as shown in Figure 14.5. Initially, rolled sheets are cut and stacked over each other. The surface of the sheets is treated to increase their roughness and enhance bonding. A rolling reduction equivalent to the thickness of a single sheet is usually provided at elevated temperatures. The strips join together during the process. The accumulative rolled sheet that is formed by this technique is cut again and stacked for subsequent passes. The process is repeated several times to obtain a desirable microstructure as well as the required properties.

The deformation temperature is an important aspect of ARB. Higher temperatures will assist recrystallization and grain growth, thereby reducing grain refinement [37]. Contrarily, low temperatures may lead to insufficient ductility and poor bond strength.

When two sheets are accumulatively roll bonded, the reduction in thickness is precisely 50% per pass. During ARB, plane strain deformation is incorporated; therefore, the increase in the sheet width is negligible. According to Saito et al. [36], under such conditions, the thickness of the parent sheet after n cycles can be given by:

$$t = \frac{t_o}{2^n}$$
(14.4)

Where t_o is the thickness of the parent sheet. The total reduction r_t after n cycles can be given as:

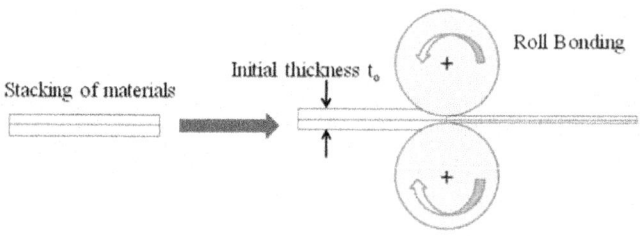

FIGURE 14.5 Schematic diagram of accumulative roll bonding technique

$$r_t = \frac{t}{t_o} = 1 - \frac{1}{2^n} \tag{14.5}$$

According to the von Mises criterion, the equivalent plastic strain ε is:

$$\varepsilon = \left\{ \frac{2}{\sqrt{3}} \ln\left(\frac{1}{2}\right) \right\} \times n = 0.8n \tag{14.6}$$

Successful ARB requires large rolling reductions. It is valuable to note that failure may occur by fracture during the process [38].

14.3.2.1 Multi-Axial Forging

MAF is a compressive deformation technique that is carried out along all three ortho-normal axes (X, Y, and Z) simultaneously to complete one cycle, as shown in Figure 14.6. This process is repeated to obtain refined microstructures. Salischev [39] originally developed this process and named it 'ABC forging'.

The process involves changes in strain path after each pass and is very effective in obtaining UFG or NS grains [40]. In 2009, Suwas et al. filed a patent on the MAF of interstitial free (IF) steel [41]. The process imposed the plane strain compression along all the sample axes concurrently to obtain UFG IF steel with enhanced strength. Similar work was also published later with a magnesium alloy [42].

Recently, Biswas et al. [17] filled an Indian patent on the development of nano-structured Ti with the highest strength, ductility, and formability together via combining multi-axial plane strain forging with the cold-rolling process. A transmission

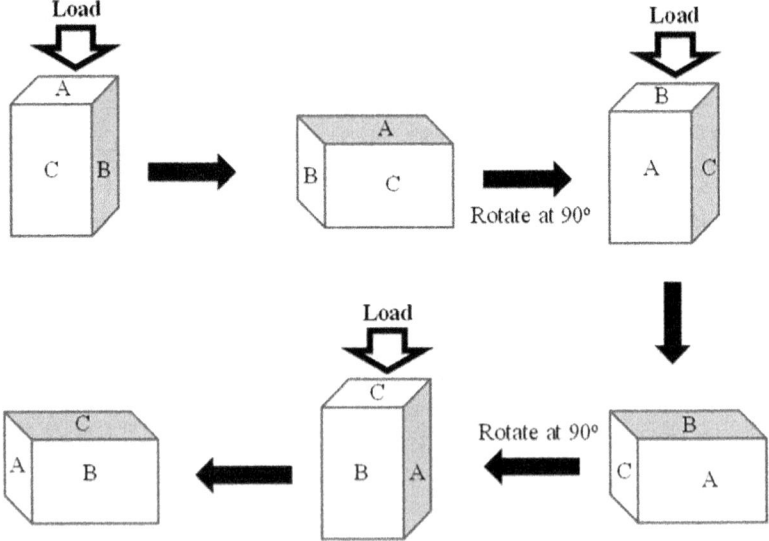

FIGURE 14.6 Schematic diagram of multi-axial forging technique (one cycle)

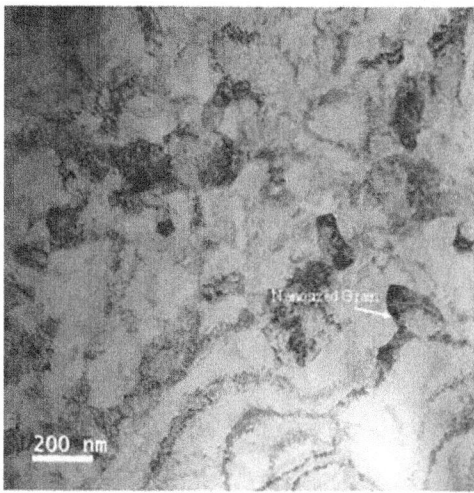

FIGURE 14.7 TEM image of nanostructured titanium obtained by multi-pass plane strain forging

electron microscope image of nanostructured Ti processed in this way is shown in Figure 14.7.

14.4   FORMATION OF OTHER NANOSTRUCTURES BY SPD

All SPD techniques are suitable to convert a coarse-grained poly-crystal structure into bulk nanostructured polycrystalline grain metals/alloys. Apart from this, the formation of other nanostructures should be taken into account, such as various dislocation features, dislocation walls, dense dislocation density structures, special grain boundary features, distribution of secondary particles, and nano twins. These structures may enhance the mechanical properties of materials. Generally, we can see that there are four types of structure formation in SPD-processed materials (Figure 14.8).

Nano twins: A high percentage of nano twins leads to the enhancement of strength but has an adverse effect on the formability of the NS materials. In contrast, long nano twins transmitted into neighboring grains lead to increased ductility to some extent. To encourage the formation of these twins, some conditions need to be satisfied: (1) low stacking fault energy (SFE), (2) low deformation temperature, and (3) low strain rate. It should always be taken into account that the optimum grain size range in any material is favorable for deformation twinning [43]. Optimum grain size can be estimated by the equation:

$$\frac{d_m}{\ln\left(\frac{\sqrt{2}d_m}{a}\right)} = \frac{(9.69 - \nu)}{253.66(1 - \nu)}\left(\frac{Ga^2}{\gamma}\right) \tag{14.7}$$

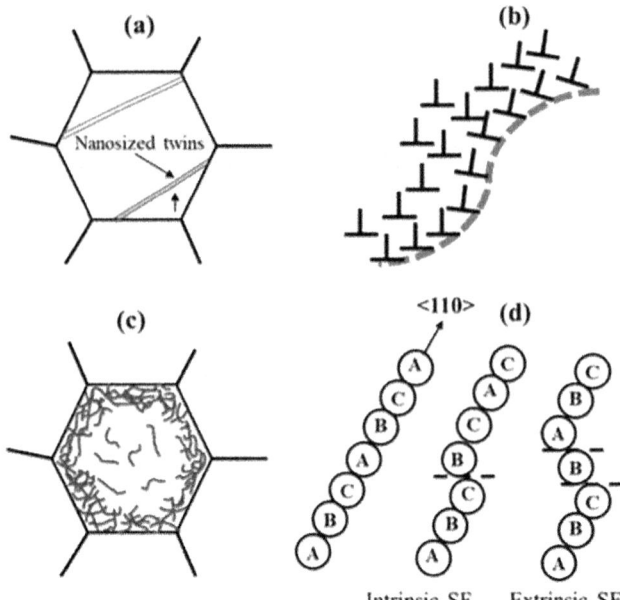

FIGURE 14.8 (a) Nano twin formation, (b) nonequilibrium grain boundaries with extrinsic grain boundary dislocation arrays, (c) intergranular cell formation, (d) stacking fault formation

where γ is the SFE, a is lattice parameter, ν is the Poisson's ratio, G is the shear modulus, and d_m is the optimum grain size of the materials.

Taking nanocrystalline face-centered cubic (FCC) materials into account, it can be seen experimentally that deformation twinning is a key mechanism for deformation [44]. It can also be seen that even in medium to high SFE nanocrystalline FCC material the deformation mechanism is twinning [44, 45, 46], even if it is not observed in coarse-grained FCC material.

Segregation of clusters: Owing to the SPD process, impurities form and/or alloying elements segregate at the grain boundary [47, 48]. This can be directly shown with 3D atomic probe tomography. The segregations at the grain boundary are about 3–5 nm in size and affect the dislocation motion, which bestows extra strength to the alloy. It can be seen the increment of strength is approx. 40% in Al alloys. [47, 49].

Grain boundaries and other defects: Because of the SPD process, dislocation density is observed at the grain boundary, as can be seen in the UFG of an Al–3%Mg alloy formed by the HPT process. This grain boundary is nonequilibrium in nature with lattice distortions 5–7 nm in width near the boundaries. This nonequilibrium grain boundary has a major effect on the mechanical behavior of the materials [16, 50].

Nanosized second phase particles: ECAP of the Al 6061 alloy leads to the fragmentation of the second phase particles or intermetallics into 10–20 nm pieces: their distribution in UFG has been observed [15, 51]. These nanoparticles can be the result of dynamic aging and, due to their size and excessive density, may block dislocation causing higher strength and also improving ductility [52, 53, 54].

14.5 PROPERTIES OF NANOSTRUCTURED SPD MATERIALS

NS materials have distinctive physical and mechanical properties that arise from the novel microstructure. These properties make the material more suitable for various specific application areas.

14.5.1 STRENGTH AND DUCTILITY

The strength of coarse to ultra-fine grained materials depends on the Hall–Petch relationship, $\sigma = \sigma_0 + kd^{-\frac{1}{2}}$, where the k and σ_0 are constants, d is the average number fraction grain size, and σ is the strength. In contrast, when the average grain size is lower than 20 nm, the above relation is no longer valid [55]. NS materials developed by the SPD process normally have high dislocation densities, enhanced random HAGB density, and other features related to imposing a large plastic strain. A finer grain size and a large number of defect densities lead to higher strength in comparison with a coarse-grained structure. [56, 57]. The reason behind this enhanced strength and strain hardenability is related to a decrease in distance between source and sink regions in the microstructure. This small distance required relatively higher stress compared with coarse-grain. Often, the twin boundaries or twin dislocation and grain boundary provide the high resistance. Grain refinement not only affects the strength but also gives rise to improvement in corrosion and fatigue strength [58]. In most cases, the enhanced strength due to grain refinement has an adverse effect on the ductility of nanostructured polycrystalline materials [59].

In contrast to the conventional process, SPD processing significantly reduces ductility. Horita et al. compared the processing routes of 3004 Al alloy by ECAP and rolling [60], observing an enhancement in strength with increasing strain imposed by ECAP or rolling. However, after imposing 1.16 strain on the material in ECAP, the ductility dropped from 32% to 14%. No further reduction was shown in subsequent passes. Cold rolling diminishes ductility by imposing a similar strain, but in this case the ductility continues to decrease, although at a slower rate, with further passes. To have the specialized combination of higher strength and enhanced ductility for structural applications, a finer grain size is not always needed. It can be seen that material with a mixture of UFG and NS grain size or a bimodal grain size distribution with an optimum amount of dislocation density has a better strength and ductility combination. The processing methods and routes always have a direct effect on the mechanical, electrical/electronic, and chemical properties of materials. Much research has indicated that, under certain conditions, SPD-processed NS materials show much better ductility.

Valiev and co-workers processed a copper nanostructure by up to 16 passes of ECAP with the application of back-pressure. The final sample has the same ductility as its coarse-grain counterpart, but the yield strength is many times higher [61].

Researchers are always seeking a combination of high strength and enhanced ductility for functional and structural applications.

Using the MAF method, the grain size of Ti6Al4V alloy (titanium alloy with 6 wt% aluminium and 4 wt% vanadium) reduces from 40 μm to 2 μm in the first pass and becomes 100 nm on the fourth pass. An enormous change of yield strength and ultimate tensile strength can be observed at the first pass. After the fourth pass, a drastic change in microstructure and mechanical properties is seen. The hardness slightly changes with subsequent passes [62]. The author suggested that the reason was flow softening via dynamic recovery at higher strains. A similar phenomenon was also seen in nickel 200 alloy [63].

ARB is also a very useful method for nanostructuring. Takata et al. investigated the properties of NS copper processed by up to eight cycles of ARB. The drastic enhancement of YS compared with the coarse-grained structure, with very little loss of ductility, suggests that the addition of trace elements such as phosphorus can increase the mechanical properties of copper [64]. Normally the ductility of a material is commanded by two factors: (1) work hardening and (2) strain rate sensitivity. A high rate of work hardening can affect the necking process to increase the ductility of materials. Accumulation of dislocations hardens the material and hinders further deformation. In NS the density of dislocations is at an extreme level so further increase is impossible. In this case, a low annealing temperature may enhance ductility without compromising strength [57, 65, 66].

14.5.2 CORROSION

The grain boundary region is the zone where the atoms are not arranged regularly: they have a higher chemical energy compared with those in the grain interior. Therefore, they react fastest. This chemical activity can be controlled by engineering the grain boundary character or forming a coincident site lattice boundary via coupled SPD and heat-treatment processes. In the field of corrosion, the research on NS material is not well established. However, many researchers have compared the effect of UFG and NS with their coarse-grained equivalents and found some enhancement in corrosion resistance [67]. However, we cannot ignore the fact that the enhancement may vary with the materials and processing routes. Many scientists raise the issue of corrosion resistance properties of these NS/UFGs. Vinogradov et al. [68] found the structure of polarization curves and the corrosive attack type remains constant in UFGs and coarse-grained materials. UFG Ti processed by SPD showed more desirable corrosion resistance in comparison with its coarse-grained counterpart in an acidic environment [69]. This may be due to the instant formation of a titanium dioxide film, which occurs because of surface defects like dislocation and grain boundaries.

The grain size reduction and accumulation of the dislocation inside the NS grain balances the energy, equivalent to that at the grain boundary. Therefore homogenous corrosion takes place more in NS grain materials than coarse-grained materials. In addition, the concentration of impurities is also found to be higher at the boundaries of grains. However, defects residing in the grains of NS titanium lead to diffusion of energy throughout the material, causing homogeneous corrosion. The existence of a large proportion of grain boundaries, as well as a high frequency of defects in the grain interiors, results in a more orderly spatial arrangement of impurities in NS titanium.

Magnesium and its alloys [70] gained great attention due to its light weight and viable biodegradable properties. Hot rolling of magnesium alloys has been found to be more potent in reducing biological corrosion rate than ECAP-mediated deformation [71]. Ben Hamu et al. [72] concluded that ECAP-propitiated deformations cause a reduction in the size of grains, and dislocations in the interiors as well as the boundaries of grains. No such dislocations are found in hot-rolled magnesium and magnesium alloys. These deformations/defects are responsible for enhancing biocorrosion rates in ECAP-treated materials; in lieu of this, it has been suggested to treat these materials thermally. In order to mitigate problems associated with ECAP processing and to generate microstructures favoring fatigue strength as well as resistivity against biological corrosion, more efficient, standardized studies should be promoted. These studies could represent SPD as a potent option in generating new implants having biodegradable properties.

14.6 APPLICATIONS

Owing to their superior properties in contrast with their coarse-grained counterparts, NS materials have grabbed the attention of the production industries. They also have the potential to enhance the quality of life. For example, Ti materials are very prominent for bioimplants: SPD-processed NS Ti has a high fatigue strength, which will increase the use of Ti in the biomedical field. NS Ti is also used in dental implants, which need a high fatigue strength because cyclic loading related to chewing crosses the fatigue limit of coarse-grained counterparts. Through nanostructuring and enhancing fatigue strength, it is possible to make a small-diameter implant for small teeth. High cycle fatigue is mainly related to strength and low cycle fatigue is related to ductility. SPD-processed NS has some improvement in high cycle fatigue but a shorter low cycle fatigue life [73]. However, it can be improved by moderate annealing without much loss of strength. Hardness is also improved in NS in comparison with its coarse-grained structure [74].

In sports instruments such as golf clubs, the shape and surface finishing/machining make products costlier. SPD-processed materials may be precluded from further surface finishing, which reduces the cost of manufacturing. The same trends are also seen in forged materials that are normally used in automobile and aerospace applications. In forged SPD materials, the heat-treatment temperature may be reduced by up to 50% [75]. These applications show the vista of NS materials processed by SPD methods. The early advantages of these SPD-processed materials are good machinability and formability at a good cost in comparison to other mechanical working methods. This makes NS materials very promising in biomedical applications, aerospace, defense, and heavy structural applications. Their production needs to be increased to an industrial level for the large-scale industries.

14.7 SUMMARY

It is widely accepted that SPD is a good process to obtain UFG/NS grains in all types of metallic materials. The competency of this method is well appreciated with reference to FCC, body-centered cubic, and hexagonal close packed metals/alloys, although there remains a need to understand the mechanism of grain refinement and the solidity of the NFG/NS grain structure and the implication of the grains for various material properties. There have been comparatively few studies on nanostructuring by SPD processes.

Of the SPD methods, ECAP is the most popular and well established. The possibility of augmenting this process is still unexplored. Alternative SPD processes such as ARB and MAF are candidates for a commercially feasible process. The improvement of mechanical properties by SPD is the main feature of these technologies. Other properties that can be improved make the materials more appealing for different applications such as hydrogen storage or bio-implantation. In short, there is plenty of study related to the development of SPD processes yet to be explored.

REFERENCES

1. Valiev, RZ., 1996. Ultrafine-grained materials prepared by severe plastic deformation. *Annales de Chimie. Science des Materiaux*, 21:369.
2. Gleiter, H. 1989. Nanocrystalline materials. *Progress in Material Science* 33: 223–330.
3. Biswas, S., S. S. Dhinwal, and S. Suwas. 2010. Room-temperature equal channel angular extrusion of pure magnesium. *Acta Materialia* 58:3247–3261.
4. Edalati, K., and Z. Horita. 2016. A review on high-pressure torsion (HPT) from 1935 to 1988. *Material Science and Engineering: A* 652:325–352.
5. Tsuji, N., Y. Saito, S-H Lee et al. 2003. ARB (Accumulative Roll-Bonding) and other new Techniques to Produce Bulk Ultrafine Grained Materials. *Advance Engineering Materials* 5:338–344.
6. Suwas, S., S. Biswas, A. Bhowmik et al. 2010. *A method to process interstitial-free, (IF) steels by adapting multi-axial, Indian Patent.*
7. Biswas, S., Satyam Suwas. 2012. Evolution of sub-micron grain size and weak texture in magnesium alloy Mg–3Al–0.4Mn by a modified multi-axial forging process. *Scripta Materialia* 66:89–92.
8. Jianyu Huang, Yuntian T. Zhu, David J. Alexander et al. 2004. Development of repetitive corrugation and straightening, *Material Science and Engineering: A* 371:35–39.
9. Valiev R.Z., N.A. Krasilnikov, and N. K. Tsenev. 1991. Plastic deformation of alloys with submicron-grained structure. *Material Science and Engineering: A* 137:35–40.
10. Kuchibhatla, S.V.N.T., A.S. Karakoti, D. Bera et al. 2007. One dimensional nanostructured materials. *Progress Materials Science* 52:699–913.
11. Dervin, S., D. D. Dionysiou, and Suresh C. Pillai. 2016. 2D nanostructures for water purification: graphene and beyond. *Nanoscale* 8:15115–15131.
12. Sun, Y., N. Liu, and Y. Cui. 2016. Promises and challenges of nanomaterials for lithium-based rechargeable batteries. *Nature Energy* 1: 16071.
13. Shen, Q., L. Jiang, H. Zhang et al. 2008. Three-dimensional Dendritic Pt Nanostructures: Sonoelectrochemical Synthesis and Electrochemical Applications. *Journal of Physical Chemistry C* 112:16385–16392.
14. Valiev, R.Z., Y. Estrin, Z. Horita et al. 2006. Producing bulk ultrafine-grained materials by severe plastic deformation. *JOM* 58:33–39.
15. Sauvage, X., G. Wilde, S.V. Divinski et al. 2012. Grain boundaries in ultrafine grained materials processed by severe plastic deformation and related phenomena. *Materials Science and Engineering: A* 540:1–12.

16. Valiev, R., R. K. Islamgaliev, I.V. Alexandrov. 2000. Bulk nanostructured materials from severe plastic deformation. *Progress in materials science* 45:103–189.

17. Biswas, S., D. K. Chouhan, A. K. Singh et al. 2019. *Indian Patent* Application no. *201931045383.*

18. Karimi, M., Mohammad Reza Toroghinejad, and Jan Dutkiewicz. 2016. Nanostructure formation during accumulative roll bonding of commercial purity titanium. *Materials Characterization* 122:98–103.

19. Yu, H. L., C. Lu, A. K. Tieu et al. 2014. Fabrication of Nanostructured Aluminum Sheets Using Four-Layer Accumulative Roll Bonding. *Materials and Manufacturing Processes* 29:448–453.

20. Valiev R.Z., A.V. Korznikov, R.R. Mulyukov. 1993. Structure and properties of ultra-fine-grained materials. *Materials Science and Engineering: A* 168:141–148.

21. Valiev, R.Z., N.A. Krasilnikov, N.K. Tsenev.1991. Plastic deformation of alloys with submicron-grained structure. *Materials Science and Engineering: A* 137:35–40

22. Reihanian M., R. Ebrahimi, N. Tsuji et al. 2008. Analysis of the mechanical properties and deformation behavior of nanostructured commercially pure Al processed by equal channel angular pressing (ECAP). *Materials Science and Engineering: A,* 473:189–194.

23. Segal V.M. 1995. Materials processing by simple shear. *Materials Science and Engineering: A* 197:157–164.

24. Kim H.S. 2002. Finite element analysis of deformation behaviour of metals during equal channel multi-angular pressing. *Materials Science and Engineering: A* 328:317–323.

25. Iwahashi, Y., J. Wang, Z. Horita et al. 1996. Principle of Equal-Channel Angular Pressing for the Processing of Ultra-Fine Grained Materials. *Scripta Materialia* 35:143–146.

26. Furukawa M., Z. Horita, and T.G. Langdon. 2002. Factors influencing the shearing patterns in equal-channel angular pressing. *Materials Science and Engineering: A* 332:97–109.

27. Lapovok R., Y. Estrin. M.V. Popov et al. 2008. Enhanced superplasticity of magnesium alloy AZ31 obtained through equal-channel angular pressing with back-pressure. *Journal Material Science* 43:7372–7378.

28. Stolyarov V.V., Y.T. Zhu, T.C. Lowe et al. 1999. Two-step SPD processing of ultrafine-grained titanium. *Nanostructured Mater* 11:947–954.

29. Beyerlein I.J., and L.S. Tóth. 2009. Texture evolution in equal-channel angular extrusion. *Progress in Material Science* 54:427–510

30. Lapovok R., D. Tomus, J. Mang et al. 2009. Evolution of nanoscale porosity during equal-channel angular pressing of titanium. *Acta Materialia* 57:2909–2918.

31. Bridgman, P.P.P.W.1935. Effects of high shearing stress combined with high hydrostatic pressure. *Physical Review* 48:825–847.

32. Orlov, D., P.P. Bhattacharjee, Y. Todaka et al. 2009. Texture evolution in pure aluminum subjected to monotonous and reversal straining in high-pressure torsion, *Scripta Materialia* 60:893–896.

33. Valiev, R.Z., M.Y. Murashkin, A. Kilmametov et al. 2010. Unusual super-ductility at room temperature in an ultrafine-grained aluminum alloy. *Journal of Material Science* 45:4718–4724.

34. Khatibi, G., J. Horky, B. Weiss et al. 2010. High cycle fatigue behaviour of copper deformed by high pressure torsion. *International Journal of Fatigue* 32:269–278.

35. Xu, C., Z. Horita, and T.G. Langdon. 2007. The evolution of homogeneity in processing by high-pressure torsion. *Acta Materialia* 55:203–212.

36. Saito, Y., H. Utsunomiya, N. Tsuji et al. 1999. Novel ultra-high straining process for bulk materials—development of the accumulative roll-bonding (ARB) process. *Acta Materialia* 47:579–583.

37. McQueen, H.J. and J. J. Jonas. 1975. Recovery and recrystallization during high temperature deformation in Treatise on Materials Science & Technology. *Elsevier* 6:393–493.
38. Tsuji, N., Y. Minamino, Y. Koizumi et al. 2002. Fabrication of ultrafine grained metallic materials by accumulative roll-bonding, in *International Symposium on Processing and Fabrication of Advanced Materials XI*: 320–334.
39. Salishchev, Gennady A., Sergey V. Zherebtsov, Oleg R. Valiakhmetov et al. 2004. Development of submicrocrystalline titanium alloys using "abc" isothermal forging. *Materials Science Forum* 447:459–464.
40. Bhowmik, A., S. Biswas, S.S. Dhinwal et al. 2011. Microstructure and Texture Evolution in Interstitial-Free (IF) Steel Processed by Multi-Axial Forging. *Materials Science Forum* 702–703:774–777.
41. Suwas, S., S. Biswas, A. Bhowmik et al. 2010. *A method to process interstitial-free (IF) steels by adapting multi-axial.*
42. Biswas, S., and S. Suwas. 2012. Evolution of sub-micron grain size and weak texture in magnesium alloy Mg-3Al-0.4Mn by a modified multi-axial forging process. *Scripta Materialia* 66:89–92.
43. Zhu Y.T., X.Z. Liao, and X.L. Wu. 2012. Deformation twinning in nanocrystalline materials. *Progress in Materials Science* 57:1–62.
44. Liao X.Z., S.G. Srinivasan, Y.H. Zhao et al. 2004. Formation mechanism of wide stacking faults in nanocrystalline Al. *Applied Physics Letter* 84:3564–3566.
45. Liao X.Z., F. Zhou, E.J. Lavernia et al. 2003. Deformation twins in nanocrystalline Al *Applied Physics Letter* 83:5062–5064.
46. Liao, X. Z., F. Zhou, E. J. Lavernia et al. 2003. Deformation mechanism in nanocrystalline Al: Partial dislocation slip. *Applied Physics Letters* 83: 632–634.
47. Nurislamova, Gulnaz, Xavier Sauvage, Maxim Murashkin et al. 2008. Nanostructure and related mechanical properties of an Al–Mg–Si alloy processed by severe plastic deformation. *Philosophical Magazine Letters* 88: 459–466.
48. Sha G., Y.B. Wang, X.Z. Liao et al. 2009. Influence of equal-channel angular pressing on precipitation in an Al–Zn–Mg–Cu alloy. *Acta Materialia* 57:3123–3132.
49. Liddicoat, P., X. Liao, Y. Zhao et al. 2010. Nanostructural hierarchy increases the strength of Al alloys. *Nature Communication* 1:63.
50. Valiev R. 2009. Nanostructuring of metallic materials by spd processing for advanced properties. *International Journal of Materials Research* 100:757–761.
51. Valiev R.Z., M.Y. Murashkin, E.V. Bobruk et al. 2009. Grain refinement and mechanical behavior of the Al alloy, subjected to the new SPD technique. *Materials Transactions* 50:87–91.
52. Zhao Y.H., X.Z. Liao, S. Cheng et al. 2006. Simultaneously increasing the ductility and strength of nanostructured alloys. *Advanced Materials* 18:2280–2283.
53. Valiev R.Z., N.A. Enikeev, M.Y. Murashkin et al. 2010. On the origin of the extremely high strength of ultrafine-grained Al alloys produced by severe plastic deformation. *Scripta Materialia* 63:949–952.
54. Zhao Y.H., Y.T. Zhu, and E.J. Lavernia. 2010. Strategies for improving tensile ductility of bulk nanostructured materials. *Advanced Engineering Materials* 12:769–778.
55. Chokshi, A.H., A. Rosen, J. Karch et al. 1989. On validity of Hall Petch relationship in nanocrystalline materials. *Scripta Matellurgica* 23:1679–1684.
56. Huang, J. Y., Y. T. Zhu, H. Jiang et al. 2001. Microstructures and dislocation configurations in nanostructured Cu processed by repetitive corrugation and straightening. *Acta Materialia* 49:1497–1505.
57. Zhu, Y. T., J. Y. Huang, Jenő Gubicza et al. 2003. Nanostructures in Ti processed by severe plastic deformation. *Journal of Materials Research* 18: 1908–1917.

58. Hashimoto, S. 2005. Latest Frontiers of Nanomaterials Research-Perspective on ECAP Method. *Koyo Engineering Journal*, 167:2.

59. Meyers M. A., A. Mishra, and D. J. Benson. 2006. Mechanical properties of nanocrystalline materials. *Progress in Materials Science* 51:427–556.

60. Horita, Z., T. Fujinami, M. Nemoto et al. 2000. Equal-channel angular pressing of commercial aluminum alloys: grain refinement, thermal stability and tensile properties. *Metallurgical and Materials Transactions A* 31: 691–701.

61. Valiev, R., I. Alexandrov, Y. Zhu et al. 2002. Paradox of Strength and Ductility in Metals Processed by Severe Plastic Deformation. *Journal of Materials Research*, 17:5–8.

62. Ghanbari, Behzad Fallah, Hossein Arabi, S. Mehdi Abbasi et al. 2016. Manufacturing of nanostructured Ti-6Al-4V alloy via closed-die isothermal multi-axial-temperature forging: microstructure and mechanical properties. *The International Journal of Advanced Manufacturing Technology* 87:755–763.

63. Hussain, Zahid, Fahad A. Al-Mufadi, Sivasankaran Subbarayan et al. 2018. Microstructure and mechanical properties investigation on nanostructured Nickel 200 alloy using multi-axial forging. *Materials Science and Engineering: A* 712: 772–779.

64. Takata, Naoki, Seong-Hee Lee, Cha-Yong Lim et al. 2007. Nanostructured bulk copper fabricated by accumulative roll bonding. *Journal of Nanoscience and Nanotechnology* 7: 3985–3989.

65. Stolyarov, Vladimir V., Yuntian T. Zhu, Terry C. Lowe et al. 2001. Microstructure and properties of pure Ti processed by ECAP and cold extrusion. *Materials Science and Engineering: A* 303: 82–89.

66. Stolyarov, Vladimir V., Y. Theodore Zhu, Igor V. Alexandrov et al. 2003. Grain refinement and properties of pure Ti processed by warm ECAP and cold rolling. *Materials Science and Engineering: A* 343: 43–50.

67. Kim, H. S., S.J. Yoo, J.W. Ahn et al. 2011. Ultrafine grained titanium sheets with high strength and high corrosion resistance. *Materials Science and Engineering: A*, 528:8479–8485.

68. Vinogradov, A., T. Mimaki, S. Hashimoto et al. 1999. On the corrosion behaviour of ultra-fine grain copper. *Scripta Materialia* 41: 319–326.

69. Balyanov, A., J. Kutnyakova, N. A. Amirkhanova et al. 2004. Corrosion resistance of ultra fine-grained Ti. *Scripta Materialia* 51: 225–229.

70. Biswas, Somjeet, Sudeep K. Sahoo, Devesh K. Chouhan, et al. 2019. Microstructure, Texture Evolution and Dynamic Recrystallization in Magnesium. *Reference Module in Materials Science and Materials Engineering*.

71. Babu, P.K. Ajeet, Akshay S. Nilawar, Pankaj Vishvakarma et al. 2013. Corrosion Behavior of Ultra Fine Grain Pure Magnesium for Automotive Applications. *SAE International Journal of Materials and Manufacturing* 6:99–104.

72. Hamu, G. Ben, D. Eliezer, and L. Wagner. 2009. The relation between severe plastic deformation microstructure and corrosion behavior of AZ31 magnesium alloy. *Journal of alloys and compounds* 468:222–229.

73. Vinogradov, A. and S. Hashimoto. 2003. Fatigue of Severely Deformed Metals. *Advanced Engineering Materials* 5:351–358.

74. Stolyarov, V. V., L. Sh Shuster, M. Sh Migranov et al. 2004. Reduction of friction coefficient of ultrafine-grained CP titanium. *Materials Science and Engineering: A* 371:313–317.

75. Zhu Y.T., T. C. Lowe, and T. G. Langdon. 2004. Performance and applications of nanostructured materials produced by severe plastic deformation, *Scripta Materialia* Volume 51:825–830.

Index

Page numbers in *italic* indicate figures. Page numbers in **bold** indicate tables.